Electromagnetic Metamaterials

Scrivener Publishing
100 Cummings Center, Suite 541J
Beverly, MA 01915-6106

Publishers at Scrivener
Martin Scrivener (martin@scrivenerpublishing.com)
Phillip Carmical (pcarmical@scrivenerpublishing.com)

Electromagnetic Metamaterials

Properties and Applications

Edited by

Inamuddin

Department of Applied Chemistry, Zakir Husain College of Engineering and Technology, Aligarh Muslim University, Aligarh, India

and

Tariq Altalhi

Department of Chemistry, College of Science, Taif University, Taif, Saudi Arabia

Scrivener Publishing

WILEY

This edition first published 2023 by John Wiley & Sons, Inc., 111 River Street, Hoboken, NJ 07030, USA and Scrivener Publishing LLC, 100 Cummings Center, Suite 541J, Beverly, MA 01915, USA

For more information about Scrivener publications please visit www.scrivenerpublishing.com.

Wiley Global Headquarters
111 River Street, Hoboken, NJ 07030, USA

For details of our global editorial offices, customer services, and more information about Wiley products visit us at www.wiley.com.

Library of Congress Cataloging-in-Publication Data

ISBN 978-1-394-16622-0

Cover image: Pixabay.Com
Cover design by Russell Richardson

Set in size of 11pt and Minion Pro by Manila Typesetting Company, Makati, Philippines

Printed in the USA

10 9 8 7 6 5 4 3 2 1

Contents

Preface

In recent years, metamaterials have become a hot topic in the scientific community due to their remarkable electromagnetic properties. Metamaterials have the ability to alter electromagnetic and acoustic waves in ways that bulk materials cannot. The metamaterials have a wide range of potential applications, including remote aerospace applications, medical appliances, sensor detectors and monitoring devices of infrastructure, crowd handling, smart solar panels, radomes, high-gain antennas lenses, high-frequency communication on the battlefield, ultrasonic detectors, and structures to shield earthquakes. A wide range of disciplines is involved in metamaterial research, including electromagnetics and electrical engineering, classical optics, microwave, and antenna engineering, solid-state physics, material sciences, and optoelectronics. This book presents an overview of metamaterials' current state of development in several domains of application.

Chapter 1 focuses on applications of metamaterials in the terahertz range, especially summarizing the performance attributes such as the gain, bandwidth, polarization, and isolation of the antenna by integrating metamaterials and various types of metamaterial-based absorbers. It also discusses their ability to manipulate and control electromagnetic waves.

Chapter 2 discusses the fundamentals of chiral metamaterial (CMM), its properties, constructions, applications, and recent advancements in the modern era. CMMs have many advantages, like design flexibility, excellent customized properties, giant optical activity, tuneability, etc., which makes them suitable for applications in imaging, bio-sensing, polarization manipulation, absorption, and other fields.

Chapter 3 reviews the theory of Metamaterial Absorbers (MMAs) and their applications in bio- and chemical sensing in mid-IR frequencies. The theoretical background on the design of MMAs is given in detail. In addition, bio- and chemical detection approaches based on Surface-Enhanced Infrared Spectroscopy (SEIRA) and refractive index change are discussed.

Chapter 4 provides a deeper understanding of double-positive medium metamaterials, their inherent features, and their various applications in sensors, photonic devices, etc. In addition, different types of metamaterials are discussed and compared with double-positive medium metamaterials, highlighting their merits over widely discussed double-negative medium metamaterials.

Chapter 5 discusses various types of photonic metamaterials and their application in life. Ways to alter the properties of this man-made material by changing its composition is also explained. Application of metamaterials in various areas, such as the health care industry, optical field, and aerospace, are all detailed. It signifies the composition, properties, and application of metamaterials.

Chapter 6 discusses the diverse topic of metamaterials with the support of theoretical models. Each theory employs a unique viewpoint to describe the same device. The main focus here is to communicate the drawbacks and benefits of available models. Additionally, the unusual advances proposed in metamaterials are discussed in detail.

Chapter 7 discusses the limitless potential of metamaterials to trigger a wide variety of applications, including band gaps, cloaking devices, electromagnetic, transformation elastodynamics, etc. Considering the recent proliferation of metamaterials and the growing interest in the associated research, three important milestones are also discussed in this chapter.

Chapter 8 discusses the basic concept of metamaterial cloaking and invisibility, and its design and simulation are explained. Types of cloaking, such as optical cloaking, acoustic cloaking, thermal cloaking, elastic cloaking, and mass diffusion cloaking, multifunctional cloaking, light diffusion cloaking, are explained. Also reviewed are various techniques of metamaterial cloaking, which includes scattering line cancellation, transmission line technique, coordinate transformation technique, and others.

Chapter 9 discusses different kinds of metamaterials, such as electromagnetic metamaterial, double-negative metamaterials, chiral metamaterials, and semiconductor metamaterials. Fundamental equations of metamaterials and the development of single negative metamaterials are explained along with their application.

Chapter 10 discusses negative-index metamaterials. Various basic concepts and theories of metamaterials, as well as scientific importance, are covered in detail. A primary focus is on the different aspects of NIMs and their potential applications in different domains, which allows access to new dimensions of material response.

Chapter 11 discusses the various kinds of electromagnetic metamaterials, their properties, uses, and types. The importance of an in-depth study of metamaterials and their potential applications in numerous aspects of life is explored. Moreover, several traditional metamaterials that are tunable using multiple design techniques are discussed.

Chapter 12 discusses the plasmonic materials and their fundamentals, such as negative refractive index and negative permeability. Surface plasmon polariton and localized surface plasmon are also explained. Different types of plasmonic materials, including graphene-based plasmonic metamaterials, nanorod plasmonic metamaterials, plasmonic meta-surfaces, self assemble plasmonic metamaterials, non-linear plasmonic materials, 2D plasmonic metamaterials are presented in detail. Furthermore, the chapter delves into the applications of these plasmonic metamaterials in nanochemistry, biosensing, photovoltaics, filter, planner ring resonator, and optical computing.

Chapter 13 explains the nonlinear effects of the metamaterial. Types of nonlinear metamaterials such as ferrite-based metamaterials, plasmonic metamaterials, dielectric materials, and some tunable nonlinear metamaterials are discussed. Applications of nonlinear metamaterials in Spling Ring Resonators (SRR) and an overall overview of nonlinear metamaterials are provided.

Chapter 14 discusses tunable metamaterials. The chapter also highlights the substantial developments that have been made in the fabrication and design of these materials.

Chapter 15 discusses metamaterials and their types. Further, the role of metamaterials in sound filtering is discussed with a focus on acoustics metamaterials. Additionally, phononic crystals are discussed in detail, along with their applications. Later, the fabrication and assembly of metamaterials used for sound filtering are discussed.

Chapter 16 presents concepts of metamaterial, radar technology, Radar Cross-Section Reduction (RCS), and different techniques of RCS reduction. Finally, the chapter concludes with future outlooks based on the current progress and advancements in metamaterials for radar cross-section reduction.

Our thanks go to Wiley and Scrivener Publishing for their continuous support and guidance.

Inamuddin
Department of Applied Chemistry, Zakir Husain College of Engineering and Technology, Aligarh Muslim University, Aligarh, India
Tariq Altalhi
Department of Chemistry, College of Science, Taif University, Taif, Saudi Arabia

1

Metamaterial-Based Antenna and Absorbers in THz Range

M. R. Nigil[1] and R. Thiruvengadathan[1,2]*

[1]Department of Electronics and Communication Engineering, Amrita School of Engineering, Coimbatore, Amrita Vishwa Vidyapeetham, Coimbatore, Tamil Nadu, India
[2]Mechanical Engineering, Department of Engineering and Technology, Southern Utah University, Cedar City, Utah, USA

Abstract

In the past decade, there has been exponential growth in the number of users of the internet, with consumers using both wireless and wired means to achieve connectivity. This rapid growth demands a wider spectrum of operations. But fabricating devices for operating at higher frequencies is difficult due to various impeding factors like power loss, reduced electromagnetic properties, and material sensitivity with increasing frequencies. In particular, there are no naturally occurring materials that can be employed to operate and exploit the terahertz (THz) region. Hence, there is a need to develop material for THz applications. Thus, this chapter presents a detailed review of the development of a class of material called metamaterial and its applications, especially in the THz regime. These are engineered materials capable of being tuned to provide desired electromagnetic properties that are not found naturally. Deployment of these materials in appropriate proportions in antennas and absorbers is found to provide significant improvements in the gain, bandwidth, and absorption coefficient. It is also found to reduce mutual coupling and enables the miniaturization of the devices.

Keywords: Metamaterials, metamaterial antennas, metamaterial absorbers, terahertz region, metamaterial design approach, tuneable metamaterials

**Corresponding author*: rajagopalanthiruveng@suu.edu

Inamuddin and Tariq Altalhi (eds.) Electromagnetic Metamaterials: Properties and Applications, (1–32) © 2023 Scrivener Publishing LLC

1.1 Introduction

The growth in technology requires a lot of new inventions along with improvements in existing devices and systems. Constant growth in wireless communication around the world demands a wider spectrum of operation. With growing frequency ranges, the possibility of fabricating devices for higher frequencies is hindered by many factors like power loss, lack of natural electromagnetic material, and material sensitivity at higher frequencies. One of the viable solutions that have gained attraction in recent years are artificially engineered materials called metamaterials. Metamaterials are materials that can be controlled to manipulate electromagnetic waves. Metamaterials have exciting application possibilities in high-frequency ranges. These materials are used in the design of antennas, absorbers, switches, etc. Implementing metamaterials in high frequencies, especially in the terahertz region, ensures wireless communication with enhanced carrier frequency, greater data rates, and channels with high capacity. Sections 1.1 and 1.2 of the chapter, summarize the issues in the terahertz region, the basics of metamaterials, the various types of metamaterials and their design approaches. Section 1.3 outlines the application of metamaterials in absorbers and then provides a more generalized summary of performance improvements achieved in antennas.

1.1.1 Terahertz Region

The terahertz spectrum is defined in the range between 0.1 and 10 THz, corresponding to wavelengths of 0.03 and 3 mm. This spectral region bridges the gap between electronics and photonics and is called the terahertz gap. THz radiation is non-destructive, non-invasive, and non-ionizing, so has no health risks [1]. Due to severe power losses at higher frequencies, scaling microwave electronics to the THz range is challenging and scaling infrared technologies, such as photonic band-gap materials, is problematic due to material sensitivity at THz frequencies [1]. Various novel hybrid approaches have been suggested to develop THz systems. One such approach is engineered materials called electromagnetic metamaterials.

1.1.2 Metamaterials

Metamaterials are artificial materials composed of subwavelength resonating structures that can modify light waves. In ordinary materials, the electromagnetic properties are obtained from atoms or molecules, but in the case of metamaterials, the desired electromagnetic property is obtained from the resonating subwavelength structures. Walser developed the term "metamaterial" to describe artificial three-dimensional periodic composites that produce a combination of more than two electromagnetic responses that are not present in nature. Hence, these are also called engineered materials [2]. Metamaterials can be constructed to operate at the desired frequency and have a broad range of electromagnetic properties that do not occur in nature. As a result, the word "meta" was coined, which means "beyond materials".

1.1.3 Classification of Metamaterials

Victor Veselago was the first scientist to suggest a categorization of metamaterials based on the permittivity and permeability of homogenous materials. Figure 1.1 shows a schematic of the classification of metamaterials

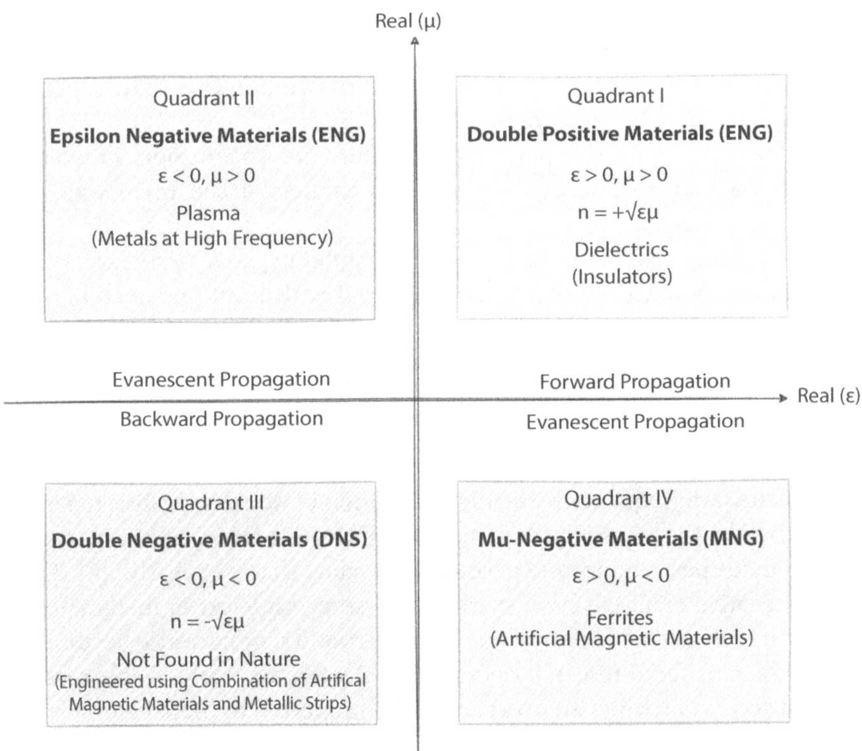

Figure 1.1 Classification of metamaterials based on permittivity and permeability. Reproduced with permission from Hossain *et al.* Decagonal c-shaped CSRR textile-based metamaterial for microwave applications, Computers, Materials & Continua, vol. 71, no.1, pp. 1677–1693, 2022. Copyright @2022 Hossain *et al.*

based on permittivity and permeability [3]. When both permittivity and permeability are negative, Snell's law, the Cerenkov effect, and Doppler shift, phenomena deviate and are reversed [4].

$$n = \pm\sqrt{\varepsilon_r \mu_r} \tag{1.1}$$

ε_r - relative permittivity of the material
μ_r - relative permeability of the material
n - refractive index

The relationship between the refractive index and the permittivity and permeability is established by Equation 1.1.

Quadrant I—Dielectrics are often found in the domain where both real electric permittivity and magnetic permeability are positive.

Quadrant II—Radiation-plasma interactions occur well beyond the infrared region and produce negative permittivity. For example, metamaterials like thin-wire lattices can dilute plasma clouds and reduce plasma frequency, leading to negative permittivity [5].

Quadrant III—Negative permeability values are uncommon and can be obtained via magnetic resonances in ferromagnets at sub-microwave frequencies (terahertz). For example, upon interaction with magnetic waves, pairs of split ring resonators and thin wires exhibit negative permeability, [6].

Quadrant IV—A double negative material with both permeability and permittivity less than zero can be achieved by controlling the electric and magnetic responses of the material. Such materials are seldom naturally observed because the frequency range at which the negative permeability and negative permittivity are realized is not only very narrow but also far away from each other [7].

At microwave frequencies, the first realization of double negative materials was realized by the combined activities of both cut wires and split ring resonators to provide permittivity and permeability values that are negative [8]. These double-negative materials have sparked widespread interest in metamaterials because of their odd and controllable behaviour, and they have also become the focus of metamaterial research operating in the terahertz and visible regions. The next section provides an overview of the properties of metamaterials.

1.1.3.1 Epsilon-Negative Metamaterials

Epsilon-negative metamaterials (ENG) are materials with ε negative and μ positive. ENG metamaterials are obtained by exposing metallic mesh of thin wires that show high pass filter properties to the application of a parallel electric field. Below the plasma frequency, the cylindrical arrays show negative permeability. These metal structures are organized into periodic patterns as in Figure 1.2(a). Figure 1.2(b) provides the single unit cell structure of the thin wire, and Figures 1.2(c) and (d) provide the effective permittivity graph and LC equivalent circuit of the thin wire metamaterial [4].

$$\varepsilon_{eff} = 1 - \frac{\omega^2{}_p}{\omega(\omega + j\Gamma)} \tag{1.2}$$

ω - propagation frequency of the electromagnetic wave
ω_{ep} - electric plasma frequency
Γ - reflection coefficient

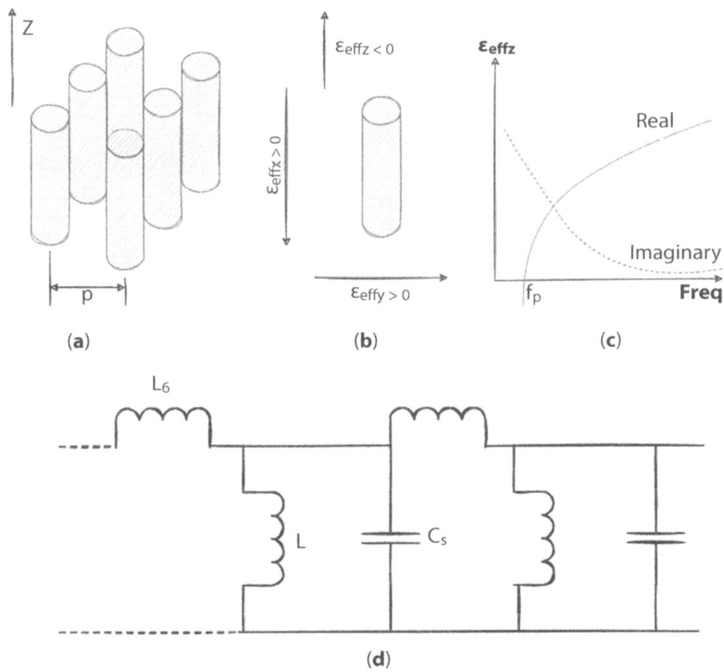

Figure 1.2 (a) Array of thin conducting metallic wires, (b) single unit cell, (c) effective permittivity plot of metallic wire array and (d) equivalent transmission circuit model. Reproduced with permission from K. Wojciech Jan and C. Thanh Nghia, in Metamaterials and Metasurfaces, ed. C.-F. Josep, IntechOpen, Rijeka, 2018, Copyright@2018 Wojciech Jan *et al.*

From Equation 1.2 the permittivity is negative for propagation frequencies lower than the plasma frequency (ω_p) and exactly at ω_p, the permittivity becomes zero. Interested readers can refer to published works for the detailed calculation of effective plasma frequency [9–11].

1.1.3.2 Mu-Negative Metamaterials

Mu-negative metamaterials (MNG) are materials with ε positive and μ negative. The Split Ring Resonator (SRR) has been the most preferred structure for mu-negative (MNG) material. The SRR unit cell is made of two circular metallic rings separated by a small gap. As shown in Figure 1.3(a), each ring has a slit on each side that is 180 degrees apart. From Figure 1.3(b), it can be inferred that the gap between rings serves as a capacitor, and the rings serve as inductors when current is passed in accordance with Faraday's law of electromagnetic induction. As a result, the two rings work together to form an LC resonance circuit.

$$\mu_{eff} = \mu'_{eff} - j\mu'_{eff} = 1 - \frac{\omega_{mp}^2 - \omega_0^2}{\omega^2 - \omega_0^2 - j\gamma\omega} \tag{1.3}$$

ω - frequency of the signal
ω_{mp} - magnetic plasma frequency
ω_0 - magnetic plasma divergence frequency
γ - losses

The geometry of the material determines the frequencies f_0 and f_{mp}. The effective permeability plot is shown in Figure 1.3(c) and the equivalent resonance circuit is shown in Figure 1.3(d).

1.1.3.3 Double-Negative Metamaterials

Negative refractive index material (NIM), or DNG material, are materials with both ε negative and μ negative. The DNG metamaterial capabilities can be attained by merging the thin wire-based ENG structure with the SRR-based MNG structure, as shown in Figure 1.4. Both ε negative and μ negative are satisfied by the combination of SRR and wire medium that serves as an artificial dielectric medium. Additionally, to take advantage of both sides of the dielectric layer, 2D metamaterials have been constructed by etching SRR on one side and a planar strip on the other.

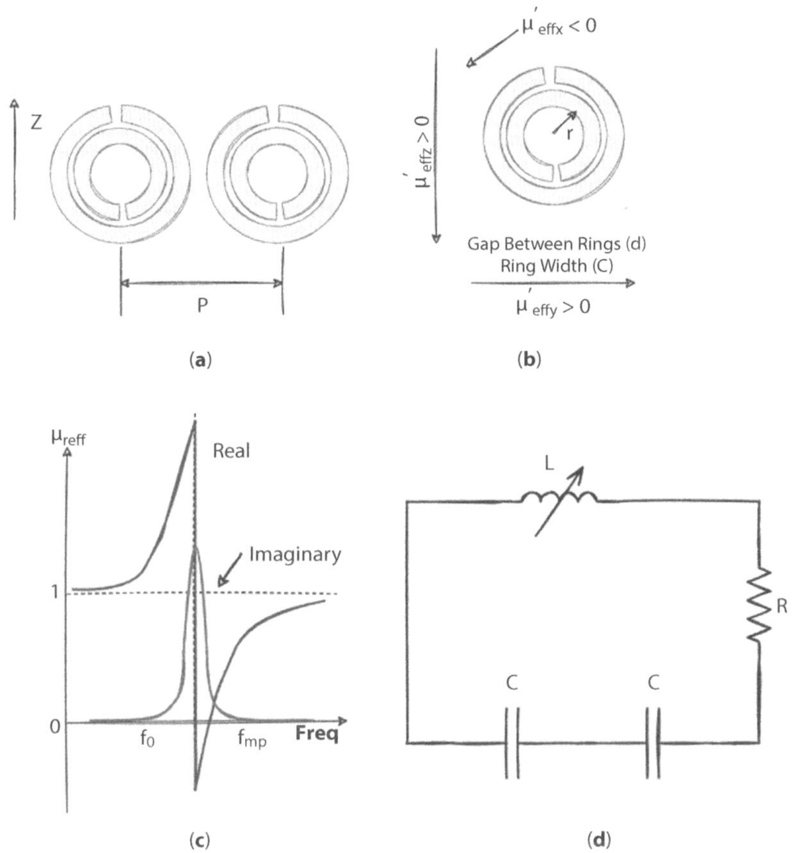

Figure 1.3 (a) Array of SRR, (b) single SRR unit cell, (c) effective permeability plot of SRR array and (d) equivalent transmission circuit model. Reproduced with permission from K. Wojciech Jan and C. Thanh Nghia, in Metamaterials and Metasurfaces, ed. C.-F. Josep, IntechOpen, Rijeka, 2018, Copyright@2018 IntechOpen Limited [4].

DNG metamaterials are highly frequency dependent due to their strong resonance behaviour, causing the refractive index to be rewritten as,

$$n = n_{eff}(\omega) = \sqrt{\varepsilon_{eff}(\omega)\mu_{eff}(\omega)} \tag{1.4}$$

Equation 1.4 shows the dependency of effective permittivity and permeability on frequency (ω), which in turn makes the refractive index a

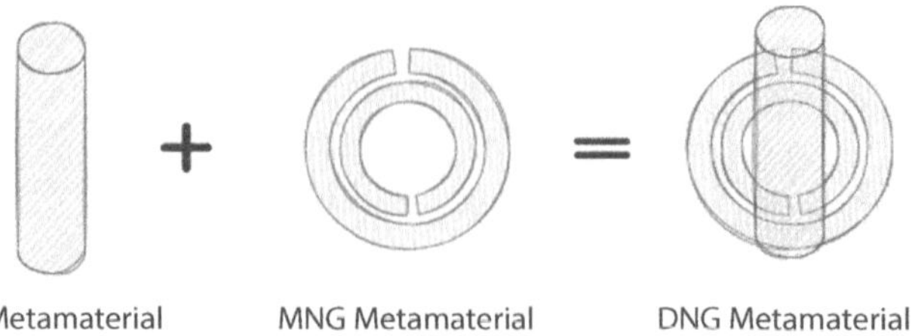

Figure 1.4 Combination of thin metallic strip wire and SRR to form double negative material.

function of frequency. Effective material permittivity and permeability can be obtained by Drude-Lorentz dispersion models [2]. Equation 1.5 and Equation 1.6 are plotted in Figure 1.5 [12].

$$\varepsilon_{eff}(\omega)=1-\frac{\omega^2{}_{ep}-\omega^2{}_{eo}}{\omega^2-\omega^2{}_{eo}-j\omega\gamma_c} \tag{1.5}$$

$$\mu_{eff}(\omega)=1-\frac{F\omega^2}{\omega^2-\omega^2{}_{mo}-j\omega\Gamma} \tag{1.6}$$

ω_{ep} - electric plasma frequency
ω_{mp} - magnetic plasma frequency
ω_{eo} - electric resonance frequency
ω_{mo} - magnetic resonance frequencies
γ_c - collision frequency
F - amplitude factor
Γ - damping factor

Both the negative sections of the ENG ($\omega_{eo} < \omega < \omega_{ep}$) and MNG ($\omega_{mo} < \omega < \omega_{mp}$) structures must coincide to create the DNG materials. For the calculation of the amplitude and damping factor, the readers are suggested to refer to [13].

These double-negative materials have gained significant traction among researchers in recent years. These double-negative materials are also known as left-handed materials (LH). These LH can be modelled as transmission line (TL) models enabling easy physical realization. In practical cases, these materials have left-handed material characteristics in some frequency

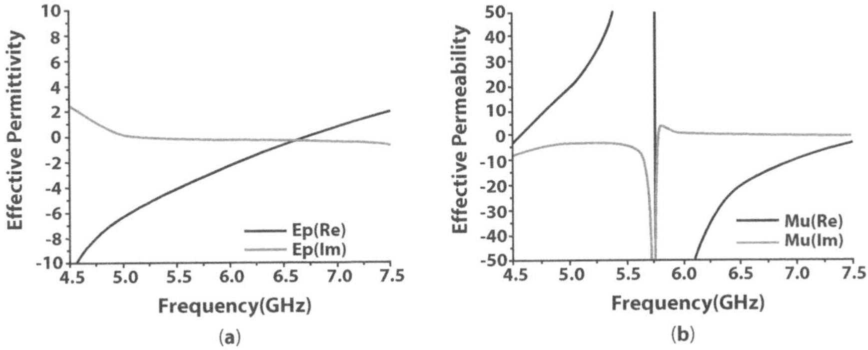

Figure 1.5 Plots of effective permittivity and permeability of NIM materials as a function of frequency. Reproduced with permission from S. S. Islam *et al.*, An Object-Independent ENZ Metamaterial-Based Wideband Electromagnetic Cloak, Scientific Reports, 2016, 6, 33624. Copyright © 2016 Springer Nature [12].

ranges and right-handed material properties (natural electromagnetic materials) in others. These materials are called CRLH (Composite Right/Left Hand) materials. This term was coined by Caloz *et al.* [14]. In the next section, various design approaches of CRLHs are summarized.

1.2 Design Approach

The performance of metamaterials may be influenced by several factors. Metals and dielectrics, which are prevalent in metamaterials, play a significant role in the dissipation of energy. There are three design approaches widely employed to achieve the desired metamaterial characteristics.

1.2.1 Resonant Approach

In the resonant approach, the LH material combines two distinct materials to form its unit cell. One material has negative permittivity and the other has positive permittivity. This approach is derived from the application of SRR [15]. Metamaterial designed using a resonant approach reacts to external factors like a magnetic field, electric field, and voltage and exhibits LH material characteristics in the vicinity of the resonant frequency. An SRR subjected to an axial magnetic field exhibits positive effective permeability above the narrow band at the quasi-static resonant frequency and negative effective permeability below the narrow band at the quasi-static resonant frequency.

A material loaded with an array of SRR usually forms a negative refractive medium exhibiting extreme stopband characteristics. The closer the SRR rings, the higher the coupling and the better the stop-band characteristics. Due to its ability to reject frequencies near the resonant frequency, the SRR approach is used in the fabrication of filters, frequency-selective surfaces, etc. Metamaterials formed using a resonant approach cannot be used for wider frequency ranges [11].

1.2.2 Non-Resonant Approach

The non-resonant approach is also called the transmission line (TL) approach. It is the dual of the RH transmission line theory [16]. In this case, the capacitors are connected in series and the inductors are placed in parallel. Like the resonant approach, the material designed exhibits effective negative permittivity and negative permeability at certain frequency ranges. The RH materials act as low-pass filters, whereas the LH materials act as high-pass filters [15]. Homogenous CRLH materials are not naturally occurring but can be created by cascading multiple CRLH unit cells such that their dimension is sufficiently small and both effective permeability and effective permittivity can be calculated. Unlike the resonant approach, the TL approach can be applied to a wide range of frequencies with materials that are significantly smaller. But, TL-based materials have relatively high insertion loss that can be overcome by using novel unit cells like ForeS [17] and S-spiral unit cells [18].

1.2.3 Hybrid Approach

The hybrid approach combines both the resonant approach and the non-resonant approach. Here the unit cell contains SRR & CSRR (resonant approach) on one side and gaps and shunted stubs (TL approach) on the other side. The SRR and gaps provide negative permeability, and CSRR and shunted stubs produce negative permittivity around the resonant frequency. The material formed usually has band-pass characteristics. The metamaterial unit cells formed using this approach are not pure LH material due to the pass band characteristics. In this approach, LH material can be designed using more than two materials, such that third material can be used to control and tune the effective permittivity and permeability of the material [18].

1.3 Applications

Metamaterials can be integrated with a wide range of devices like antennas, absorbers, switches etc. This chapter will be focused on summarizing the metamaterial application to improve antenna performance and provide a summary of various types of metamaterial-based absorbers.

1.3.1 Metamaterial Absorbers

Metamaterials can absorb electromagnetic radiation such as light, giving them greater adaptability and effectiveness than traditional absorbers. These electromagnetic metamaterials are used to create a high electromagnetic absorber as well as to manipulate the loss components of effective parameters. The scarcity of strong terahertz properties in natural materials has provided a significant impetus for driving the research area of terahertz metamaterials. This section is a brief introduction to various metamaterial-based absorbers.

The two most important parameters in the absorption mechanism are the reflected power S_{11} and transmitted power S_{21}. Both the reflected power and transmitted power should be minimized simultaneously, to achieve maximum absorptivity (A). The absorptivity is given by Equation 1.7.

$$A = 1 - |S_{11}|^2 - |S_{21}|^2 \tag{1.7}$$

Most of the absorber designs are backed by complete metal lamination, thereby ensuring zero transmission ($S_{21} = 0$). The design is made in such a way that the input impedance is matched exactly with the free space impedance S_{21}.

Metamaterial absorbers can be used to regulate effective electromagnetic parameters. Metamaterial absorbers can be easily fabricated and scaled from microwave, terahertz, and infrared, to optical regions. There are two types of absorbers: dielectric-based absorbers and magnetic absorbers. Dielectric-based absorbers are good at absorbing electric fields, while magnetic absorbers absorb magnetic fields. To understand the working of absorbers, consider a simple dielectric absorber acting as an insulator placed in an electric field. The electric field generates static electric charges that give rise to dielectric polarization. This induced dielectric polarization causes an internal electric field. When an EM wave is incident on the surface of the dielectric, the wave gets attenuated and energy loss occurs because of thermal energy due to the phase cancellation between the incident electric field and the internal electric field. Correlating to the transmitted power S_{21}, the lower the transmission coefficient, the better the attenuation property of the dielectric medium. So, it can be concluded that the amount of absorptivity (A) indirectly depends on the frequency, electrical permittivity, and magnetic permeability. Metamaterial absorbers are ultra-thin. The first metamaterial structure was reported by Landy *et al.* in 2008 [19]. Metasurface-based absorbers can be classified into two types: passive absorbers and active absorbers. Based on the bandwidth and ranges of operating frequency, there are various types of passive absorbers such as passive single/multi-band metamaterial absorbers [20], passive bandwidth-enhanced broadband absorbers [21] and broadband absorbers [22].

Conventional passive absorbers offer no flexibility after fabrication. Active absorbers can be re-used by controlling them with external stimuli. External stimuli can be achieved magnetically, mechanically, or electronically. Electronically controlled absorbers use active components, such as PIN diodes (PN junction diodes with an intrinsic layer) and varactor diodes. Those components offer high tuning speed, wideband tuning, compact size, and low cost. There are several types of active absorbers; they are switchable absorbers-reflectors, switchable absorbers, and tuneable absorbers. These switchable absorbers exhibit two bands of operation depending on the external stimuli [23].

1.3.1.1 Switchable Absorbers-Reflectors

A switchable absorber-reflector exhibits absorption and reflection responses in two different conditions, regulated by external control. External control can be achieved by two fabrication techniques that are used to provide switching operations; they are PIN diodes and varactor diodes.

A DC voltage source needs to be connected across the PIN diode, such that the diode can be ON or OFF, by regulating the external voltage levels. [24] Note that the biasing of a single metamaterial is easy but, in a metasurface absorber structure, unit cells are arranged in a periodic pattern. The most crucial factors in the case of metamaterial absorbers are the biasing networks that generate internal electric or magnetic fields, causing dielectric or magnetic polarization. Figure 1.6(a) shows the incorporation of PIN diodes in the metasurface unit cell and Figure 1.6(b) shows the equivalent electrical model of the structure.

The absorbers are further classified into two types based on their band of operation; they are narrow band and broadband absorbers. Broadband absorbers operate over a wide range of frequencies, and narrow-band absorbers, as the name suggests, operate over narrow frequency ranges.

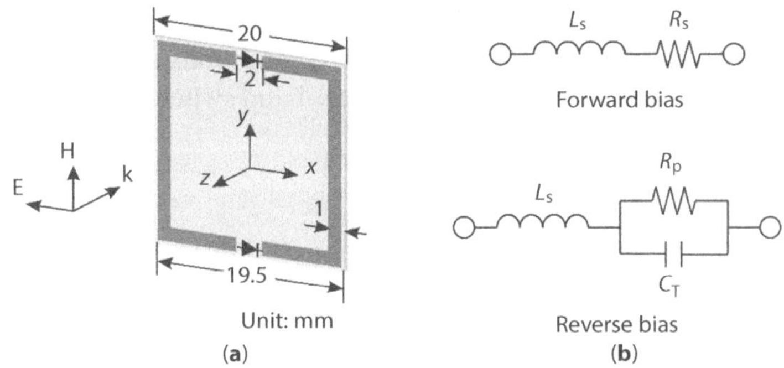

Figure 1.6 (a) Metasurface with two PIN diode connected in-between the gaps. (b) equivalent transmission line model of metamaterial for the ON and OFF states of the PIN diode. Reproduced with permission from Chaimool *et al.*, Design of a PIN Diode-Based Reconfigurable Metasurface Antenna for Beam Switching Applications, International Journal of Antennas and Propagation, 2019, 7216324. Copyright © 2019 Sarawuth Chaimool *et al.* [67].

An absorber is said to be polarization insensitive if its absorption property at different polarization angles is the same. Based on the range of operation and polarization, the switchable absorber-reflectors are classified into polarization-insensitive switchable absorber-reflectors, broadband polarization-sensitive switchable absorber-reflectors, and broadband polarization-insensitive switchable absorber-reflectors. By modeling the metasurface absorbers as transmission models and extending the structure by mounting multiple lumped resistors in the structure, one can increase the range of operation and the switching can occur at multiple bands over the range.

1.3.1.2 Switchable Absorbers

A switchable absorber exhibits a response that can be switched between two frequency bands by an external stimulus. PIN diodes are mostly used for generating the switching operation [25]. The absorption is happening in both states, but the frequency is changing due to the diode action. Like switchable absorber-reflectors, switchable absorbers are classified based on bandwidth and polarization sensitivity. They are polarization-independent single-band switchable absorbers based on a circular split ring and cross-dipole geometry, and polarization-independent dual-band switchable absorbers. In the circular split ring geometry, the diodes are in series connection, so there will be a large supply of voltage that is required to turn ON the diodes, but in the case of cross-dipole geometry the PIN diodes are in parallel connection, so the overall supply voltage is small [26]. Combining the circular split ring and cross dipole designs can give resonance conditions at four frequencies. These is called polarization-independent dual-band switchable absorbers [27].

1.3.1.3 *Tuneable Absorbers*

A tuneable absorber exhibits a variable absorption frequency that can be regulated by an external bias voltage. The structure has an infinite number of states. Most of the tuneable absorbers are realized by varactor diodes. A varactor diode, while operated under reverse voltage, exhibits tuneable junction capacitance, represented by Equation 1.8 [28].

$$C_J = \frac{C_{JB}}{\left(1+\dfrac{V_R}{V_J}\right)^M} + C_P \tag{1.8}$$

The varactor-modelled tuneable absorbers are polarization sensitive in nature. Depending on the placement of the varactor diode, a polarization-independent single-band tunable absorber can be achieved. Note that, unlike PIN diodes, the varactor diodes operate under reverse bias voltage, and the varactor diodes should be placed in a parallel combination to reduce the voltage requirement. Embedding the polarization-independent single-band tuneable absorber with an internal biasing network provides an increase in the tuneable frequency. Using the biasing network, complexity can be reduced and the structure can be expanded to dual or multi-band tunable absorbers [29].

1.3.2 Metamaterial Antenna

Metamaterial antennas are antennas wherein metamaterials are utilized to increase the antenna system's performance. A metamaterial coating is applied to improve the radiation and increase the radiating power. When utilized in an antenna system, metamaterials with negative permeability may allow for qualities like high directivity, superior operational frequency, and increased radiated power. This increases bandwidth range by using metamaterials as a superstrate over traditional patch antennas. High-directivity antennas are frequently made from materials with a refractive index of zero. Using the unusual properties of metamaterials, antennas with compact size and improved performance can be obtained.

Primarily, unit cells are made to interact with the antenna structure to obtain the desired antenna performance. There are many types of metamaterial incorporation techniques that are implemented to achieve the desired antenna performance. Listed below are a few:

1. Metamaterial Loading: Metamaterials made of periodic subwavelength unit cells loaded into antennas.
2. Metamaterial Inspired Antenna: In this approach, the meta resonators are loaded into the structure of the antenna. The loading may involve two or multiple meta-resonator unit cells.
3. Metasurface Loading: Artificial thin layers made of subwavelength periodic elements in a uniform or non-uniform pattern are arranged to form planar 2D surfaces.
4. Resonant Dispersion: Fractal geometries of RH material and LH material-based unit cells are incorporated into the proximity of the antenna.

Sections 1.3.2.1 to 1.3.2.4 provide a brief description of various improvements achieved in antenna performance by loading metamaterial. Though there are numerous benefits to incorporating metamaterials, this chapter will only summarize the reduction in antenna size, improvements in gain and bandwidth, generating circularly polarized electromagnetic waves via antenna structures from the linearly polarized antenna, and finally achieving isolation between antenna arrays.

1.3.2.1 Miniaturization

(1) Metamaterial Loading
Growing technology demands compact antennas with more accurate resonant conditions. Electrically small antennas (ESA) with such requirements are hard to achieve because as the antenna size decreases, the operating wavelength also decreases, causing an imbalance in the impedance [30]. One of the possible solutions is integrating the antenna with periodic subwavelength metamaterial unit cells. Various metamaterial techniques include ENG/MNG materials, high permittivity or permeability shells, photonic crystals, and near-field resonators [31–36].

The primary goal of antenna miniaturization is to make the antennas' physical dimensions independent of the loading. According to S.K. Sharma *et al.*, this can be achieved by using a meta resonator such as a zeroth-order resonator (ZOR) [37]. The ZOR provides a zero-phase constant and an infinite wavelength at frequencies other than zero. This enables antenna miniaturization just by tuning the distributed lumped elements like capacitors or inductors. These ZORs are made of ENG/MNG or CRLH TL models [38].

(2) Metamaterial Inspired Antenna
As mentioned above, the unit cells are structured in an array configuration to show ENG, MNG, or DNG characteristics. In contrast to the previous technique of metamaterial loading, where the metamaterials are loaded near the device, in this technique, metamaterial-inspired antennas, the metamaterials are used as a fabrication component. In this method, the unit cells are engraved into the antenna patch, ground or in the substrate, making them a more integral part of the antenna fabrication material [5]. These integrated metamaterials act as resonant structures and, by controlling the resonating components, size reduction, increased gain, and good impedance matching are reported in R. K. Saraswat *et al.* [39].

(3) Metasurface Loading
There are various types of metasurfaces: high impedance surface (HIS), reactive impedance surface (RIS), frequency selective surface (FSS), and electromagnetic bandgap (EBG).

HIS is a periodic arrangement of sub-wavelength structures atop a dielectric medium with a metal ground plane on the other side [40]. HIS materials suppress surface waves at other frequencies and have zero to low phase reflection (θ) at specific frequencies. At zero-phase reflection, the surface becomes an artificial magnetic material (AMM). This AMM functions as an artificial ground, doubling the fields while reducing the antenna thickness. It is suggested that refer to [40] for the formula of phase reflection.

RIS is like HIS but can store magnetic energy or electric energy and reflects power like ideal magnetic or electric conductors [41]. RIS structures reduce the size by improving the propagation constant, leading to a decreased wavelength at the resonant frequency [42]. RIS also provides isolation between the antenna and dielectric, thereby decreasing impedance mismatch [43]. Both FSS and EBG metasurfaces operate like RIS and HIS.

(4) Resonant Dispersion
As mentioned above, CRLH metamaterial unit cells are loaded for antenna miniaturization. These CRLH unit cells act as internal load and adjust the capacitance and inductance values of the CRLH TL model, contributing to compact-size antennas.

An important point to note from (1) to (4) is that all the size reduction is achieved by altering the reactive elements of the TL model of the metamaterial unit cell. The adjustment of the reactance, inductance, and capacitance of the metamaterials makes the antenna resonate at a lower frequency than its intended frequency of operation, resulting in reduced antenna size.

(5) Metamaterial for Antenna in Terahertz Region

So far, the general idea of metamaterial inclusion in antennas has been discussed briefly. Now, let us extend the discussion to terahertz region application by considering an antenna with dimensions of 131.5 μm x 155 μm x 10 μm and a patch with dimensions of 71.5 μm x 95 μm on dielectric made of quartz with a relative permittivity of 3.78 and tanδ (loss tangent) of order 10-3. The antenna resonates at 0.97 THz without the inclusion of metamaterials with an S11 of -28.76 dB and a fractional bandwidth of around 4.12% [44].

The most commonly used metamaterial is circular SRR. It is used to study the improvements achieved in the patch antenna proposed. The SRR used as loading has a 20 μm outer radius and a 15 μm inner radius with a 5 μm split width and a 5 μm ring thickness. For the calculation of the permittivity and permeability of SRR, refer to [45, 46]. The SRR is incorporated into the dielectric of the terahertz antenna.

The SRR resonance frequency is adjusted by controlling the capacitance (C) due to the split in the rings and the inductance (L) in the two rings due to Faraday's law. The structure of SRR is shown in Figure 1.3a. Note that inductance is varied by varying the dimension of the ring and capacitance is varied by altering the split size. By modeling SRR as an LC resonant element and controlling the inductance and capacitance, it results in negative permeability or even zero refractive indexes at or near the resonant frequency [47]. The proposed antenna with SRR loading on the substrate showed an increase in the resonant frequency from 0.97 THz to 1.02 THz because the designed SRR metamaterial resonates at 1.02 THz. Also, substrate size was reduced from 131 x 155 μm^2 to 128.5 x 150 μm^2 patch size was reduced from 71.5 x 95 μm^2 to 68.5 x 90 μm^2. As mentioned above, considerable size reduction can be achieved just by adjusting the capacitance and reactance of the metamaterials.

1.3.2.2 Gain and Bandwidth Improvement

In recent years, along with the downsizing of antennas, there has been a huge demand for antennas with a wider bandwidth of operation and improved gain in wireless communication. Compact antennas suffer from low gain and narrow bandwidth due to weak radiation characteristics. Although different methods and strategies are suggested across the literature, considering various other methods, the most cost-efficient and least complex method is introducing metamaterials.

(1) Metamaterial Loading

The core idea behind achieving improved gain and bandwidth by using metamaterials is to control the electromagnetic waves at the surface by modifying the antenna structure to store less power and radiate more power. So, to achieve high radiated power, the metamaterial unit cells loaded into the antenna should decrease the reactance of the antenna.

For example, Yuan *et al.* proposed an antenna with a planar metamaterial structure incorporated into the ground plane, offering discontinuities such that the antenna stores less energy and radiates more power, providing an increased gain and wider bandwidth of operation [48]. A similar technique is suggested by Urul *et al.* using DNG metamaterial in conventional microstrip antennas [49]. DNG metamaterial reduces impedance mismatch and reduces the reactance of the antenna structure. This allows the antenna to radiate high power, thereby achieving improved gain (4.36 to 7.45 dB) and bandwidth.

(2) Metamaterial Inspired Antenna

Metamaterial deployment in the fabrication process has significantly improved the performance of planar antennas. The most deployed metamaterial structures are the SRR and complementary SRR (CSRR). Like metamaterial loading, this technique also focuses on controlling the electromagnetic waves at the surface and increasing the radiated power.

For example, Patel *et al.* proposed an antenna structure deployed with corrugated and noncorrugated SRR [50]. The proposed microstrip antenna with SRR achieved an improved gain from 6.2 dB to 7dB and improved bandwidth from 230 MHz to 430 MHz.

(3) Metasurface Loading

Unlike metamaterial loading and metamaterial-inspired antennas, where the metamaterial just provides improvements by changing the surface wave characteristics of the antenna, here the metasurfaces act as an aperture for the antenna, radiating or receiving EM waves from or into the feed, playing an integral part of the antenna function. But, like the previous techniques, metasurfaces also regulate surface waves to increase radiated power.

For example, consider the well-known metasurface EBG in antenna design. EBG acts as a transmission or reflection surface and suppresses surface waves so that high gain can be achieved. P. Sambandam *et al.* proposed a monopole antenna whose radiating patch is partially integrated with EBG, which has a significant increase in bandwidth (3.1-10.6 GHz) and a peak gain of 6.25 dB [51].

(4) Resonant Dispersion
CRLH metamaterial can also be used to improve antenna performance. The presence of a zeroth-order resonator improves the bandwidth of the antenna when the CRLH unit cell is loaded. With the right tweaking of the parallel and series inductance and capacitance of the LC equivalent circuit of the CRLH transmission line model, antenna bandwidth can be increased. Similarly, integrating SRR and CSRR with CRLH can boost the gain of the antenna [52–58].

Incorporating metamaterials into antenna structures decreases the quality factor (Q) of the resonating antenna structure. It is well known that the bandwidth of an antenna is inversely proportional to the quality factor. This justifies the core idea of achieving high gain and enhanced bandwidth because, with a reduced quality factor, the antenna radiates more power, thereby increasing the gain.

For the readers convenience, various improvements like gain and bandwidth enhancements and miniaturization of antennas achieved by the inclusion of various metamaterials in the antenna structure are summarized in Table 1.1.

(5) Metamaterial for Antenna in Terahertz Region
The metamaterial loading technique can be extended to the terahertz region without any difficulty because THz waves have shorter wavelengths and require miniaturized antennas to transmit. With the inclusion of metamaterials, conventional antennas can be made to resonate at lower frequencies than their intended frequency, and vice versa, because the resonant frequency is inversely proportional to capacitance and inductance.

Considering the same patch antenna with SRR inclusion discussed previously, the gain of the antenna increases from 1.44 dB to 5.75 dB. It can be seen that metamaterial loading helps the antenna to radiate maximum power in the desired direction. As a result, the directivity and gain are increased because the SRR loaded in the substrate acts as a reflecting surface and suppresses the antenna's surface wave. By suppressing the surface waves, the S11 (return loss) of the antenna increased from -28.75 dB

Table 1.1 Influence of metamaterials on antenna size, gain, and bandwidth of antenna.

Ref	**Metamaterial technique**	**Operation frequency (GHz)**	**Size reduction (%)**	**S11 BW (%)**	**Gain (dB)**
L.-Y. Liu *et al.* 2015 [30]	ENG-TL metamaterial with circular shaped closed complementary resonator	1.16, 2.32, 3.58	-	8.54, 10.25, 35.75	1.59, 2.1, 4.97
Gupta *et al.* 2019 [68]	Split ring resonator (SRR)	2.78, 6.02	-	3.59, 25.41	-
Rao *et al.* 2018 [69]	Complementary SRR	1.8	48	-	7
Gudibandi *et al.* 2020 [36]	HRI (metasurface loading)	2.5	45.7	-	-1.4
Hong *et al.* 2019 [70]	FSS (metasurface loading)	1.59, 1.96	-	-	-
Gupta *et al.*2017 [35]	CRLH (Resonant method)	1.62, 2.78	51.9	2.46, 1.07	1.05, 2.59

to -65.26 dB and the VSWR ratio also decreased from 1.07 to 1.00. In the metamaterial loaded patch antenna considered, there was no enhancement in bandwidth reported. The fractional bandwidth remained constant at 4.12 %. Note that the bandwidth can also be increased by decreasing the quality factor as mentioned above.

1.3.2.3 Circular Polarization

The polarization of an antenna is a crucial parameter in determining the performance of the antenna. Depending on the requirements, the antenna is either linearly polarized or circularly polarized. Most of the fabricated antennas are linearly polarized. In a linear polarized (LP) EM wave, the fields and direction of propagation are all mutually perpendicular. In a circular polarized (CP) EM wave, the electric field lags or leads the magnetic field by 90 degrees and appears to be rotating. If the wave rotates in the anti-clockwise direction, it is said to be right-hand circular polarization (RHCP) and left-hand circular polarization (LHCP) if it rotates in the clockwise direction.

CP provides several advantages, including improved absorption and reflectivity; reduced phasing difficulties; fading and multipath interferences; protection against weather changes; and providing flexibility in antenna orientation. Two parameters are used for determining circular polarization: (1) axial ratio (AR) and (2) surface current distribution. For an ideal circularly polarized antenna, the AR should be unity or 0 dB. This is not applicable in practical situations; therefore, as a rule of thumb, for a good CP antenna, the AR should range from 1 to 2.5. Various techniques and arrangements are reported in the literature to achieve CP. This section details the role of metamaterials in achieving circular polarization.

(1) Metamaterial Loading

As mentioned above, the important factor in achieving circular polarization is creating the phase difference. This phase difference can be achieved by incorporating metamaterial structures in association with an antenna. These metamaterials feature resonance qualities, which allow the antenna's inductance and capacitance to be tuned. As a result, an antenna loaded with a metamaterial structure produces a variation in effective reactance and modifies the surface current distribution, which aids in the achievement of circular polarization in antennas. It should be noted that RHCP or LHCP can be produced by appropriately orienting the metamaterial unit cell.

(2) Metamaterial Inspired Antenna
Circular polarization is depending on the geometry of the antenna and the elements attached to it. M. Venkateshwara Rao *et al.* postulated, by utilizing ring resonators like SRR or CSRR metamaterials on antennas the effective inductances and capacitances can be altered such the antenna becomes circularly polarized [59].

(3) Metasurface Loading
Materials like HIS, RIS, EBG, FSS, etc., can also be used to achieve circular polarization. For example, J. Chatterjee *et al.* [60] verified that using RIS between the radiating plane and the ground plane reduces mutual coupling and increases current distribution consistently. At a certain frequency, the RIS surface inductance cancels out with a capacitive effect on the radiating patch, resulting in the antenna radiating a circularly polarized wave with a wider AR bandwidth.

(4) Resonant Dispersion
A CRLH made of a meander (zigzag) line produces the capacitance effect, and a short-circuited stub produces the inductance effect. As previously stated, the polarization characteristics of an antenna are governed by its surface current. According to L. Y. Liu *et al.* [61], an antenna with two uneven meander lines placed between antenna patches acts as a parallel inductance and a pair of slots etched on the antenna patch acts as a series capacitance. The current distribution is disturbed by the newly added capacitance and inductance, which result in two orthogonal operational modes with a phase difference creating a circular polarized antenna.

(5) Metamaterial for Antenna in Terahertz Region
Creating a circularly polarized antenna structure using SRR loading relies heavily on Faraday's law of electromagnetic induction. Faraday's law states that a current will be induced in a conductor which is exposed to a changing magnetic field and will be in a direction such that it opposes the changing magnetic field that produces it. Implementing SRR in the substrate of the patch antenna under study uses the converse of Faraday's law; that is, by altering the current distribution, the antenna operating at 1.02 THz can produce circularly polarized electromagnetic waves by creating a phase difference of 90 degrees between electric and magnetic fields. Multipath effects can be reduced with this circularly polarized terahertz patch antenna [62].

1.3.2.4 *Isolation*

Mutual coupling between adjacent antennas causes a reduction in diverse performance parameters such as antenna gain and reflection coefficient. The mutual coupling also affects the transmission capacity of the wireless channel and decreases the multiplexing efficiency.

In a MIMO system, if the antennas are placed less than λ /4, high mutual coupling occurs between adjacent antennas, causing performance degradation. Among various methods reported in the literature, metamaterials are the most viable solution for suppressing mutual coupling. Metamaterial structures are used in MIMO systems to break up the flow of current and serve as a decoupling component between adjacent antennas, resulting in isolation. Controlling the magnetic field inside the exciting antenna enables maximum isolation.

(1) Metamaterial Loading
Each metamaterial unit cell can be modelled into its equivalent TL model. This is the important analysis and design technique that has come in handy throughout the application section. The same is the case for achieving isolation. Metamaterial unit cells that are modelled in terms of LC components can also act as filters with band-stop characteristics. This band-stop characteristic of metamaterials has proved to reduce coupling between adjacent antennas. For example, R. Mark *et al.* proposed that MIMO antenna systems with circular-shaped near-zero permittivity and permeability metamaterial, offer decoupling between adjacent antennas and have a maximum isolation of 41 dB [63].

(2) Metamaterial Inspired Antenna
The CSRR structure incorporated into the antenna design significantly reduces mutual coupling. For example, V. Najafi *et al.* [64] proposed that Vivaldi antennas with CSRR placed between antenna elements, due to their band stop characteristics, reduce mutual coupling and increase isolation between two Vivaldi antenna arrays.

(3) Metasurface Loading
The Metasurface EBG structure is made of a series of subwavelength structures. At the operating frequency, these structures act as a band-stop filter, confining the surface waves within the antenna. MIMO systems with split EBG structures achieve a higher order of isolation, according to X. Tan *et al.* [65].

Table 1.2 Metamaterial approaches to achieve enhanced isolation and respective isolation achieved in dB.

Reference	Metamaterial approach for increased isolation	Isolation achieved (dB)
Shabbir *et al.* 2020 [71]	ENG metamaterial loading	>20
Reddy *et al.* 2021 [72]	Metamaterial loading	>25
Irene *et al.* 2018 [73]	SRR metamaterial inspired antenna	>20
Ghosh *et al.* 2020 [74]	HRI (metasurface loading)	10
Zhu *et al.* 2017 [75]	FSS (metasurface loading)	>16
Thakur *et al.* 2020 [76]	CRLH (resonant method)	35

(4) Resonant Dispersion
A unique approach is suggested by A. A. Ibrahim *et al.* [66]. The MIMO system with a CRLH structure made of a meander line and short-circuited stub achieved ZOR by altering the dimension of the CRLH unit cell such that the structure provided reverse currents at the resonant frequency. These reverse currents tend to excite the antenna, thereby increasing the isolation between multiple antennas packed closely together.

(5) Metamaterial for Antenna in Terahertz Region
Though isolation is not explicitly verified for the patch antenna under study, it is important to note that isolation can be achieved between an array of antennas even in the terahertz region by utilizing the band stop characteristic of the SRR loaded into the substrate of the antenna. Table 1.2 provides various metamaterial techniques used and the isolation achieved by including them in the antenna structure.

1.4 Conclusion

This chapter provided an overview of the implications of fabricating devices in the terahertz region and suggests a novel artificial engineering material to overcome the constraint. Such materials are also called "metamaterials". Depending on the nature of the permittivity and permeability, the metamaterials are classified into epsilon-negative metamaterials, mu-negative-metamaterials, and double-negative materials.

Using the unit cell of ENG, MNG or DNG metamaterials as transmission models and altering the effective reactance and dimension after incorporation with antenna structures provided various improvements in the antenna performance, such as high gain and enhanced bandwidth, generation of circular polarization, making the antenna independent of orientation, thereby reducing cross polarization loss; increased isolation between adjacent antennas; and finally, structure improvement includes miniaturization of the antenna. Described below are the following properties exploited to achieve the improvements.

1. In the miniaturization of the antenna, the reactance of the metamaterials is tuned such that the antenna resonates at a frequency lower than the intended frequency.
2. Gain is enhanced by increasing the radiated power by controlling the electromagnetic surface waves using metamaterials.
3. Bandwidth is increased by incorporating metamaterials, which reduce the quality factor, which in turn increases the bandwidth and the radiation efficiency.
4. Circular polarization is achieved by adjusting the reactive elements of the metamaterials such that the current distribution is perturbed, providing a phase difference of 90 degrees.
5. Isolation is achieved by using the band stop characteristics of the metamaterial, which in tweaking offers zero permittivity and permeability at the frequency of operation of the antenna.

Also, the application section provides a short overview of various metamaterial/metasurface-based absorbers fabricated with the incorporation of PIN diodes and varactor diodes for the generation of dielectric and magnetic polarization. In the case of absorbers in the terahertz region, the main property exploited is described below.

Integration of PIN diodes and varactor diodes in the metasurface and exciting them using a biasing network produces static electric charges. These static charges provide an internal electric and magnetic field. When a terahertz wave is an incident on the metasurface, the electric field is attenuated and dissipated in the form of thermal energy due to destructive interference between the internal electric/magnetic field and the incident electric/magnetic field. Various types of metasurface absorbers, such as switchable absorbers-reflectors, switchable absorbers, and tuneable reflectors, were also briefed based on their polarization sensitivity and operation bandwidth. Also, as far as the application in the terahertz region is concerned, the metamaterial inclusion can be extended without any difficulty because metamaterials are one of the best solutions to overcome the difficulty in realizing devices in the terahertz region. The metamaterials can also be used for a wide range of applications, including absorbers, phase shifters, waveguides, and in stealth technology. Readers are encouraged to explore various other applications of metamaterials.

References

1. Xu, C., Ren, Z., Wei, J., Lee, C., Reconfigurable terahertz metamaterials: From fundamental principles to advanced 6G applications. *iScience*, 25, 103799, 2022.
2. Withayachumnankul, W. and Abbott, D., Metamaterials in the terahertz regime. *IEEE Photonics J.*, 1, 99, 2009.
3. Hossain, K., Sabapathy, T., Jusoh, M., Soh, P.-J., Al-Bawri, S.-S., Osman, M.-N., Rahim, H.-A., Torrungrueng, D., Akkaraekthalin, P., Decagonal C-shaped CSRR textile-based metamaterial for microwave applications. *Comput. Mater. Contin.*, 71, 1677, 2022.
4. Krzysztofik, W.J. and Cao, T.N., Metamaterials in application to improve antenna parameters, in: *Metamaterials and Metasurfaces*, J.C. Ferrer (Ed.), pp. 63–85, IntechOpen, London, 2018.
5. Pendry, J.B., Holden, A.J., Robbins, D.J., Stewart, W.J., Magnetism from conductors and enhanced nonlinear phenomena. *IEEE Trans. Microw. Theory Tech.*, 47, 2075, 1999.
6. Powell, D.A., Shadrivov, I.V., Kivshar, Y.S., Cut-wire-pair structures as two-dimensional magnetic metamaterials. *Opt. Express*, 16, 15185, 2008.
7. Pendry, J.B. and Smith, D.R., Reversing light with negative refraction. *Phys. Today*, 57, 37, 2004.
8. Smith, D.R., Padilla, W.J., Vier, D.C., Nasser, S.C., Schultz, S., Composite medium with simultaneously negative permeability and permittivity. *Phys. Rev. Lett.*, 84, 4184, 2000.
9. Belov, P.A., Tretyakov, S.A., Viitanen, A.J., Dispersion and reflection properties of artificial media formed by regular lattices of ideally conducting wires. *J. Electromagn. Waves Appl.*, 16, 1153, 2002.
10. Maslovski, S.I., Tretyakov, S.A., Belov, P.A., Wire media with negative effective permittivity: A quasi-static model. *Microw. Opt. Technol. Lett.*, 35, 47, 2002.
11. Sarychev, A.K. and Shalaev, V.M., *Electrodynamics of metamaterials*, pp. 1–247, World Scientific Publishing Co., New Jersey (USA), 2007.
12. Islam, S.S., Faruque, M.R.I., Islam, M.T., An object-independent ENZ metamaterial-based wideband electromagnetic cloak. *Sci. Rep.*, 6, 33624, 2016.

13. Rani, R., Kaur, P., Verma, N., Metamaterials and their applications in patch antenna: A review. *Int. J. Hybrid Inf. Technol.*, 8, 199–212, 2015.
14. Caloz, C. and Itoh, T., Application of the transmission line theory of left-handed (LH) materials to the realization of a microstrip "LH line", in: *IEEE Antennas and Propagation Society International Symposium*, pp. 412–415, 2002.
15. Eleftheriades, G.V., Siddiqui, O., Iyer, A.K., Transmission line models for negative refractive index media and associated implementations without excess resonators. *IEEE Microw. Wirel. Compon. Lett.*, 13, 51, 2003.
16. Marqués, R., Martel, J., Mesa, F., Medina, F., Left-handed-media simulation and transmission of EM Waves in subwavelength split-ring-resonator-loaded metallic waveguides. *Phys. Rev. Lett.*, 89, 183901, 2002.
17. Jokanovicé, B. and Bengin, V., Novel left-handed transmission lines based on grounded spirals. *Microw. Opt. Technol. Lett.*, 49, 2561, 2007.
18. Bonache, J., Martín, F., Gil, I., García, J., Marqués, R., Sorolla, M., Microstrip bandpass filters with wide bandwidth and compact dimensions. *Microw. Opt. Technol. Lett.*, 46, 343, 2005.
19. Landy, N.I., Sajuyigbe, S., Mock, J.J., Smith, D.R., Padilla, W.J., Perfect metamaterial absorber. *Phys. Rev. Lett.*, 100, 207402, 2008.
20. Li, H., Yuan, L.H., Zhou, B., Shen, X.P., Cheng, Q., Cui, T.J., Ultrathin multiband gigahertz metamaterial absorbers. *J. Appl. Phys.*, 110, 014909, 2011.
21. Bhattacharyya, S., Ghosh, S., Srivastava, K.V., Equivalent circuit model of an ultra-thin polarization-independent triple band metamaterial absorber. *AIP Adv.*, 4, 097127, 2014.
22. Ghosh, S., Bhattacharyya, S., Chaurasiya, D., Srivastava, K.V., An ultra wideband ultrathin metamaterial absorber based on circular split rings. *IEEE Antennas Wirel. Propag. Lett.*, 14, 1172, 2015.
23. Glybovski, S.B., Tretyakov, S.A., Belov, P.A., Kivshar, Y.S., Simovski, C.R., Metasurfaces: From microwaves to visible. *Phys. Rep.*, 634, 1, 2016.
24. Yoo, M. and Lim, S., Active metasurface for controlling reflection and absorption properties. *Appl. Phys. Express*, 7, 112204, 2014.
25. Zhu, B., Huang, C., Feng, Y., Zhao, J., Jiang, T., Dual band switchable metamaterial electromagnetic absorber. *Prog. Electromagn. Res. Lett.*, 24, 121–129, 2010.
26. Hakla, N., Ghosh, S., Srivastava, K.V., Shukla, A., A dual-band conformal metamaterial absorber for curved surface, in: *2016 URSI International Symposium on Electromagnetic Theory, EMTS 2016*, Institute of Electrical and Electronics Engineers Inc., pp. 771–774, 2016.
27. Ghosh, S. and Srivastava, K.V., Broadband polarization-insensitive tunable frequency selective surface for wideband shielding. *IEEE Trans. Electromagn. Compat.*, 60, 166, 2018.
28. Ma, B., Liu, S., Kong, X., Jiang, Y., Xu, J., Yang, H., A novel wide-band tunable metamaterial absorber based on varactor diode/graphene. *Optik*, 127, 3039, 2016.

29. Ghosh, S. and Srivastava, K.V., Polarisation-independent tunable absorber with embedded biasing network. *Electron. Lett.*, 53, 1176, 2017.
30. Liu, L.Y. and Wang, B.Z., A broadband and electrically small planar monopole employing metamaterial transmission line. *IEEE Antennas Wirel. Propag. Lett.*, 14, 1018, 2015.
31. Volakis, J.L., Mumcu, G., Sertel, K., Chen, C.C., Lee, M., Kramer, B., Psychoudakis, D., Kiziltas, G., Antenna miniaturization using magnetic-photonic and degenerate band-edge crystals. *IEEE Antennas Propag. Mag.*, 48, 12, 2006.
32. Ziolkowski, R.W. and Erentok, A., Metamaterial-based efficient electrically small antennas. *IEEE Trans. Antennas Propag.*, 54, 2113, 2006.
33. Irci, E., Sertel, K., Volakis, J.L., Antenna miniaturization for vehicular platforms using printed coupled lines emulating magnetic photonic crystals. *Metamaterials*, 4, 127, 2010.
34. Ko, S.T. and Lee, J.H., Miniaturized ENG ZOR antenna with high permeability material, in: *2010 IEEE International Symposium on Antennas and Propagation and CNC-USNC/URSI Radio Science Meeting - Leading the Wave, AP-S/URSI*, Toronto, 2010.
35. Gupta, A. and Chaudhary, R.K., A miniaturized dual-band ZOR antenna using epsilon negative transmission line loading. *Int. J. Microw. Wirel. Technol.*, 9, 1735, 2017.
36. Gudibandi, B.R., Murugan, H.A., Dhamodharan, S.K., Miniaturization of monopole antenna using high refractive index metamaterial loading. *Int. J. RF Microw. Comput. Aided Eng.*, 30, e22163, 2020.
37. Sharma, S.K., Gupta, A., Chaudhary, R.K., Epsilon negative CPW-fed zeroth-order resonating antenna with backed ground plane for extended bandwidth and miniaturization. *IEEE Trans. Antennas Propag.*, 63, 5197, 2015.
38. Majedi, M.S. and Attari, A.R., Dual-band resonance antennas using epsilon negative transmission line. *IET Microw. Antennas Propag.*, 7, 259, 2013.
39. Saraswat, R.K. and Kumar, M., A quad band metamaterial miniaturized antenna for wireless applications with gain enhancement. *Wireless Pers. Commun.*, 114, 3595, 2020.
40. Sievenpiper, D., Zhang, L., Broas, R.F., Alexöpolous, N.G., Yablonovitch, E., High-impedance electromagnetic surfaces with a forbidden frequency band. *IEEE Trans. Microw. Theory Tech.*, 47, 2059, 1999.
41. Zhang, Y., Von Hagen, J., Younis, M., Fischer, C., Wiesbeck, W., Planar artificial magnetic conductors and patch antennas. *IEEE Trans. Antennas Propag.*, 51, 2704, 2003.
42. Wu, J. and Sarabandi, K., Reactive impedance surface TM mode slow wave for patch antenna miniaturization [AMTA corner]. *IEEE Antennas Propag. Mag.*, 56, 279, 2014.
43. Jagtap, S., Chaudhari, A., Chaskar, N., Kharche, S., Gupta, R.K., A wideband microstrip array design using RIS and PRS layers. *IEEE Antennas Wirel. Propag. Lett.*, 17, 509, 2018.

44. Devapriya, A.T. and Robinson, S., Investigation on metamaterial antenna for terahertz applications. *J. Microw. Optoelectron. Electromag. Appl.*, 18, 377, 2019.
45. Chen, X., Grzegorczyk, T.M., Wu, B.I., Jr., Kong, J.A., Robust method to retrieve the constitutive effective parameters of metamaterials. *Phys. Rev. E Stat. Nonlin. Soft Matter Phys.*, 70, 016608, 2004.
46. Selvaraju, R., Jamaluddin, M.H., Kamarudin, M.R., Nasir, J., Dahri, M.H., Complementary split ring resonator for isolation enhancement in 5G communication antenna array. *Prog. Electromagn. Res. C*, 83, 217, 2018.
47. Ma, F., Lin, Y.S., Zhang, X., Lee, C., Tunable multiband terahertz metamaterials using a reconfigurable electric split-ring resonator array. *Light Sci. Appl.*, 3, e171, 2014.
48. Yuan, B., Zheng, Y.H., Zhang, X.H., You, B., Luo, G.Q., A bandwidth and gain enhancement for microstrip antenna based on metamaterial. *Microw. Opt. Technol. Lett.*, 59, 3088, 2017.
49. Urul, B., Gain enhancement of microstrip antenna with a novel DNG material. *Microw. Opt. Technol. Lett.*, 62, 1824, 2020.
50. Patel, S.K. and Argyropoulos, C., Enhanced bandwidth and gain of compact microstrip antennas loaded with multiple corrugated split ring resonators. *J. Electromagn. Waves Appl.*, 30, 945, 2016.
51. Sambandam, P., Kanagasabai, M., Ramadoss, S., Natarajan, R., Alsath, M.G.N., Shanmuganathan, S., Sindhadevi, M., Palaniswamy, S.K., Compact monopole antenna backed with fork-slotted EBG for wearable applications. *IEEE Antennas Wirel. Propag. Lett.*, 19, 228, 2020.
52. Ali, A., Mitra, A., Aïssa, B., Metamaterials and metasurfaces: A review from the perspectives of materials, mechanisms and advanced metadevices. *Nanomaterials*, 12, 1027, 2022.
53. Chen, P.W. and Chen, F.C., Asymmetric Coplanar Waveguide (ACPW) Zeroth-Order Resonant (ZOR) antenna with high efficiency and bandwidth enhancement. *IEEE Antennas Wirel. Propag. Lett.*, 11, 527, 2012.
54. Chi, P.L. and Shih, Y.S., Compact and bandwidth-enhanced zeroth-order resonant antenna. *IEEE Antennas Wirel. Propag. Lett.*, 14, 285, 2015.
55. Jang, T., Choi, J., Lim, S., Compact coplanar waveguide (CPW)-fed zeroth-order resonant antennas with extended bandwidth and high efficiency on vialess single layer. *IEEE Trans. Antennas Propag.*, 59, 363, 2011.
56. Ji, J.K., Kim, G.H., Seong, W.M., Bandwidth enhancement of metamaterial antennas based on composite right/left-handed transmission line. *IEEE Antennas Wirel. Propag. Lett.*, 9, 36, 2010.
57. Kumar, S. and Kumari, R., Bandwidth enhanced coplanar waveguide-fed composite right/left handed antenna loaded with resonating rings. *IET Microw. Antennas Propag.*, 13, 2134, 2019.
58. Niu, B.J., Feng, Q.Y., Shu, P.L., Epsilon negative zeroth- and first-order resonant antennas with extended bandwidth and high efficiency. *IEEE Trans. Antennas Propag.*, 61, 5878, 2013.

59. Rao, M., Madhav, B.T.P., Anilkumar, T., Nadh, B., Metamaterial inspired quad band circularly polarized antenna for WLAN/ISM/Bluetooth/WiMAX and satellite communication applications. *AEU Int. J. Electron. Commun.*, 97, 229, 2018.
60. Chatterjee, J., Mohan, A., Dixit, V., Broadband circularly polarized h-shaped patch antenna using reactive impedance surface. *IEEE Antennas Wirel. Propag. Lett.*, 17, 625, 2018.
61. Liu, L.Y. and Wang, B.Z., Compact broadband linearly and circularly polarized antenna based on CRLH transmission line, in: *Asia-Pacific Microwave Conference (APMC)*, pp. 1–3, 2016.
62. Dong, Y., Toyao, H., Itoh, T., Compact circularly-polarized patch antenna loaded with metamaterial structures. *IEEE Trans. Antennas Propag.*, 59, 4329, 2011.
63. Mark, R., Singh, H.V., Mandal, K., Das, S., Mutual coupling reduction using near-zero ϵ and μ metamaterial-based superstrate for an MIMO application. *IET Microw. Antennas Propag.*, 14, 479, 2020.
64. Najafy, V. and Bemani, M., Mutual-coupling reduction in triple-band MIMO antennas for WLAN using CSRRs. *Int. J. Microw. Wirel. Technol.*, 12, 762, 2020.
65. Tan, X., Wang, W., Wu, Y., Liu, Y., Kishk, A.A., Enhancing isolation in dual-band meander-line multiple antenna by employing split EBG structure. *IEEE Trans. Antennas Propag.*, 67, 2769, 2019.
66. Ibrahim, A.A. and Abdalla, M.A., CRLH MIMO antenna with reversal configuration. *AEU Int. J. Electron. Commun.*, 70, 1134, 2016.
67. Chaimool, S., Hongnara, T., Rakluea, C., Akkaraekthalin, P., Zhao, Y., Design of a PIN diode-based reconfigurable metasurface antenna for beam switching applications. *Int. J. Antennas Propag.*, 2019, 7216324, 2019.
68. Gupta, A. and Singh, R., A miniaturized elliptically shaped split ring resonator antenna with dual-band characteristics, in: *Advances in Signal Processing and Communication, Select Proceedings of ICSC*, S. Manhas, B.S. Rawat, A. Trivedi, V. Karwal (Eds.), Springer Verlag, pp. 37–44, 2019.
69. Rao, P.H., Sajin, J.S., Kudesia, K., Miniaturisation of switched beam array antenna using phase delay properties of CSRR-loaded transmission line. *IET Microw. Antennas Propag.*, 12, 1960, 2018.
70. Hong, T., Wang, M., Peng, K., Gong, S., Ultrathin and miniaturized frequency selective surface with closely located dual resonance. *IEEE Antennas Wirel. Propag. Lett.*, 18, 1288, 2019.
71. Shabbir, T., Saleem, R., Al-Bawri, S.S., Shafique, M.F., Islam, M.T., Eight-port metamaterial loaded UWB-MIMO antenna system for 3D system-in-package applications. *IEEE Access*, 8, 106982, 2020.
72. Reddy, M.H., Sheela, D., Parbot, V.K., Sharma, A., A compact metamaterial inspired UWB-MIMO fractal antenna with reduced mutual coupling. *Microsyst. Technol.*, 27, 1971, 2021.

73. Irene, G. and Rajesh, A., A dual-polarized UWB–MIMO antenna with IEEE 802.11ac band-notched characteristics using split-ring resonator. *J. Comput. Electron.*, 17, 1090, 2018.
74. Ghosh, J. and Mitra, D., A technique for reduction of mutual coupling by steering surface wave propagation. *Microw. Opt. Technol. Lett.*, 62, 1957, 2020.
75. Zhu, X., Yang, X., Song, Q., Lui, B., Compact UWB-MIMO antenna with metamaterial FSS decoupling structure. *Eurasip. J. Wirel. Commun. Netw.*, 2017, 115, 2017.
76. Thakur, E., Jaglan, N., Gupta, S.D., Design of compact triple band-notched UWB MIMO antenna with TVC-EBG structure. *J. Electromagn. Waves Appl.*, 34, 1601, 2020.

2

Chiral Metamaterials

Wasefa Begum[1], Monohar Hossain Mondal[2], Ujjwal Mandal[1*] and Bidyut Saha[1†]

[1]Homogeneous Catalysis Laboratory, Department of Chemistry, The University of Burdwan, Burdwan, WB, India
[2]Chemical Sciences Laboratory, Government General Degree College, Singur, Hooghly, WB, India

Abstract
Chiral metamaterials (MTMs) are a special type of composite material having nonsuperimposable mirror image structures which leads to strong cross-coupling interaction between electric field and magnetic field showing unusual or giant optical properties in them. In the last few decades, chiral metamaterials have drawn the attention of researchers because of their excellent advantages over natural chiral materials, such as their design flexibility, ability to achieve excellent customized properties, giant optical activity, tunability or switchability, and so on. The control or manipulation of optical activity, basically electromagnetic properties are fundamental to many applications which invoke the further development of such composite materials. Ability to achieve attractive customized properties, tuneability, and design flexibility; make them suitable for significant applications in medical and bio sectors, bio-sensing, absorber, optical filters, antenna, display, photography, etc. In this write-up, we have tried to outline the fundamentals of chiral metamaterial, its properties, constructions, applications, and recent advances in the modern era.

Keywords: Chirality, circular dichroism, metamaterial, optical activity, reconfigurable, sensing

**Corresponding author*: umandal@chem.buruniv.ac.in
†*Corresponding author*: bsaha@chem.buruniv.ac.in

Inamuddin and Tariq Altalhi (eds.) Electromagnetic Metamaterials: Properties and Applications, (33–52) © 2023 Scrivener Publishing LLC

2.1 Introduction

Asymmetry prevails everywhere in this universe, from the living world to galaxies, in biology and ecology, from physiology to anatomy, from micro-components to macrocomponents in science and technology, and so on. Asymmetry arises due to the violation or absence of symmetry elements and chiral materials are asymmetric materials having no plane of symmetry; basically, construct nonsuperimposable mirror images possessing optical activity. In general, the capability of manipulating the plane of polarization of EM (electromagnetic waves) is called optical activity. In a chiral medium, optical activity is nothing but the consequence of cross-coupling between the electric field and magnetic field. Due to weak electromagnetic coupling, naturally occurring chiral materials like sugars, amino acids, quartz, etc. exhibit weak optical activity [1]. In 1811, Dominique Arago first discovered the optical activity of chiral materials [2]. Optical activity has great importance in various fields of science like optoelectronics, molecular biology, display and imaging application, analytical chemistry, and many more [3, 4].

Manipulating and monitoring electromagnetic properties like circular dichroism and optical activity is the basis of diverse utilization, such as optical data storage, display, wireless communication, internet, imaging, etc. [5]. Advancement in these fields of technology demands structured or composite materials with exotic electromagnetic properties. Over the last few decades, the interest of researchers is rising in composite materials having unusual and wonderful optical activities and these composite materials are commonly known as "metamaterials." These metamaterials are mainly artificially constructed materials, engineered to induce several customized properties, which are absent in natural materials [6]. They are mainly designed based on their application. Metamaterials can exhibit negative refractive index, permeability, and permittivity, and the properties and features of metamaterials depend on their structure and construction rather than their component. Due to having the ability to achieve attractive customized properties, tuneability, and design flexibility, they have significant applications in aerospace, biosensor, absorber, optical filters, antennae, energy harvesters, etc. [7].

Chiral metamaterials (CMM) are a special type of class-bearing structures which possess nonsuperimposable mirror images. As mirror symmetry is absent, strong cross-coupling occurs between the magnetic field and electric field, which results in an unusual or giant optical property in chiral metamaterials. Chiral metamaterials show extremely strong optical activity and several unique features like negative refractive index [8, 9], the anti-Doppler effect [10], negative permittivity and permeability, etc. Being a metamaterial, they have also flexibility in design fabrication and tunability as well as the ability to gain customized characteristics as per requirement. Moreover, the control of optical activity in chiral metamaterial is possible [11]. All these attractive features make them suitable in various application fields, such as bio-detector, absorber, antenna, imaging, display, smart devices, medical and bio sectors, communication technology, and optical fibers [7, 12, 13]. In this review, we will discuss chiral metamaterial, their properties, constructions, applications, and recent advances in different segments.

2.2 Fundamentals of Chiral Metamaterials and Optical Activity Control

Circular dichroism (CD) and Optical activity (OA) are the most important electromagnetic properties manifested as an optical response in any chiral medium. Optical activity is the capability of manipulating the plane of polarization of EM (electromagnetic waves) in a chiral medium, which is the consequence of cross-coupling between magnetic and electric fields, whereas CD is the difference in the absorption of circularly polarized (RCP & LCP) light (Figure 2.1a,b) [14, 15]. Circular dichroism mainly depends on the transmission of light. Thus, it may be expressed in terms of ellipticity, which is a function of the relative difference in transmission. Furthermore, circular dichroism plays an important role in material absorption. Differences in the absorption of RCP and LCP can

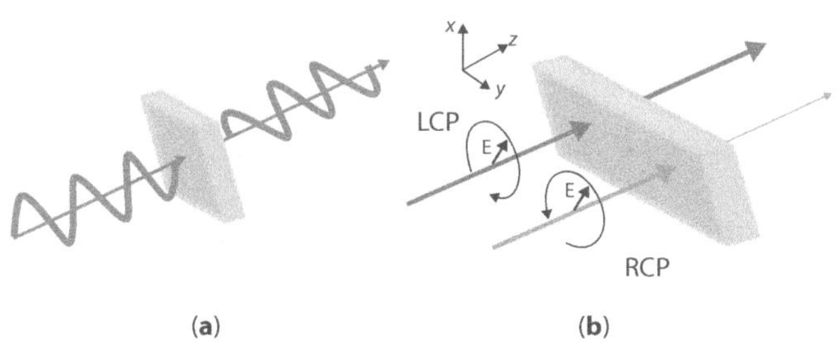

Figure 2.1 a) Optical activity:In chiral medium, the electric field vector of a linearly polarized wave rotating about theaxis which is parallel with respect to its axis of propagation (+z axis). 1. b) Circular dichroism: Difference in absorption of right and left circularly polarized light resulting different transmissions for LCP and RCP to each other. Figure 2.1a and 2.1b is reproduced from [11].

be considerably increased by resonance effect and circular dichroism in composite materials [11, 16].

Resonance in chiral metamaterial is another important element that favors achieving a giant optical or chiral response in chiral media. Strong resonance in constituent chiral meta-atoms arises due to excitation by external electromagnetic waves and as a result, strong polarization occurs. Thus, resonance in chiral metamaterial helps to enhance or manipulate optical activity, circular dichroism cum ellipticity [17, 18].

The major attraction of chiral metamaterial is its capability to exhibit a negative refractive index and this is also the prime reason which invokes its further development and research. The negative refractive index is known to arise from the oppositely directed phase velocity and group velocity and also propagation of energies and phase of the wave are oppositely directed. As earlier spoken, the optical activity of chiral MTMs can be exceptionally enhanced which implies that the direction of the electric field produced by linearly polarized light may be rotated or manipulated within a very small distance while passing through a chiral medium [11, 19, 20]. In this regard, a large number of chiral MTM designs to achieve giant optical chirality have been proposed which will be discussed in segment 3.

In the practical field of an application controlling the dispersion and bandwidth of chiral material are very significant which becomes a real challenge as chiral MTMs generally depend on resonances to achieve large optical chirality, and also strong CD (circular dichroism) within a limited frequency range. Dispersion of the chiral metamaterials is also influenced by the resonant characteristics of the materials, and so there arises a necessity to compromise between broad bandwidth and enhanced circular dichroism [11]. Hannam and his co-researchers reported a spectrally flat and experimentally large optical activity with little ellipticity by utilizing a meta-surface, comprised of pairs of twisted meta-atoms and their complements [21].

2.3 Construction of Chiral Metamaterial

From the basic idea, it can be said that chiral metamaterial can be constructed by interrupting the mirror symmetry of a three-dimensional arranged material. We have tried to represent the units of chiral metamaterial having different structural features. Several chiral unit cells are arranged to obtain giant chiral metamaterial [Table 2.1] having different structures including a Y-shaped structure [14], Slavic symbol [22, 23],

Table 2.1 Types of chiral metamaterial of different unit cell.

Sl. No.	Types of chiral metamaterial of different unit cell	References
1.	Y-shaped structure	[14]
2.	Slavic symbol	[22, 23]
3.	helix structure	[24]
4.	multilayered arc structures	[25]
5.	twisted split ring	[26, 27]
6.	twisted cross structure	[28, 29]
7.	Multilayergammadion structure	[4]
8.	Twisted SRR U-shaped structure	[40]

helix structure [24], multilayered arc structures [25], twisted split ring [26, 27], twisted cross structure [28, 29]. Due to greater chirality in periodical unit cells, cross-coupling involving electric and magnetic fields becomes stronger. This coupling interaction is capable of being externally controlled by adjusting the size, shape, or type of the chiral unit cells to get exotic electromagnetic properties e.g., highly efficient absorption [30, 31], conversion of broadband polarization [32, 33], optical imaging which break through the diffraction limit [34, 35].

Figure 2.2 represents various types of chiral metamaterial units. Figure 2.2a, depicts a bilayer chiral metamaterial unit having a cross-wire design with a twisted angle ϕ which gives excellent OA as well as a negative refractive index [19]. The structure will possess chiral character if the angle (twisted) ϕ equals to other than the multiple of $\pi/2$ where n is an integer. In the cross-wire pairs, the cross-coupling interaction occurs at chiral resonance possessing potent chirality around the resonating frequencies. The resulting strong chirality splits the refractive indices which results in a negative refractive index of one circularly polarized wave [36].

Figure 2.2b represents multilayered gammadion structured chiral metamaterial in parallel planes with a definite twist angle with nearby planes. This multi-layered version of chiral metamaterial shows a negative refractive index, negative permeability, giant optical activity, negative permittivity moreover, enhanced polarization rotation and circular dichroism than bi-layered structure. This extraordinary performance makes it compatible to use in ultrathin polarization rotators and circular polarizers in various practical fields [4].

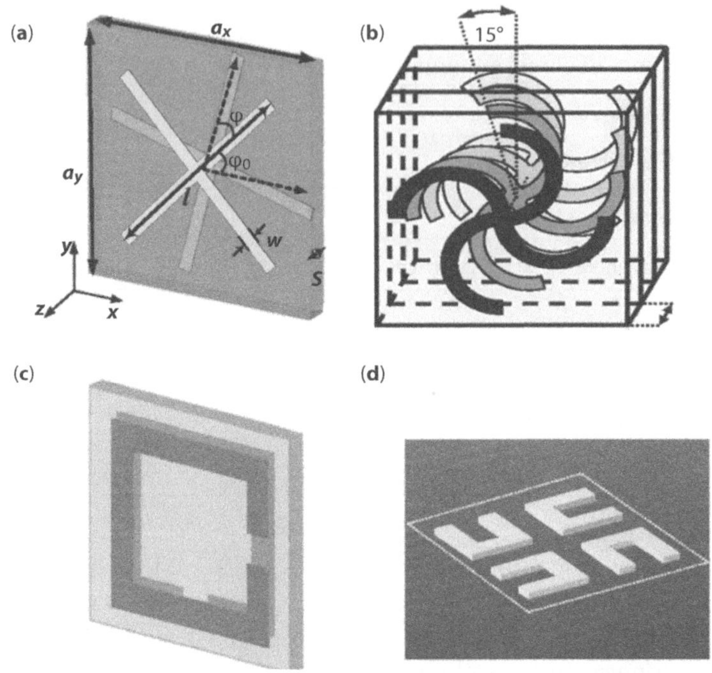

Figure 2.2 Representation of chiral metamaterial-based unit cells. (a) Twisted cross wire structure [19]; (b) multilayer gammadion structure; (c) Twisted split ring structure; (d) Twisted SRR structure of U-shaped; (a) is replicated with consent from [19]. ©American Physical Society, 2009; (b) is replicated with consent from [4]. ©American Physical Society, 2009; (c) is replicated with consent from [13]; (d) is replicated with consent from [40].

Ye and He *et al.* reported two-layered chiral MTMs having enantiomeric patterns with the four-fold rotational axis of symmetry to acquire 90° polarization transformation [37]. Another study on twisted bi-layered U-shaped SRR was reported to design 90° polarization transformation [38]. Huang and his co-workers proposed a polarization rotator, consisting of two pairs of twisted bi-layered SRRs with a twist angle equal to 90° exhibits broadband optical activity (Figure 2.2c) [39].

Another Chiral metamaterial composed of a bi-layered Y-shaped metallic structure with a twisted angle ϕ and the angle between the parallel branches remained at 120°. At the lower resonance frequency, the incident linearly polarized wave is transformed to LCP whereas at a higher resonance frequency the wave is converted into a cross-polarized wave. The planar structure makes it appropriate for application in multi-polarization antenna systems [14].

With further development, other studies were reported regarding the construction of chiral metamaterials. Figure 2.2d depicts chiral metamaterial constructed by basic units built of two pairs of twisted U-shaped metallic structures with a rotation angle equal to 90° with neighbors. These types of CMMs have multiple layers having 90° neighboring twist angles and generally are applied for 90° polarization transformation [40].

Some other chiral metamaterials do not contain C4 symmetry and they were reported to act as the highly efficient polarizer. The efficacy of polarization conversion can be enhanced with proper designing of the structural parameters. In a study, a horn antenna that is dual circularly polarized was proposed in the Ku band which was formulated on chiral MTMs. The CMM unit cells are composed of a bi-layered metallic arc structure which is twisted by an angle θ_2 with the top layer and bottom layer. This CMM can realize circular dichroism at two different working frequencies. The chiral metamaterial decomposes the linearly incident wave into two differently circularly polarized waves with only 0.6 dB degradation of the gain of the horn antenna around the two working frequencies [41].

In another study, a dual-band chiral metamaterial was reported to be composed of two pairs of arcs of different sizes with increased operating frequency. It was also studied that increase in the number of arcs on both sides of the dielectric lamina results in higher resonant frequency. Figure 2.3a depicts the multiband circular polarizer for converting linearly polarized incident waves into differently circular polarized waves having lower degradation and different rotations at four different resonances. Figure 2.3b depicts the surface current distribution of CMM which resonates at a low value. This chiral MTM can be developed to be used potentially in

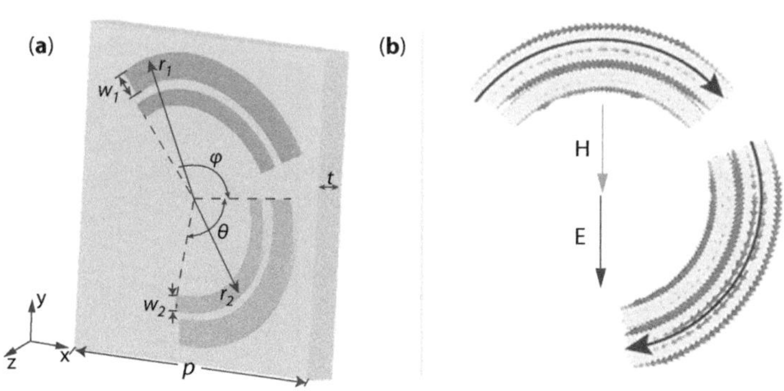

Figure 2.3 a) Unit cell with two pairs of twisted arcs; b) Surface current distribution at the first resonances of the CMM. a) & b) are replicated with consent from [42]. ©AIP Publishing LLC, 2012.

circularly polarized antenna systems [42, 43]. The combination method of chirality with anisotropy results in enhancement in polarization conversion of CMMs. The twisted arc CMM inspired the construction of more CMM for nonlinear optical devices [44].

2.4 Applications

In recent days, the advancement of communication technology demands multi-polarization states of electromagnetic waves and in this regard, chiral metamaterial shows promising activities. Chiral metamaterials having twisted cross structure [19]; SRR structure of U-shaped [45]; multilayered twisted gammadion structure [4] can be used to achieve linear to circular polarization transformation, whereas some works are focused on 90° polarization transformation.

2.4.1 Chiral Metamaterials in the Chiral Sensing

Generally, chiral molecules show different biochemical properties with their mirror images. The electromagnetic fields of chiral molecules interact in different manners with chiral material or matter for left and right circularly polarized waves, and this effect of chiral metamaterials can be used as a platform for chiral sensing due to possessing strong optical activity around the vicinity of metamaterials [46]. Figure 2.4a depicts chiral metamaterial with gammadion structure which has efficient resonance near 800 nm wavelength and shows strong circular dichroism signals due to

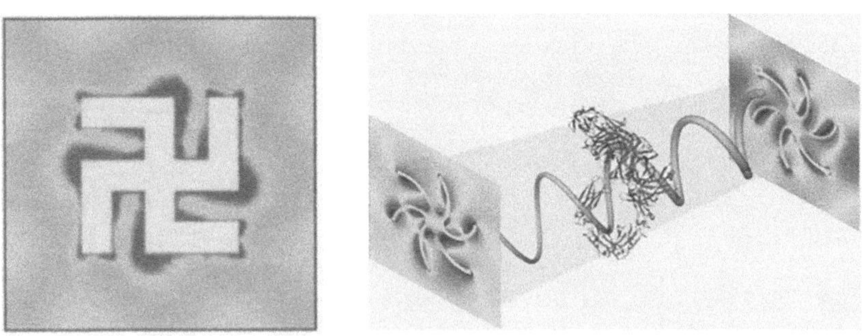

(a) (b)

Figure 2.4 Enhancements of Optical chirality in the surrounding area of different CMM: (a) gammadion like structure [47], (b) shuriken structure [49], a) is replicated with consent from [47], © 2010, Nature Publishing Group; b) is replicated with consent from [49], ©2018, American Chemical Society.

its chiral structure. The optical chirality surrounding the nearby field of the gammadion CMM structure is sufficiently large for strong coupling with chiral molecules and this strong coupling results in a significantly large shift [47]. In Figure 2.4b the shuriken-like structured metamaterial was employed for the characterization of protein interface structures [48] as well as the structural order of bio-interface [49]. This type of structure has two resonances nearly located at wavelengths about 710 and 740 nm. The coupling of different modes becomes asymmetric when the arrangement of chiral molecules takes place on the surface of the shuriken and this asymmetric interaction is the pivotal point that is generally used for the chiral sensing of different types of biomolecules [48, 49]. Schfering and Giessen have reported a few facts about the construction of platforms based on chiral MTMs and from their study we know that greatly twisted planar configuration without any sharp corners can be utilized for getting an uninterrupted area of strong optical chirality [50]. In addition to that, a compact 3D chiral structure needed to be used to obtain a large difference in optical chirality. Despite having many advantages, chiral MTM sensing platforms have some drawbacks too, and the limitations arise due to the contamination between strong and relatively weak CD signals, which are generated by the platforms and chiral molecules, respectively [51, 52].

Generally, a chiral metamaterial platform gives strong CD signals even when the chiral molecule is absent. When platforms are occupied with chiral molecules, both of them contribute to the experimentally obtained CD signals. So, it is difficult to discriminate the signals through the MTM platforms which intensify comparatively weaker molecular circular dichroism signals. The signs of molecular Circular Dichroism signals are very significant for chiral sensing since they include stereo-chemical data of that particular chiral molecule. So, there is a chance of contamination of the stereo-chemical data due to the strong optical chirality of the metamaterial platform [46].

2.4.2 Reconfigurable Chiral Metamaterial

The key part for achieving intelligent electromagnetic devices is the proper or desired tuning of electromagnetic characteristics of the constituent chiral metamaterial. In recent days, the most challenging task is the construction of desired chiral metamaterial and in this regard, many active materials, such as semiconductors [53], microelectromechanical system (MEMS) [54], phase change materials [55], etc. have been induced to the construction method of CMMs to achieve controllable electromagnetic features and reconfigurable chirality. Active devices like positive-intrinsic-negative (PIN) diodes and varactors have been also introduced in the microwave region for the construction of active CMMs [56]. The unit cell of this type of reconfigurable CMM in the microwave region consists of a metallic structure having rectangular and cross apertures in the top and bottom layers. Four PIN diodes are loaded in the cross aperture and separated into two parts which work oppositely so that by altering the working states of the PIN diodes, the handedness of the material can be altered [56].

Despite the widespread use of active devices in metamaterial in microwave range for tuning the features, they are not compatible with metamaterial of higher frequency because of their millimeter-scale volume [57, 58]. Active devices like semiconductors, MEMS, and phase change materials are used in metamaterial having higher operating frequency than terahertz to construct reconfigurable CMMs. Yin and his co-researchers have reported the construction of reconfigurable CMM by introducing a layer of $Ge_3Sb_2Te_6$, a phase change material between the two-stacked nanorods [59]. In this case, a higher spectral shift of circular dichroism was observed to reach 18% with thermal control while the refractive index of the phase material changed from 3.5 + 0.01i to 6.5 + 0.06i in heating conditions (160°C) in the mid-IR region. Kenanakis and his co-workers reported different chiral structures based on a two-layer conductor design to achieve switchable giant optical activity by altering metallic parts of the structures by Si which is photo-conducting and can be transformed from insulating to conducting condition with the help of the photoexcitation process [60].

In another study, Zhang *et al.* reported silicon pads incorporated a unit cell of the three-dimensional chiral metal molecule. The photo-induced switching of chirality manipulates the optical activity of the molecules resulting in the reversion of circular dichroism between right and left circularly polarized waves [61]. A planer CMM is represented in Figure 2.5

Figure 2.5 Schematic representation of unit cell of the reconfigurable dynamic bi-layered conjugate gammadion-shaped gold resonators chiral metamaterials with semiconductor substrate. This is reprinted with permission [62]. Copyright American Physical Society, 2012.

with a conjugated two-layered metal structure consisting of gammadion-shaped gold resonators. This chiral metamaterial was fabricated on silicon-on-sapphire (SOS) wafer to get higher and more dynamic tunability, and superb optical activity was realized through near-infrared excitation of photo-carriers in the intrinsic silicon islands [62].

2.4.3 Chiral Metamaterial Absorber

The basis of circular dichroism and optical activity is the discrepancy in refraction indices of RCP and LCP revealed from the theoretical analysis of the chiral medium. The origin of Circular dichroism is the difference of refraction indices for circular polarization in the image parts that result in different extents of absorption between the circular polarizations when traveling through the optically active medium. So, chiral metamaterial may be used to absorb definite circular polarization and explore various applications in the field of imaging, detecting, display, communication system, sensor devices, biology detection, stealth, and so on [13].

Li and his co-workers reported a chiral metamaterial absorber exhibiting large circular dichroism in the near-IR region. Figure 2.6a depicts the schematic structure of the CMM composed of the periodic metallic z-shaped antenna on top of (PMMA) poly (methyl methacrylate) spacer and Ag metal backplane. The circularly polarized wave impinged into it normally resulting in a discrepancy in circular absorption. It was reported

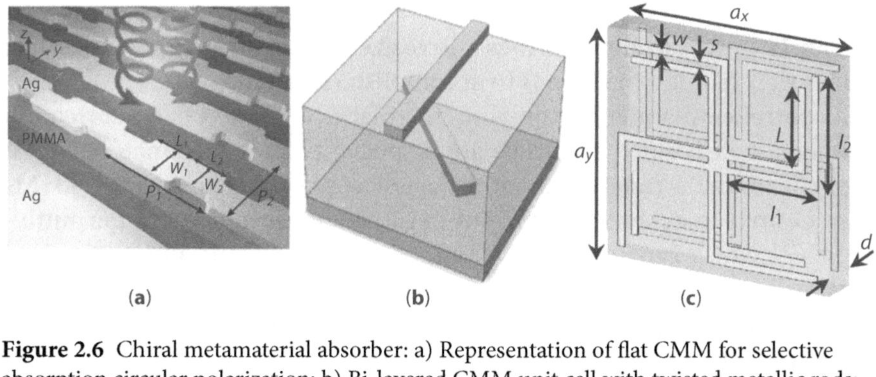

Figure 2.6 Chiral metamaterial absorber: a) Representation of flat CMM for selective absorption circular polarization; b) Bi-layered CMM unit cell with twisted metallic rods; c) Schematic design of unit cell of CMM with dual band absorption; (a) is replicated with consent from [63]. ©2015, Nature Publishing Group; (b) is replicated with consent from [64]. ©2016,American Chemical Society; (c) is replicated with consent from [65]. ©2013, Progress in Electromagnetics Research Symposium.

that the Left-handed CMM perfectly absorbed LCP light at the resonant frequency but reflected 90% of the Right-handed circularly polarized light and vice-versa in the case of right-handed CMM [63]. The combination of these chiral metamaterial absorbers with active materials like semiconductors results in tunable absorption activities of the material.

Wang and his co-researchers reported a chiral metamaterial (Figure 2.6b) consisting of the chiral unit having a b-layered twisted metallic rod. They observed that at the wavelength of 8.1μm the metamaterial perfectly absorbed the incident RCP light while reflecting almost of the LCP light. Hence selective perfect absorption can be achieved for CP (circularly polarized) light and its handedness is preserved over a large range of incident angles [64].

Chiral metamaterials are suitable not only for selective absorption but also for multiband absorption because of their multiple modes of cross-interaction within the magnetic and electric field. Figure 2.6c depicts a chiral metamaterial absorber with a conjugated gammadion patterned unit cell composed of two metallic (copper) double arms. This is reported that the chiral metamaterial acts as radar-absorbing material and multiband absorption of the linearly polarized incident waves is achieved almost perfectly. Extra introduced branches can lead to an increment in the bands of resonant frequency chiral MTM. The absorption spectra consist of two separate peaks and the absorption is considerably well at the incident angle (0-30)° [65].

2.4.4 Applications of Chiral Metamaterial as Multifunctional Sensors

Scientists have executed various microsized artificial intermediates like LHMs (left-handed materials) to accomplish electromagnetic (EM) waves in a controlled manner by applying microfabrication and nanofabrication methods. Left-handed materials are a special type of metamaterials that depends on the arrangements of the classical plasmonic resonator. LHMs have desired to achieve concurrent negative permeability and permittivity values which provide negative refraction. These artificially synthesized metamaterials can show excellent properties unlike their natural congeners [66, 67]. Artificially produced split-ring resonators (SRR) are superior to conventional approaches in generating effective magnetic responses from optical to microwave regions in various types of metamaterials. Many recent findings are capable to establish magnetic resonances by varying SRR configurations [68–72]. The SRR topology systems are consists of coupled SRR unit cells (should be more than one) which convey mutual capacitance and inductance at the resonance frequency following their positions [73]. Magneto-electric coupling is defined as an electric field inducing magnetic polarization which causes a magnetic field inducing electrical polarization in a chiral medium [74]. Studies on metamaterials have stretched to an advanced level because of the excellent progress in the fabrication and designing of microscale and nanoscale level metamaterials. The effective relationship between sensing technologies and metamaterials has directed modern science to develop advanced sensing technologies using such modern-day materials. Various environmental parameters can alter resonance frequencies, for this reason, scientists have developed metamaterial-based devices containing SRRs for sensing data of environmental parameters [75].

2.4.4.1 Applications of Chiral Metamaterial as Temperature, Humidity, and Moisture Sensors

Karaaslan and Bakir *et al.* [75] in their investigations have elaborately explained various sensing aptitudes such as humidity sensing, density sensing, and temperature sensing with the help of Chiral MTM sensors based on split ring resonators with double split (SRDS). For sensing the change in temperature, humidity, and density, sensor layers are occupied by various materials which have variable relative permittivity (εi) and the dimensions of sensor layers should be kept fixed for measuring the mentioned environmental parameters. They have reported that chiral SRDS-based density sensors are utilized in realizing the effects of change in the density of calcium chloride in silica gel material. In the same study, they have also reported that Chiral MTM structures based on SRDS are very much efficient for sensing marrowbone temperature, which plays a significant role in diagnostic purposes. Electromagnetic waves can cure various notorious diseases; this therapeutic property of EM was established by recording the heat, radiated from tissues due to the absorption of energy from EM. Generally, biological tissues and skin which exhibit conductivity properties, absorb the energy of EM to a higher extent than that of muscles and fat. So, we may conclude that Chiral devices exhibit diverse utilization in various fields e.g., humidity sensor, density sensor for $CaCl_2$, temperature sensor for marrowbone, etc. Moreover, the SRDS enclosure might be applied in near future for terahertz applications instead of using sensors based on metamaterial which is specially used for sensing chiral structures [75].

2.5 Conclusion and Future Perspective

During the last few years, the rising interest in various chiral metamaterial designs and their applications invokes researchers for its further development. In this regard, it should be mentioned that the progress is already reflected in the operating frequencies of chiral metamaterials which it has increased from microwave to optical range. Advancement in construction technology and design of the CMMs also has accelerated its development towards the smart era. All these betterments make CMMs suitable candidates for application in the field of imaging, bio-sensing, polarization manipulation, absorption, and many more [13]. The control or manipulation of optical activity and circular dichroism, basically electromagnetic properties is the fundamental of many applications which have drawn the attention of researchers for its further development in smart devices. By changing conformation, CMM can indicate the change in moisture which can be utilized as a potential tool for weather detection. Furthermore, it also has great application in the sensing of change in heat and humidity which invokes its utilization in climate monitoring studies [75]. Chiral metamaterials also can act as a great platform for chiral sensing and as the majority of the biological tissues and molecules are chiral it can be expected for CMMs to have more utilities in the field of biological sensing and detection applications [46]. Moreover, a combination of CMMs with active devices leads to achieving switchable chirality and giant optical activity.

However, more progress in the fabrication technology of CMMs is still needed for its higher efficiency and stronger chirality for its further development and utilities. Besides that, polarization manipulation in the case of smart detection as well as communication systems invokes more development in reconfigurable CMMs which is still a challenging task. Furthermore, control of circular dichroism and optical activity in the visible wavelength range also needs improvement as there are limited designs that can be achieved experimentally as well as they exhibit incomplete manipulation of material optical properties [11].

Acknowledgment

The authors acknowledge The University of Burdwan for providing infrastructural facilities. The authors declare no conflict of interest.

References

1. Kim, T.T., Sang, S.O., Park, H.S., Zhao, R., Kim, S.H., Choi, W., Min, B., Hess, O., Optical activity enhanced by strong intermolecular coupling in planar chiral metamaterials. *Sci. Rep.*, *4*, 5864, 2014.
2. Barron, L.D., *Molecular light scattering and optical activity*, Cambridge University Press, Cambridge, UK, 2004.
3. Svirko, Y., Zheludev, N., Osipov, M., Layered chiral metallic microstructures with inductive coupling. *Appl. Phys. Lett.*, 78, 498, 2001.
4. Plum, E., Zhou, J., Dong, J., Fedotov, V.A., Koschny, T., Soukoulis, C.M., Zheludev, N.I., Metamaterial with negative index due to chirality. *Phys. Rev. B*, 79, 035407, 2009.
5. Halim, B.I. and Boutejdar, A., Design and improvement a novel microstrip antenna using array of composite right/left handed transmission line (CRLH-TL) technique for multiband applications, in: *2019 IEEE International Symposium on Phased Array System & Technology (PAST)*, IEEE, pp. 1–5, 2019.
6. Chohan, J.S. and Singh, R., Thermosetting polymer application as metamaterials, in: *Reference Module in Materials Science and Materials Engineering*, Elsevier, Amsterdam, The Netherlands, 2021.
7. Kumar, R., Kumar, M., Chohan, J.S., Kumar, S., Overview on metamaterial: History, types and applications. *Mater. Today: Proc.*, 56, 3016, 2022.
8. Smith, D.R., Pendry, J.B., Wiltshire, M.C.K., Metamaterials and negative refractive index. *Science*, 305, 788, 2004.
9. Soukoulis, C.M., Linden, S., Wegener, M., Negative refractive index at optical wavelengths. *Science*, 315, 47, 2007.
10. Veselago, V.G., The electrodynamics of substances with simultaneously negative values of ε and μ. *Phys-USP*, 10, 509, 1968.
11. Oh, S.S. and Hess, O., Chiral metamaterials: Enhancement and control of optical activity and circular dichroism. *Nano Converg.*, 2, 1, 2015.
12. Rogacheva, A.V., Fedotov, V.A., Schwanecke, A.S., Zheludev, N.I., Giant gyrotropy due to electromagnetic-field coupling in a bilayered chiral structure. *Phys. Rev. Lett.*, 97, 177401, 2006.
13. Ma, X., Pu, M., Li, X., Guo, Y., Gao, P., Luo, X., Meta-chirality: Fundamentals, construction and applications. *Nanomaterials*, 7, 116, 2017.
14. Ma, X., Huang, C., Pu, M., Pan, W., Wang, Y., Luo, X., Circular dichroism and optical rotation in twisted Y-shaped chiral metamaterial. *Appl. Phys. Express*, 6, 022001, 2013.
15. Stojanović, D.B., Gligorić, G., Beličev, P.P., Belić, M.R., Hadžievski, L., Circular polarization selective metamaterial absorber in terahertz frequency range. *IEEE J. Sel. Top. Quantum Electron.*, 27, 1, 2020.
16. Jing, Z., Bai, Y., Feng, X., Wang, T., Ullah, H., Zhang, Z., Tunable circular dichroism of composite metamaterial on the basis of the phase transition of VO_2. *EPL- Europhys. Lett.*, 125, 35001, 2019.

17. Liu, Y. and Zhang, X., Metamaterials: A new frontier of science and technology. *Chem. Soc. Rev.*, 40, 2494, 2011.
18. Banerjee, S., Pal, B.P., Chowdhury, D., Resonance phenomena in electromagnetic metamaterials for the terahertz domain: A review. *J. Electromagn. Waves Appl.*, 34, 1314, 2020.
19. Zhou, J., Dong, J., Wang, B., Koschny, T., Kafesaki, M., Soukoulis, C.M., Negative refractive index due to chirality. *Phys. Rev. B*, 79, 121104, 2009.
20. Pendry, J.B., A chiral route to negative refraction. *Science*, 306, 1353, 2004.
21. Hannam, K., Powell, D.A., Shadrivov, I.V., Kivshar, Y.S., Broadband chiral metamaterials with large optical activity. *Phys. Rev. B*, 89, 125105, 2014.
22. Zhao, R., Zhang, L., Zhou, J., Koschny, T., Soukoulis, C.M., Conjugated gammadion chiral metamaterial with uniaxial optical activity and negative refractive index. *Phys. Rev. B*, 83, 035105, 2011.
23. Huang, W., Zhang, Y., Tang, X., Cai, L.S., Zhao, J., Zhou, L., Wang, Q., Huang, C., Zhu, Y., Optical properties of a planar metamaterial with chiral symmetry breaking. *Opt. Lett.*, 36, 3359, 2011.
24. Esposito, M., Tasco, V., Todisco, F., Cuscunà, M., Benedetti, A., Sanvitto, D., Passaseo, A., Triple-helical nano wires by tomographic rotatory growth for chiral photonics. *Nat. Commun.*, 6, 6484, 2015.
25. Cui, Y., Kang, L., Lan, S., Rodrigues, S., Cai, W., Giant chiral optical response from a twisted-arc metamaterial. *Nano Lett.*, 14, 1021, 2014.
26. Novitsky, A.V., Galynsky, V.M., Zhukovsky, S.V., Asymmetric transmission in planar chiral split-ring metamaterials: Microscopic Lorentz-theory approach. *Phys. Rev. B*, 86, 075138, 2012.
27. Decker, M., Zhao, R., Soukoulis, C.M., Linden, S., Wegener, M., Twisted split-ring-resonator photonic metamaterial with huge optical activity. *Opt. Lett.*, 35, 1593, 2010.
28. Kenanakis, G., Zhao, R., Stavrinidis, A., Konstantinidis, G., Katsarakis, N., Kafesaki, M., Soukoulis, C.M., Economou, E.N., Flexible chiral metamaterials in the terahertz regime: A comparative study of various designs. *Opt. Mater. Express*, 2, 1702, 2012.
29. Decker, M., Ruther, M., Kriegler, C.E., Zhou, J., Soukoulis, C.M., Linden, S., Wegener, M., Strong optical activity from twisted-cross photonic metamaterials. *Opt. Lett.*, 34, 2501, 2009.
30. Landy, N.I., Sajuyigbe, S., Mock, J.J., Smith, D.R., Padilla, W.J., Perfect metamaterial absorber. *Phys. Rev. Lett.*, 100, 207402, 2008.
31. Ye, Y.Q., Jin, Y., He, S., Omnidirectional, polarization-insensitive and broadband thin absorber in the terahertz regime. *J. Opt. Soc. Am. B: Opt. Phys.*, 27, 498, 2010.
32. Grady, N.K., Heyes, J.E., Chowdhury, D.R., Zeng, Y., Reiten, M.T., Azad, A.K., Taylor, A.J., Dalvit, D.A.R., Chen, H.T., Terahertz metamaterials for linear polarization conversion and anomalous refraction. *Science*, 340, 1304, 2013.

33. Ekinci, Y., Solak, H.H., David, C., Sigg, H., Bilayer Al wire-grids as broadband and high-performance polarizers. *Opt. Express*, 14, 2323, 2006.
34. Liu, Z., Lee, H., Xiong, Y., Sun, C., Zhang, X., Far-field optical hyperlens magnifying sub-diffraction-limited objects. *Science*, 315, 1686, 2007.
35. Fang, N., Lee, H., Sun, C., Zhang, X., Sub-diffraction-limited optical imaging with silver superlens. *Science*, 308, 534, 2005.
36. Tretyakov, S., Sihvola, A., Jylhä, L., Backward-wave regime and negative refraction in chiral composites. *Photonics Nanostructures - Fundam. Appl.*, 3, 107, 2005.
37. Ye, Y. and He, S., 90° polarization rotator using a bilayered chiral metamaterial with giant optical activity. *Appl. Phys. Lett.*, 96, 203501, 2010.
38. Mutlu, M. and Ozbay, E., A transparent 90 polarization rotator by combining chirality and electromagnetic wave tunneling. *Appl. Phys. Lett.*, 100, 051909, 2012.
39. Huang, C., Ma, X., Pu, M., Yi, G., Wang, Y., Luo, X., Dual-band 90 polarization rotator using twisted split ring resonators array. *Opt. Commun.*, 291, 345, 2013.
40. Liu, N. and Giessen, H., Three-dimensional optical metamaterials as model systems for longitudinal and transverse magnetic coupling. *Opt. Express*, 16, 21233, 2008.
41. Ma, X., Huang, C., Pan, W., Zhao, B., Cui, J., Luo, X., A dual circularly polarized horn antenna in Ku-band based on chiral metamaterial. *IEEE Trans. Antennas Propag.*, 62, 2307, 2014.
42. Ma, X., Huang, C., Pu, M., Wang, Y., Zhao, Z., Wang, C., Luo, X., Dual-band asymmetry chiral metamaterial based on planar spiral structure. *Appl. Phys. Lett.*, 101, 161901, 2012.
43. Ma, X., Huang, C., Pu, M., Hu, C., Feng, Q., Luo, X., Multi-band circular polarizer using planar spiral metamaterial structure. *Opt. Express*, 20, 16050, 2012.
44. Rodrigues, S.P., Lan, S., Kang, L., Cui, Y., Cai, W., Nonlinear imaging and spectroscopy of chiral metamaterials. *Adv. Mater.*, 26, 6157, 2014.
45. Li, Z., Zhao, R., Koschny, T., Kafesaki, M., Alici, K.B., Colak, E., Caglayan, H., Ozbay, E., Soukoulis, C.M., Chiral metamaterials with negative refractive index based on four "U" split ring resonators. *Appl. Phys. Lett.*, 97, 081901, 2010.
46. Yoo, S. and Park, Q.H., Metamaterials and chiral sensing: A review of fundamentals and applications. *Nanophotonics*, 8, 249, 2019.
47. Hendry, E., Carpy, T., Johnston, J., Popland, M., Mikhaylovskiy, R.V., Lapthorn, A.J., Kelly, S.M., Barron, L.D., Gadegaard, N., Kadodwala, M.J.N.N., Ultrasensitive detection and characterization of biomolecules using superchiral fields. *Nat. Nanotechnol.*, 5, 783, 2010.
48. Tullius, R., Platt, G.W., Khorashad, L., Gadegaard, N., Lapthorn, A.J., Rotello, V.M., Cooke, G., Barron, L.D., Govorov, A.O., Karimullah, A.S., Kadodwala,

M., Superchiral plasmonic phase sensitivity for fingerprinting of protein interface structure. *ACS Nano*, 11, 12049, 2017.
49. Kelly, C., Tullius, R., Lapthorn, A.J., Gadegaard, N., Cooke, G., Barron, L.D., Karimullah, A.S., Rotello, V.M., Kadodwala, M., Chiral plasmonic fields probe structural order of biointerfaces. *J. Am. Chem. Soc.*, 140, 8509, 2018.
50. Schäferling, M., Dregely, D., Hentschel, M., Giessen, H., Tailoring enhanced optical chirality: Design principles for chiral plasmonic nanostructures. *Phys. Rev. X*, 2, 031010, 2012.
51. Etxarri, A. and Dionne, J.A., Surface-enhanced circular dichroism spectroscopy mediated by nonchiral nanoantennas. *Phys. Rev. B*, 87, 235409, 2013.
52. Yoo, S., Cho, M., Park, Q.H., Globally enhanced chiral field generation by negative-index metamaterials. *Phys. Rev. B*, 89, 161405, 2014.
53. Lv, T.T., Zhu, Z., Shi, J.H., Guan, C.Y., Wang, Z.P., Cui, T.J., Optically controlled background-free terahertz switching in chiral metamaterial. *Opt. Lett.*, 39, 3066, 2014.
54. Kan, T., Isozaki, A., Kanda, N., Nemoto, N., Konishi, K., Takahashi, H., Gonokami, M., Matsumoto, K., Shimoyama, I., Enantiomeric switching of chiral metamaterial for terahertz polarization modulation employing vertically deformable MEMS spirals. *Nat. Commun.*, 6, 8422, 2015.
55. Wang, Q., Rogers, E.T.F., Gholipour, B., Wang, C.M., Yuan, G., Teng, J., Zheludev, N.I., Optically reconfigurable metasurfaces and photonic devices based on phase change materials. *Nat. Photonics*, 10, 60, 2016.
56. Ma, X., Pan, W., Huang, C., Pu, M., Wang, Y., Zhao, B., Cui, J., Wang, C., Luo, X., An active metamaterial for polarization manipulating. *Adv. Opt. Mater.*, 2, 945, 2014.
57. Huang, C., Pan, W., Luo, X., Low–loss circularly polarized transmit array for beam steering application. *IEEE Trans. Antennas Propag.*, 64, 4471, 2016.
58. Cui, J., Huang, C., Pan, W., Pu, M., Guo, Y., Luo, X., Dynamical manipulation of electromagnetic polarization using anisotropic meta-mirror. *Sci. Rep.*, 6, 30771, 2016.
59. Yin, X., Schäferling, M., Michel, A.K.U., Tittl, A., Wuttig, M., Taubner, T., Giessen, H., Active chiral plasmonics. *Nano Lett.*, 15, 4255, 2015.
60. Kenanakis, G., Zhao, R., Katsarakis, N., Kafesaki, M., Soukoulis, C.M., Economou, E.N., Optically controllable THz chiral metamaterials. *Opt. Express*, 22, 12149, 2014.
61. Zhang, S., Zhou, J., Park, Y.S., Rho, J., Singh, R., Nam, S., Azad, A.K., Chen, H.T., Yin, X., Taylor, A.J., Zhang, X., Photoinduced handedness switching in terahertz chiral metamolecules. *Nat. Commun.*, 3, 1, 2012.
62. Zhou, J., Chowdhury, D.R., Zhao, R., Azad, A.K., Chen, H.T., Soukoulis, C.M., Taylor, A.J., O'Hara, J.F., Terahertz chiral metamaterials with giant and dynamically tunable optical activity. *Phys. Rev. B*, 86, 035448, 2012.
63. Li, W., Coppens, Z.J., Besteiro, L.V., Wang, W., Govorov, A.O., Valentine, J., Circularly polarized light detection with hot electrons in chiral plasmonic

metamaterials, in: *Proceeding of the Frontiers in Optics*, San Jose, CA, USA, OSA Publishing, Washington, DC, USA, 2015.
64. Wang, Z., Jia, H., Yao, K., Cai, W., Chen, H., Liu, Y., Circular dichroism metamirrors with near-perfect extinction. *ACS Photonics*, 3, 2096, 2016.
65. Tehrani, K.N., Abdolali, A., Zarifi, D., Kashani, F., Application of chiral layers and metamaterials for the reduction of radar cross section. *Prog. Electromagn. Res.*, 137, 759, 2013.
66. Pendry, J.B., Negative refraction makes a perfect lens. *Phys. Rev. Lett.*, 85, 3966, 2000.
67. Pendry, J.B., Electromagnetic materials enter the negative age. *Phys. World*, 14, 47, 2001.
68. Yang, J.J., Huang, M., Sun, J., Double negative metamaterial sensor based on microring resonator. *IEEE Sens. J.*, 11, 2254, 2011.
69. Yang, J.J., Huang, M., Lan, Y., Li, Y., Microwave sensor based on a single stereo-complementary asymmetric split resonator. *Int. J. Rf. Microw. C. E.*, 22, 545, 2012.
70. Schueler, M., Mandel, C., Puentes, M., Jakoby, R., Metamaterial inspired microwave sensors. *IEEE Microw. Mag.*, 13, 57, 2012.
71. Melik, R., Unal, E., Perkgoz, N.K., Santoni, B., Kamstock, D., Puttlitz, C., Demir, H.V., Nested metamaterials for wireless strain sensing. *IEEE J. Sel. Top. Quantum Electron.*, 16, 450, 2009.
72. Cheng, Y., Nie, Y., Cheng, Z., Gong, R.Z., Dual-band circular polarizer and linear polarization transformer based on twisted split-ring structure asymmetric chiral metamaterial. *Prog. Electromagn. Res.*, 145, 263, 2014.
73. Ekmekci, E., Averitt, R.D., Sayan, G., Effects of substrate parameters on the resonance frequency of double-sided SRR structures under two different excitations. *PIERS Proceedings*, Cambridge, U.S.A, vol. 5, p. 538, 2010.
74. Kriegler, C., Bianisotropic photonic metamaterials. *IEEE J. Sel. Top. Quantum Electron.*, 16, 367, 2009.
75. Karaaslan, M. and Bakir, M., Chiral metamaterial based multifunctional sensor applications. *Prog. Electromagn. Res.*, 149, 55, 2014.

3

Metamaterial Perfect Absorbers for Biosensing Applications

Habibe Durmaz[1]* and Ahmet Murat Erturan[2]

[1]*Department of Electrical and Electronics Engineering, Karamanoglu Mehmetbey University, Karaman, Türkiye*
[2]*Department of Electrical and Electronics Engineering, Konya Technical University, Konya, Türkiye*

Abstract
Metamaterials are manufactured materials that do not occur naturally in nature and have unique features, such as a negative refractive index, exceptional transmission, and a negative Doppler effect. These features enable them for use in stealth technology, super-lenses, perfect absorption, ultrasonic sensors, etc. With its capacity to accomplish complete absorption of incoming electromagnetic radiation with subwavelength features, perfect absorbers based on these material groups have grabbed the interest of many researchers. Here, we present the fundamental theory behind the perfect absorption mechanism in metamaterials having sub-wavelength structures via interference, transmission line, and impedance matching theories. We provide an overview of metamaterial perfect absorbers with 100% absorptivity. In addition, we discuss biosensors based on metamaterial perfect absorbers, such as surface-enhanced infrared spectroscopy (SEIRA) and refractive index sensors. Finally, we conclude the book chapter by giving our opinions on what we think will contribute to the development of metamaterial perfect absorbers.

Keywords: Metamaterials, metamaterial perfect absorbers, plasmonics, plasmonic biosensors, metamaterial biosensors, SEIRA, refractive index sensors, perfect absorption

**Corresponding author*: durmazgul@gmail.com

Inamuddin and Tariq Altalhi (eds.) Electromagnetic Metamaterials: Properties and Applications, (53–84) © 2023 Scrivener Publishing LLC

3.1 Introduction

Metamaterials (MMs) are sub-wavelength composite materials comprised of dielectrics and highly conducting periodic metallic resonators. The main benefit of MM over natural materials is that it may achieve any desired operating wavelength and perfect absorption by selecting appropriate geometric parameters of the resonator and layer thicknesses. The MMs' unique electromagnetic features helped develop them to exhibit exceptional phenomena such as negative refraction [1], perfect lens [2], and electromagnetic stealth [3]. Among ongoing studies on MM-based devices, the MM perfect absorber (MMPA) fully exhibits the advantages of metamaterials. The investigation of the metamaterial-based perfect absorber (MMPA) has long been one of the favorable properties and important applications [4]. With its perfect absorption mechanism, MMPA can be used to convert energy from the incident electromagnetic wave to surface plasmon polaritons (SSPs) at full efficiency. Their small size and operation without input power allow their use in many potential applications at the different spectral ranges when properly designed structures and manufacturing [5].

The first known work on metamaterials was theoretically presented by the Russian physicist Victor Veselago in 1968 [6]. Metamaterials have been defined as artificial materials with both negative permittivity and negative permeability. Although it is known that materials with negative electrical permittivity with fine-wire array structures are based on plasma theory, it was necessary to reveal materials with negative permeability to realize Veselago's theory. The "Swiss Roll" proposal for material with negative effective magnetic permeability was first presented to the literature by Pendry in 1999 [7]. Later in 2000, Smith *et al.* introduced the first metamaterial with simultaneously negative effective electrical permittivity ($\varepsilon_{effective}$) and negative effective magnetic permeability ($\mu_{effective}$) using thin wire arrays and periodic Split Ring Resonator (SRR) arrays [8]. Along with these studies, metamaterials have become popular in many fields, such as electrical engineering, physics, microwave and antenna engineering, optoelectronics, optics, materials science, and semiconductor physics.

MMPA was first proposed by Landy *et al.* in 2008 based on the structure consisting of an electrical resonator and a cut wire operating in the microwave region [9]. Absorber surfaces ensure that the incoming electromagnetic wave is absorbed as heat and ohmic losses. MMPAs usually consist of metal-insulator-metal materials (Figure 3.1), shown in the literature with various applications. However, in recent studies in the literature, these absorbers made of all metal [10] or all-dielectric materials [11] have also been reported. MMPAs, including microwave [12, 13], infrared [14, 15], terahertz [16, 17], and visible region [18, 19] have been the subject of many research areas. Since Landy, there has been a great effort to develop MMPA for many potential application areas, such as cells [20, 21], bio/chemical sensors [21], photodetectors [22], and so on.

For perfect absorption to occur, the electric and magnetic fields of the incoming electromagnetic wave must be fully compatible with the absorber. The absorption rate is represented by the expression $A(\omega) = 1 - R(\omega) - T(\omega)$ [11]. Here, $A(\omega)$ is the absorption rate, $R = (\omega)$ and $T(\omega)$ are reflection and transmission rates, respectively. For absorption to be approximately one, transmission and reflection must be zero. To have perfect absorption, transmission and reflection must be zero. Transmission of the incoming electromagnetic wave can be minimized to zero by using an optically thick, highly conductive metallic layer at the bottom of the metamaterial. To minimize the reflection, the impedance of the free space is needed to be matched to the impedance of the MMA by suitably tailoring the geometric structure and material of resonators [23]. Since MMs have complex electrical permittivity ($\varepsilon(\omega) = \varepsilon_1 + i\varepsilon_2$)) and complex magnetic permeability ($\mu(\omega) = \mu_1 + i\mu_2$)), when $\varepsilon(\omega) = \mu(\omega)$ is satisfied with a suitable resonator geometry and material thicknesses; therefore, the normalized impedance becomes one yielding perfect absorption at a single operating frequency [23]. This chapter presents a brief review of the theory of electromagnetic MMAs and their applications in bio/chemical sensing in mid-IR frequencies. In Section 3.1.2, the theoretical background on the design of the MMAs is given. Section 3.1.3 shows the metamaterial design, working principle, and design methods. Section 3.1.4 discusses bio- and chemical detection approaches in the mid-IR regime based on Surface-Enhanced

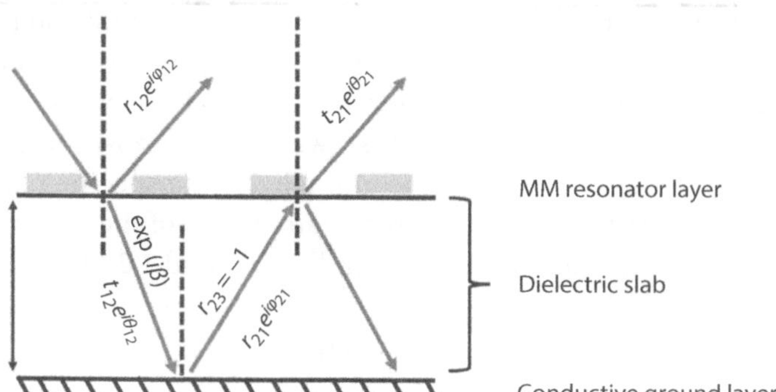

Figure 3.1 Schematic representation of multiple reflections from a metamaterial substrate.

Infrared Spectroscopy (SEIRA) and refractive index change. Finally, a conclusion is given in Section 3.1.5.

3.1.1 Theoretical Backgrounds

This section reviews the general theories that explain the physical phenomena behind the absorption mechanism in metamaterials. The first phenomenon is impedance matching with free space, where the effective permittivity and permeability of the MM can be adjusted in a way that the resulting electric and magnetic resonance satisfy the perfect absorption. With this configuration, all incoming waves are trapped and forced to be absorbed in the interface with no reflection. The second phenomenon is the interference theory in the dielectric layer of the MMA.

3.1.1.1 Impedance Matching Theory

An absorber based on metamaterial generally consists of periodic metallic structures sitting on a dielectric material and at the back coated with an optically thick metallic layer. Because of the electric and magnetic fluxes in the metal and dielectric substrate, there is a significant impedance mismatch between MMs and the surrounding environment. The electric permittivity $\varepsilon(\omega)$ and magnetic permeability $\mu(\omega)$ of corresponding MM fields are $\varepsilon(\omega) = \varepsilon_0\varepsilon_r(\omega)$ and $\mu(\omega) = \mu_0\mu_r(\omega)$, respectively, where ε_0 and μ_0 are the free space permittivity and permeability. $\varepsilon_r(\omega)$ and $\mu_r(\omega)$ are the relative permittivity and permeability of the adjacent materials, respectively. Additionally, electric permittivity and magnetic permeability have frequency-dependent real and imaginary parts as given $\varepsilon(\omega) = \varepsilon(\omega)'+i\varepsilon(\omega)''$ and $\mu(\omega) = \mu(\omega)' + i\mu(\omega)''$ [24]. Since the MMPA has a thick metallic ground plane at the back, there is zero transmittance on the other side. Therefore, we only need to deal with reflection at the front interface. When an incoming light hits on the surface of the MM, three physical events happen; reflection and scattering from the surface and transmission through the dielectric layer. To understand the perfect absorption mechanism, first, we need to know the penetration constant or skin depth (δ) of the incoming light in a specific material [25, 26]. Skin depth is the thickness where the amplitude of the electromagnetic wave decreases to 1/e (0.369), as given in Eqs. 3.1 and 3.2 [26].

$$e^{-m\delta} = 1/e \qquad (3.1)$$

$$\delta = \sqrt{\frac{2}{w\mu\sigma}} \tag{3.2}$$

Here ω is the angular frequency, μ is the permeability, and σ is the electrical conductivity. If the ground plane of the MM is thicker than the penetration depth (δ), the incoming wave cannot travel into the metal; therefore, the transmission becomes "0". Also, the reflection must be zero for perfect absorption in a metamaterial.

The refractive index can also express absorptivity since the absorption coefficient is directly related to the reflection coefficient based on the theory developed by Agustin Jean Fresnel. Fresnel's equations (Eqs. 3.3 and 3.4) describe how much of an incoming electromagnetic wave will be transmitted and how much will be reflected at an interface. The general reflectivity and reflection coefficient from the metamaterial for TE and TM polarized light according to Fresnel equations are [23, 27]:

$$R_{TE} = |r_{TE}|^2 = \left| \frac{\mu_r cos\theta - \sqrt{n^2 - sin\theta}}{\mu_r cos\theta + \sqrt{n^2 - sin\theta}} \right|^2 \tag{3.3}$$

$$R_{TM} = |r_{TM}|^2 = \left| \frac{\varepsilon_r cos\theta - \sqrt{n^2 - sin\theta}}{\varepsilon_r cos\theta + \sqrt{n^2 - sin\theta}} \right|^2 \tag{3.4}$$

where r_{TE} and r_{TM} are the reflection of TE and TM polarized waves, and $n = \sqrt{\varepsilon_r \mu_r}$ is the effective index of the MAs. When an electromagnetic wave is at the normal incident ($\theta = 0°$) at the air-MA interface, Eq. 3.5 reduces to:

$$R = \left| \frac{\sqrt{\mu_r} - \sqrt{\varepsilon_r}}{\sqrt{\mu_r} + \sqrt{\varepsilon_r}} \right|^2 = \left| \frac{Z - Z_0}{Z + Z_0} \right|^2 \tag{3.5}$$

Here $Z = \sqrt{\frac{\mu}{\varepsilon}}$ and $Z_0 = \sqrt{\frac{\mu_0}{\varepsilon_0}}$ are the impedance of the MM and free space, respectively. The absorptivity rate is given as $A(\omega) = 1 - R(\omega) - T(\omega)$. Since the metallic ground plane leads to zero transmission, $T(\omega)$ becomes zero, and absorptivity reduces to Eq. 3.6.

$$A = 1 - R = 1 - \left|\frac{Z - Z_0}{Z + Z_0}\right|^2 = 1 - \left|\frac{\sqrt{\mu_r} - \sqrt{\varepsilon_r}}{\sqrt{\mu_r} + \sqrt{\varepsilon_r}}\right|^2 \tag{3.6}$$

According to equations above show that when $Z = Z_0$ or $\mu_r = \varepsilon_r$, the impedance matching is satisfied. Therefore, absorptivity becomes "1" without an additional term for reflectivity, leading to perfect absorption. This critical condition is called impedance matching theory, where the impedance of the MM is equal to the impedance of free space. The impendence matching condition is satisfied when electric and magnetic fields are in resonance simultaneously. Having either electric or magnetic resonance at a time, the impedance matching is not satisfied. Consequently, perfect absorption cannot be achieved.

The incident electromagnetic wave in the material has an electric permittivity of the $\tilde{\varepsilon}(\omega) = \varepsilon_0 \tilde{\varepsilon}_r(\omega)$ and magnetic permeability of $\tilde{\mu}(\omega) = \mu_0 \tilde{\mu}_r(\omega)$, which are expressed by the Drude-Lorentz model as in Eqs. 3.7 and 3.8 [23, 28, 29]:

$$\tilde{\varepsilon}_r(\omega) = \varepsilon_\infty + \frac{\omega_p^2}{\omega_0^2 - \omega^2 - i\gamma_e \omega} \tag{3.7}$$

$$\tilde{\mu}_r(\omega) = \mu_\infty + \frac{\omega_{p,m}^2}{\omega_{0,m}^2 - \omega^2 - i\gamma_m \omega} \tag{3.8}$$

where the plasma frequencies are ω_p and $\omega_{p,m}$, the oscillator center frequencies are ω_0 and $\omega_{0,m}$, the damping frequencies are γ_e and γ_m, and the electric permittivity and magnetic permeability are ε_∞ and μ_∞, respectively. The electrical permittivity and magnetic permeability values reveal a time-varying electric and magnetic field with the incoming electromagnetic wave, and a Drude-Lorentz-type resonance response occurs depending on the frequency.

3.1.1.2 Interference Theory

Absorption can be due to the interference of electromagnetic waves destructively, which can be controlled by the structure's geometry. A metamaterial absorber is a coupled system where the light-matter interactions at the resonator and dielectric layer interface induce an electric field while the near field couplings result in currents flowing in opposite directions of the resonator and the bottom metal layer. The electromagnetic wave is partially reflected at the front surface with metallic structures (resonators), while the highly conductive metallic layer at the back serves as a perfect reflector with a 180° phase delay.

The incident electromagnetic wave may be scattered, reflected, induced surface electromagnetic waves, transmitted, or absorbed at the surface of the metamaterials. We can simply neglect the scattering from the surface, where the average surface roughness is much smaller than the wavelength. Then, a portion of the incoming wave is reflected at the air-metamaterial interface with a reflection coefficient $\tilde{r}_{12}(\omega)=r_{12}(\omega)e^{i\varphi_{12}(\omega)}$ and transmitted partially into the metamaterial with a transmission coefficient $\tilde{t}_{12}(\omega)=t_{12}(\omega)e^{i\varphi_{12}(\omega)}$, as shown in Figure 3.1. The transmitted wave is reflected back from the highly conductive plane with −1 reflection coefficient. During the transmission in the dielectric substrate, the propagating wave has a propagation constant $\tilde{\beta}=\beta_1+i\beta_2=\sqrt{\varepsilon_d}\,k_0 d$ in a complex form, where k_0 wavenumber in free space, d is the traveling distance (thickness of the dielectric layer), β_1 is the phase of propagation, β_2 is the measure of the absorption in the dielectric material, and $\tilde{\beta}$ is the phase delay caused by total internal reflection from the optical mirror at the back. The transmitted wave back into the dielectric layer will be partially reflected and transmitted back to the surface of the MM with a new reflection and transmission coefficient of $\tilde{r}_{21}(\omega)=r_{21}(\omega)e^{i\varphi_{21}(\omega)}$ and $\tilde{t}_{21}(\omega)=t_{21}(\omega)e^{i\varphi_{21}(\omega)}$, respectively. The back-and-forth reflection and transmission of the incident wave may interfere constructively or destructively in the dielectric layer, and the reflection coefficient can be re-stated as:

$$\tilde{r}(\omega)=\tilde{r}_{12}(\omega)\frac{\tilde{t}_{12}(\omega)\tilde{t}_{21}(\omega)e^{2i\tilde{\beta}}}{1+\tilde{r}_{21}(\omega)2^{2i\tilde{\beta}}} \tag{3.9}$$

Now we can state the absorption, $A(\omega)=1-|\tilde{r}(\omega)|^2$, by minimizing the reflection and neglecting the transmission through the slab of dielectric material that is backed with a thick metallic layer. This result is valid for a normal incident of the electromagnetic wave; when the wave is inclined at an angle θ, the wave takes a longer distance in the dielectric layer. As a result, the propagation distance can be updated as $\tilde{\beta}=\sqrt{\varepsilon_d}\,k_0 e'$, where $d'=\frac{d}{cos\theta'}$ is the updated propagation length in the dielectric material. The angle of refraction is estimated from Snell's law $\sqrt{\varepsilon_d}\,sin\theta'=sin\theta$ [30].

The physical origin of the absorption mechanism of a metamaterial absorber can be well explained based on interference theory as an alternative method.

3.1.2 Metamaterial Designs

Light-matter interactions have been widely investigated in science and engineering. Since the discovery of metamaterials around 2000, light-matter interactions have got great attention since these interactions can be achieved with artificially engineered MMs that are not found naturally. As we covered in the previous section, a MMA generally consists of a dielectric layer sandwiched between periodically arranged sub-wavelength size metallic structures and a thick metallic ground plane. These periodic sub-wavelength structures (meta-atoms) are called resonators that homogenize the incoming electromagnetic field leading to an average response (e.g., permittivity, permeability) from the materials by solving Maxwell's equations under certain boundary conditions [31]. The resonators can be in any form such as square [32], triangle [33], rectangle [34], or circle [35], or more complex structures like fractal [36], and chiral [37, 38]. The physical basis behind perfect absorption depends on the Drude-Lorentz model. For perfect absorption to occur, the equivalent impedance of the metamaterial that is related to the electrical permittivity (ε) and magnetic permeability (μ) values and the free space impedance must be perfectly matched. The electric field is generated by the resonator's geometric structure, and the magnetic field is generated by the dielectric layer; thus, the thicknesses of the sub-wavelength structures and the dielectric material have a direct impact on the resonance frequency and absorption/reflection/transmission strength [23]. The bottom metal layer works as an optical mirror, ensuring that there is no transmission on the other side of the metal layer. The middle dielectric layer is to keep and absorbs the incident electromagnetic wave there. A typical metamaterial absorber layer is shown in Figure 3.2 (a). When an incoming electromagnetic wave illuminates an MM, alternating electric and magnetic fields are induced in the dielectric layer with an anti-parallel current in between two metal layers, as illustrated in Figure 3.2 (b), resulting in the random formation of Surface Plasmon Polaritons (SPPs) along the interface of the metal-dielectric substrate. The polarization and positioning of metallic nanoparticles around the surface plasmons stimulated by incoming radiation results in the formation of Localized Surface Plasmon Polaritons (LSPP). The freedom of adjusting the geometry of the structures paves the way to design MMs from radio waves to the ultraviolet region, where the light-matter interactions occur

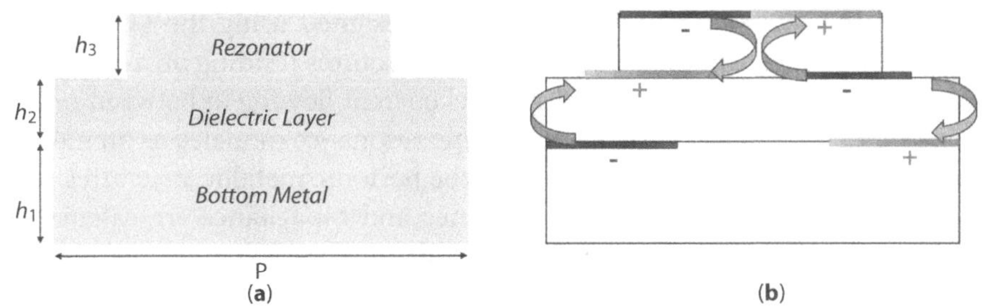

Figure 3.2 a) Schematic representation of a metamaterial absorber. b) Current flow in a metamaterial structure.

at the atomic level. The unique properties of metamaterials enable their use in a variety of applications such as biosensing [39], optical filtering [40], and optical cloak [41] as a narrow band [42], wide-band [43], dual-band [44], and multi-band [45] absorbers.

To understand the design strategies, we need to have an insight into light-matter interactions deeply. For that, the equivalent circuit method and transmission line theory are the most frequently used methods to provide absorbers' impedance matching and model the resonance frequency. In the following sub-section, these two theories will be covered in detail.

3.1.2.1 Equivalent Circuit and Impedance Matching in Metamaterial Perfect Absorbers

To understand the physical origin of the metamaterial absorbers, Maxwell's equations are needed to be solved numerically based on different techniques [46]. The structure of MM unit-cell can be in a complicated form of geometric structures that make the analytical solution of Maxwell's equation inside the MMs difficult. Hence, much commercial software has been developed for numerical solutions of Maxwell's equation to calculate various properties of MMs such as reflection, transmission, absorption, field and charge distributions. The outcome of the numerical simulation can provide valuable information for researchers to comprehend the theory of the absorption mechanism of these materials; however, the dependence of the geometrical shape and parameters, the thicknesses of the layers, the gap between structures, and so on in a unit cell is not sufficient enough. An analytical calculation is necessary to understand the systematic relation between these aforementioned parameters on the absorption mechanism. This chapter will cover two common analytical methods, namely equivalent-circuit (LC) and transmission line theory.

In LC theory, metamaterial structure is designed using the LC equivalent circuit model, where periodic metallic structures residing on a dielectric layer have a resistance R. The anti-parallel current flowing in between metal layers induced by the electric and magnetic resonance modeled as an inductor (L), and capacitors (C) is formed by the periodic metallic structures and the metallic ground plane. The inductance and capacitance are calculated analytically based on the geometrical parameters of the subwavelength structures. The estimated resonance frequency of the metamaterial can be found by using equivalent capacitance and inductance. The resonance frequency can be tuned with geometrical parameters and estimated systematically through the change in equivalent capacitor and inductance; therefore, it can be confirmed with experiments. Figure 3.3 shows an Asymmetric Electric Split-Ring Resonator (AESRR) structure and its corresponding circuit.

Lu *et al.* [48] applied an equivalent circuit (LC) model to a hybrid broadband absorber, whose unit cell is composed of a dielectric layer sandwiched between the hybrid patches antennas in rhombus and circular shapes and metallic ground plane. According to this model, the rhombus-shaped patch antenna and the metallic ground plane create a capacitor, whose capacitance and inductance are approximately given by $C \propto (\frac{D^2}{h})$ and $L \propto h$. The rhombus length is D, while the dielectric layer's thickness is h. The magnetic resonance frequency f_m is:

$$f_m = \frac{1}{2\pi\sqrt{LC/2}} \propto \frac{1}{D} \tag{3.10}$$

Equation 3.10 shows that f_m is inversely correlated only to the rhombus length, not the other geometrical parameters, such as width and the gap. This formulation cannot be used in numerical calculations; however, can give conception into MMs design and optimization. Hoque *et al.* also implemented a similar LC circuit model for a more complex structure [49].

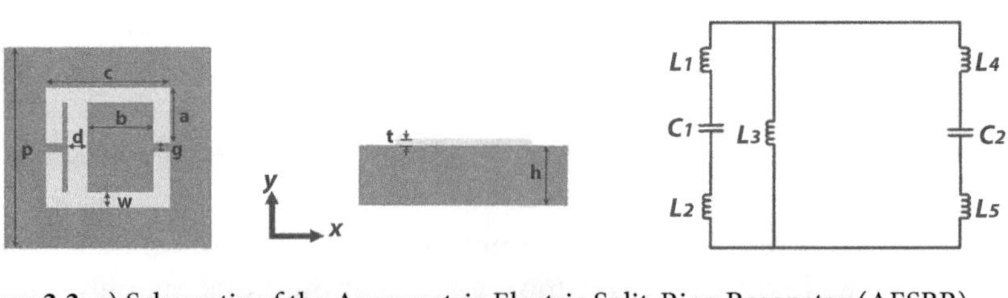

Figure 3.3 a) Schematic of the Asymmetric Electric Split-Ring Resonator (AESRR) structure. b) The equivalent circuit of AESRR. Reproduced with permission from [47].

Later, Pendry *et al.* [50] used the same analogy on a continuous cut wire pair (CWP) with a more realistic approach. The electric and magnetic resonances can be estimated based on the gap between CWP layers.

3.1.2.2 Transmission Line Theory

Transmission line theory is another method used to model impedance matching. This method is practically applied for obtaining the resonance frequency of the absorber in microwave applications since it simplifies the complex structures. In this method, the MMA is handled with two parts: the RLC circuit for the resonators and a short transmission line for the dielectric-metal layer. The resonator elements are considered as a serial and parallel RLC circuit, while the ground plane is considered as a transmission line with a short circuit in the entire circuit system. This base is expressed analytically by Eq. 3.11 [51].

$$Z_{base} = jZ_M^{TM,TE} tan\ tan(\beta h) \tag{3.11}$$

Here, $Z_M^{TM,TE}$ TM and TE denote the free space impedance for polarization, while β and h denote the wave constant propagating through the dielectric material and substrate thickness. The scattering parameter is modeled as a result;

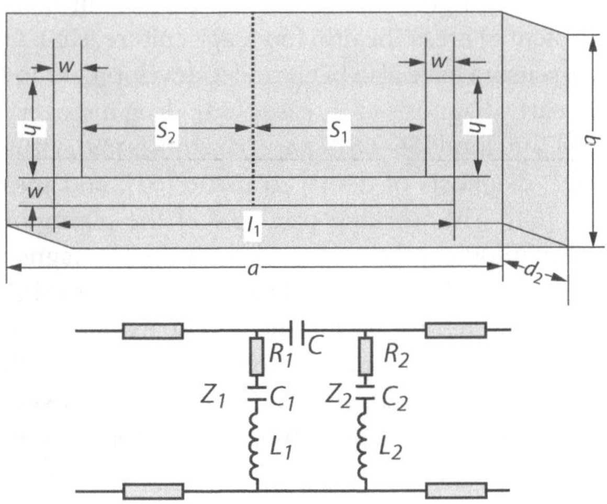

Figure 3.4 Schematic diagram of the structure unit cell and transmission line equivalent circuit of the designed structure. Reproduced with permission from [52].

$$|S_{11}(\omega)| = \left| \frac{Y_m^{TM,TE} - Y_{in}(\omega)}{Y_m^{TM,TE} + Y_{in}(\omega)} \right| \tag{3.12}$$

where $Y_m^{TM/TE}$ denotes the free space characteristic admittance for TM and TE polarization. The transmission line information for a MM is given in Figure 3.4. While the air gap of the U-shaped resonator is drawn with the capacitances C, the current path of the resonator is modeled by drawing the resistor R and inductances L [52].

3.1.3 Biosensing with Metamaterial Perfect Absorbers

A biosensor is a device integrated into a biological sensing part for analyte detection ranging from drug molecules to food and waterborne microorganisms. Early detection and precise diagnosis are highly crucial in a bio-detection system. Better sensing technologies are especially needed during epidemic outbreaks of avian and swine flu, Ebola virus, severe acute respiratory syndrome coronavirus (SARS-CoV), and the recent SARS-CoV-2 or COVID-19. It is important to develop easily accessible sensor platforms to detect these viruses' onset for the global solution. Fast, accurate, real-time diagnostic systems are urged to develop to prepare the health care system for the re-emergence of these kinds of epidemics or other common diseases such as cancer, bacterial infections, etc. To meet this need, conventional sensing methods, chromatography, electrochemical, and the enzyme-linked immunosorbent test (ELISA), have been studied extensively in the bio-detection area and many potential areas (health, food, agriculture, etc.). Over 40 years, plasmonic based-sensors have also been under development in many potential areas such as early diagnosis of diseases [53], drug discovery [54], pathogen detection [55], and food [56–58], cancer diagnosis [59], virus and bacteria detection [60, 61], diagnosis of damaged tissue [62], and measurement of blood properties [63]. The working principle of the plasmonic biosensors is to control and manipulate the subwavelength electromagnetic wave with plasmonic structures based on surface plasmon polaritons (SPPs) [64]. Since plasmonic-based structures usually take up little space and can be easily integrated into photonic circuits to operate from the terahertz to the visible frequency regime. Plasmonic biosensor systems are preferred over conventional sensor systems to overcome the disadvantage of the aforementioned methods since they offer label-free detection, high reusability, fast response time, real-time measurement, and reduced complexity and variability of the methods proposed so far, such as long incubation, separation, cleaning, and detection. However, plasmonic sensors also have some disadvantages; nonspecificity of the binding surface, misinterpretations of the data, and limited mass transportation. MMPAs have been discussed in the literature with many applications in wide and narrow bands. Broadband absorbers have generally been used in solar energy harvesting [65, 66], photoelectric sensing [67, 68], thermal emitter [69, 70], optical imaging [71, 72] and photodetection [73, 74] applications. However, narrow-band absorbers are preferred in sensing [75, 76] and spectroscopic [77, 78] applications since they can show spectral change easily [79]. In this book chapter, we only cover the Refractive Index (RI) sensor for detecting the spectral shift due to refractive index change in the surroundings of the metamaterial sensors [80] and SEIRA methods for detecting the vibrational modes of the molecules [81, 82] among others (LSPR, SER, SERS) based on metamaterial perfect absorber for real-time, sensitive, reliable, label-free, robust and noninvasive biosensing.

3.1.3.1 *Refractive Index*

A refractive index (RI) sensor, also called refractometric sensors, based on MPA is an optical technique used to detect the change in the refractive index of the surrounding environment after binding a target molecule such as a virus, protein, or toxic fluid. These RI sensors are the most favorable platform among other optical sensor systems since it is highly sensitive to the refractive index change it offers label-free detection. The electromagnetic field incoming to the sensor surface couples with the surface plasmons and propagates along the interface of metal-dielectric substrates while decaying exponentially in both substrates. The surface plasmons are the collective oscillations of free electrons at plasmonic antennas and the dielectric substrate interface. The coupling only occurs when their momentum is matched along with the interface. This decaying field is very responsive to changes in dielectric medium (RI change); therefore, it can be utilized as a real-time, precise, selective, and sensitive sensor system. The incoming radiation is absorbed in a certain resonance frequency of the plasmonic nanoantenna with and without the analyte. It is worth mentioning that the index of the medium, as well as the contact of each analyte with the antenna, produces a change in the effective refractive index, resulting in a shift in the spectral positions of the resonance frequencies in the two situations. Therefore, it gives a measure to detect the amount of analyte on the surface of the MA.

Several parameters can be used to validate the performance of the biosensors, namely limit of detection (LOD), sensitivity, Q-factor, and figure of merit (FOM). The refractive index sensitivity or selectivity $\left(S = \frac{\Delta\lambda}{\Delta n} = \frac{nm}{RIU}\right)$ is defined as the fraction of the spectral shift ($\Delta\lambda$) of the plasmonic mode before and after the target molecules to the change in the refractive index (Δn) [83]. In other words, sensitivity is a measure of a sensor to detect a specific analyte or live cell from an admixture of similar or different agents. The overall sensitivity (specificity or selectivity) is directly related to full-width half-maximum (FWHM) as Sherry *et al.* defined the term FOM, which is given by $FOM = \frac{S}{FWHM}$ [83, 84]. FOM is frequently used to measure the sensitivity of the sensor system that evaluates its precision. According to FOM, the change in the refractive index can be defined as the smallest measurable change in the refractive index. It determines the properties of the nanostructure and the properties of the optical sensors [85]. Another parameter to evaluate sensor quality is the Quality-Factor $\left(Q = \frac{\lambda_{res}}{FWHM}\right)$ of the sensor which is the fraction of the resonance wavelength over FWHM. The high Q-factor PAs are very advantageous for precise biosensing applications. The detection limit $\left(LOD = \frac{3\sigma}{S}\right)$ is also an important figure of merit in biosensing applications. It is defined as the minimum number of analytes that can cause a change in the sensor signal. σ is the background noise without an analyte in the sensor platform.

Recent studies have aimed to design biosensors based on MPAs in simple structures that have high FOM, S, and Q factors, which are crucial for sensing low concentrations of biomolecules and low-index materials with high accuracy. The detection of low-index material like Aerogel is very difficult with common detection techniques; however, it can be achievable with a high FOM system. Farhadi *et al.* showed the detection of aerogels based on the Plasmon Induced Transparency (PIT) sensor system by monitoring the shift in the spectral position of the resonance frequency with changing refractive index [86]. The sensor platform consists of a thin layer of Ag film and PMMA grating on top of an SOI wafer and backed with sub-wavelength grating (SWG) made up of SiO_2, Si grating, and sensing material, as shown in Figure 3.5(a). Compared to the homogeneous waveguides, the proposed structure detection sensitivity is increased by a factor of 6 with SWG with its high Q factor (1870), FOM (3.3×10^2), and S (173 RIU^{-1}) sensitivity. In that work, the author detected refractive index changes as low as $\Delta n = 0.002$ (around $n = 1.0$), as shown in Figure 3.5(b).

In another work conducted by Liu *et al.*, an ultra-narrow dual-band metamaterial absorber in the near-IR regime was proposed, and its biomolecular sensitivity was investigated [83]. The structure of the proposed PA consists of TiO_2 plasmonic antennas resting on the SiO_2 spacer layer

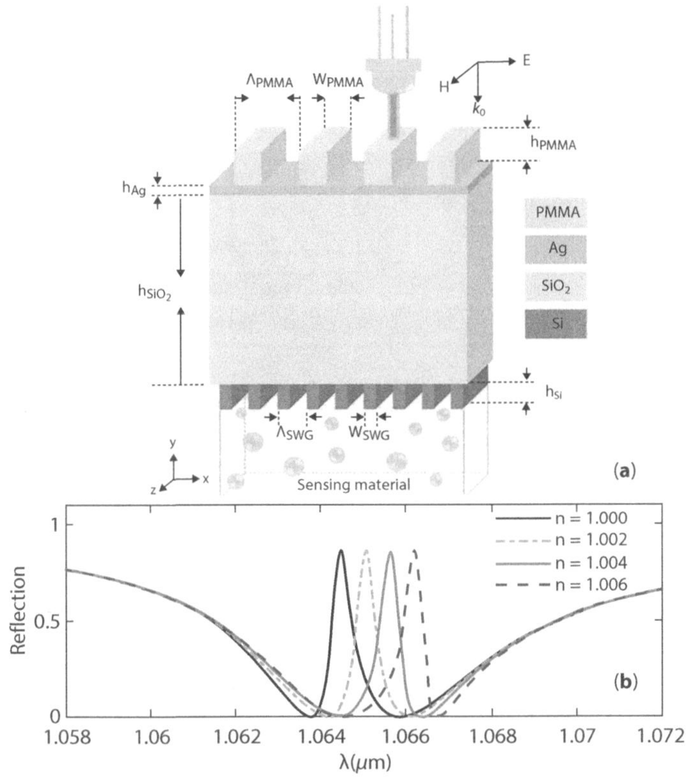

Figure 3.5 a) Schematic configuration of the proposed plasmonic system. b) Reflection spectra of the plasmonic system for refractive index change with the step of 0.005 when Silicon SWG is used. Reproduced with permission from [86].

coated with a thick gold metal film at the back, as shown in Figure 3.6 (a). The Q factor of the proposed MA had a record value of 484 in recent years. The first and the second calculated FOM values of the dual-band MA are as high as 5 RIU^{-1} and 83 RIU^{-1}, respectively. Figure 3.6 (b-c) shows the ultra-narrow double-band perfect absorber and the spectral shift concerning the change in the refractive index of the medium.

A Surface Plasmon Resonance (SPR)-based biosensor has been reported for the detection of Coronavirus (SARS-CoV-2), which has infected 500 million people worldwide and killed approximately 6 million people to date, is a newly reported virus that can be transmitted from person to person. The World Health Organization (WHO) declared the SARS-CoV-2 virus, which broke out in 2019, as a global epidemic [87]. Rapid, reliable, and precise diagnosis has become important to control the spread of this

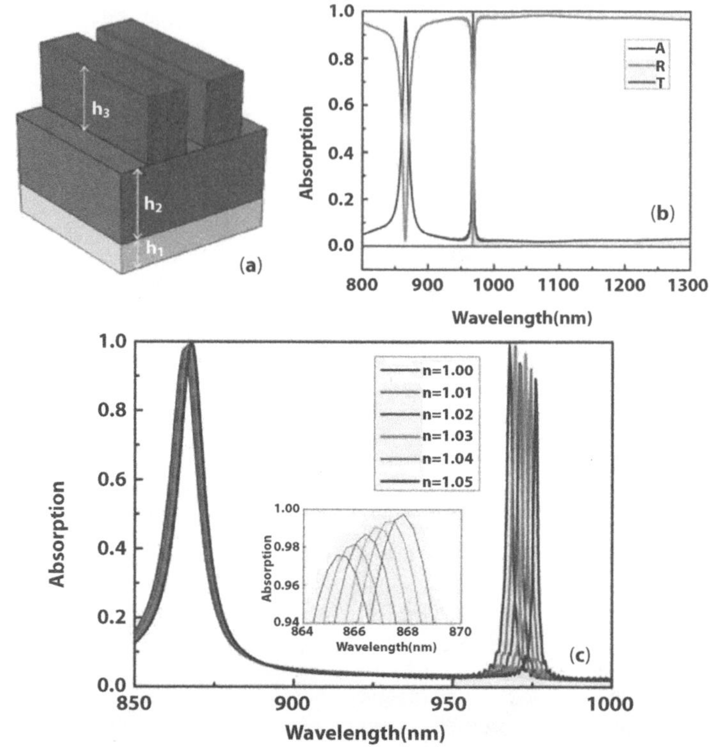

Figure 3.6 a) Three-dimensional view of the designed absorber, b) Spectra of R, T, and A, c) The numerical calculation results of the absorption spectrum when the RI of the environment changes. Reproduced with permission from [83].

disease and prepare a point of care for future pandemics like COVID-19. Akib *et al.* proposed a bio-sensor structure composed of PBS/graphene/PtSe2/Au multi-layers backed with a Bk coupling prism, as shown in Figure 3.7 (a) is designed to detect COVID-19 spike proteins, anti-spike proteins, and viral RNA [88].

In the simulation of Ref 88, the sensor platform was found to be sensitive at an SPR angle of 183.33° /RIU and SPR frequency (SPRF) of 833.33 THz/RIU for sensing COVID-19 virus spike RBD, an SPR angle of 153.85° /RIU and SPRF of 726.50 THz/RIU for detecting the anti-spike protein, and an SPR angle of 140.35° /RIU and SPRF of 500 THz/RIU for sensing the viral RNA. In the same study, the author claimed that the proposed sensor would shorten the long processing times compared to conventional sensors and reduce clinical costs by reducing the false-positive results. Figure 3.7 (b–d) shows the sensing capacity of the proposed

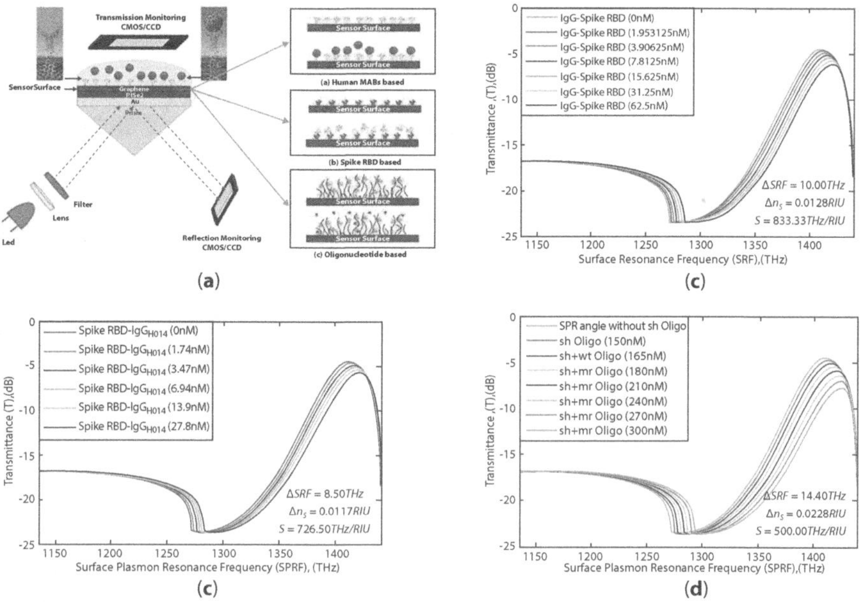

Figure 3.7 a) FDTD simulation schematic for the proposed SPR sensor. b) SPRF characteristics of Bk7/Au(50 nm)/PtSe2(2 nm)/Graphene(1.7 nm) substrates with immobilized IgG (H014 or S309) as ligand and different concentration levels of SARS-CoV-2 spike RBDs (concentration of 1.953125 nM to 62.5 nM) as the analyte. c) SPRF characteristics of Bk7/Au(50 nm)/PtSe2(2 nm)/Graphene (1.7 nm) substrates with immobilized SARS-CoV-2 spike RBDs as the ligand and different concentration levels of IgG-H014 (concentration of 1.74 nM to 27.8 nM) as the analyte. d) SPRF curve characteristics of Bk7/Au(50 nm)/PtSe2(2 nm)/Graphene (0.34 nm L) substrates without probe (sh-Oligo), with sh-Oligo and different concentrated levels of oligonucleotide (wt or mr type) binding for recognition of SARS-Cov-2 virus RNA. SPR angle and SPRF shift right due to binding with oligonucleotides. Reproduced with permission from [88].

biosensor via the IgG-Spike RBD and Spike RBD-IgGH proteins of the SARS-CoV-2 virus at different concentrations ranging from 1.9531 nM to 62.5 nM and 1.74 nM to 27.8 nM, respectively. The detection capacity and sensitivity of the proposed biosensor and SARS-CoV-2 detection ability graphs are shown.

It has been shown that biosensors proposed in recent years can perform more sensitive biosensing with their high FOM, QF (Quality Factor), sensitivity and refractive index values. Table 3.1 shows the sensitivities of the biosensors that have been proposed in recent years. According to the data obtained from 5 different structures, it is obvious that the FOM value of the proposed structure in Ref (86) is at a very high level.

Table 3.1 FOM, QF, sensitivity, and refractive index values of biosensors proposed in recent years.

FOM [RIU^{-1}]	**QF [deg/RIU]**	**Sensitivity**	**Refractive index change**	**Reference**
480	1870	3.3×10^2	2×10^{-3}	[86]
83	484	-	-	[83]
47.00	258.50	183.3	-	[88]
61.55	-	800	0.01	[89]
44.5	123.455	-	-	[90]

3.1.3.2 Surface-Enhanced Infrared Absorption

SEIRA is an analytical technique that provides label-free detection by collective excitation of surface plasmons by enhancing vibrational signals of molecular bonds and structures with incident radiation [91]. First, in 1980 Hartstein *et al.* applied the particle plasmon polaritons to IR absorption (named SEIRA for the first time) via randomly distributed silver nanoparticles with resonance frequency ranging in the visible region. Their findings showed that the vibrational bands of the underlying film were enhanced by a factor of 20 due to silver nanoparticles [92]. Later, with this discovery, the SEIRA technique was frequently used to analyze chemical [93] and biological molecules [94]. Since then, the SEIRA method has been used to identify the molecular fingerprints of the many biomolecules and chemical compounds by sensing the vibrational modes in the chemical bonds of the associated molecules. In the chemical industry and biomolecule sensing, multispectral sensing is critical for identifying the many complex biological and chemical compounds. SEIRA spectroscopy has recently become a very important tool for detecting molecular fingerprints as a sensitive, precise, robust, label-free, cost-effective, nonbulky, real-time, and noninvasive method.

The physical mechanism underlying the SEIRA mechanism is based on enhancing the absorbed molecules signal by the coupling of the incoming IR radiation with the sub-wavelength structures on the surface and the dipole interactions between adjacent structures, leading to the presence of Fano-resonance [91]. The sharp deeps in reflection spectra are attributed to the absorption of the incident electromagnetic radiation by the vibrational bands of the corresponding molecules. By looking at the spectral position of these deeps, molecules can be identified.

The resonance wavelength of the plasmonic antennas for SEIRA applications can be calculated from a half-wave dipole as follows:

$$\lambda = \frac{2L}{m} n a_1 + a_2 \tag{1.13}$$

Here m is the mode number, L is the antenna length, and n is the refractive index of the surrounding environment. The a_2 corresponds to the geometry of the structure and material parameters, while a_1 is related to the reflection phase at the edges of the nanoantenna. The perfect absorber induced by incoming radiation is expressed by Lorentzian as follows:

$$L(\omega) = \frac{a}{(\omega - \omega_0)^2 + a^2} \tag{1.14}$$

where ω_0 and a denote the resonance frequency and resonance line width, respectively. In contrast to Lorentzian, asymmetric Fano-resonances are expressed as:

$$I\alpha \frac{(F_y + \omega - \omega_0)^2}{(\omega - \omega_0)^2 + \gamma^2} \tag{1.15}$$

Here, ω_0 and denotes the resonance position and width, respectively. The vibrational mode occurring in the MPA is an effect of Fano-resonance and is represented by Eq. 3.14.

A dual-band metamaterial perfect absorber (MPA) was employed by Li *et al.* to detect poly(ethyl cyanoacrylate) (PECA), a key polymer in the glue industry [95]. The proposed MPA platform consists of three layers: MgF_2 dielectric spacer sandwiched between a thick gold substrate and gold resonators, which provided more than 90% absorption in both resonances, as shown in Figure 3.8 (a). The PECA molecules are simultaneously detected by matching their vibrational bands to the resonance frequencies of the MPA based on the SEIRA method in Figure 3.8 (b). The sensitivity of the system (an ultra-high sensitivity of 0.76%/nm) is validated by different thicknesses of the PECA, and quantitative analyses are carried out for the amount of the PECA on this chip surface.

Recent studies demonstrate that SEIRA is commonly used as a sensing tool with advantages. However, this method shows insufficient sensitivity to the molecules' distributions and nonlinear response to their concentrations [96].

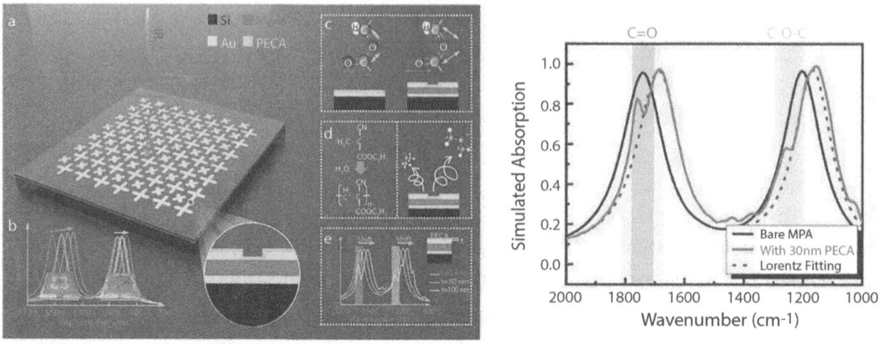

(a) (b)

Figure 3.8 Illustration of multifunctional chemical sensing platform for poly(ethyl cyanoacrylate) on-chip detection. a) 3D schematic view of the dual-resonant MPAs sensing platform. Individual resonance tuning in a dual-resonant array by adjusting the length of the corresponding antenna. b) Vibrational sensing of the PECA, *in situ* observation of ECA polymerization and precipitation, and thickness measurement of PECA by using a multifunctional MPA-based sensing platform. The simulated spectra of the dual-resonant MPA before and after were covered with 30 nm PECA film, showing the details of detecting the two vibrations of PECA simultaneously. Black curve: bare MPA, Red solid line: MPA with PECA, Blue dashed line: baseline. The PECA molecules are simultaneously detected by matching their vibrational bands (stretching band C=O at 1747.5 cm^{-1} and C-O-C at 1252.8 cm^{-1}) to the resonance frequencies of the MPA based on the SEIRA method. Reproduced with permission from [95].

Lin *et al.* proposed a metamaterial perfect absorber that is combined with a hybrid SEIRA and refractive index sensor to overcome this problem for highly sensitive and precise detection of analytes [97]. In this study, the detection of a widely used polyvinyl chloride (PVC) in the industry has been investigated with this hybrid MPA sensor platform for the accurate result since PVC is available in many products with less than 10% (since its use is restricted or prohibited in many countries). This level of concentration is not detectable with conventional methods with high accuracy. The hybrid sensor system consists of a golden split-ring resonator sitting on silica backed with a highly conductive gold plane, as demonstrated in Figure 3.9 (a). The resonator shape is designed to provide Fano resonance, and these resonances are matched with the vibrational modes of the PVC. In this study, the author chose the detect C-Cl bond at 616 to 690 cm^{-1} since it is more unique for identifying the PVC than the most common C-H bonds. Based on the SEIRA results, the reflection dip at 615 cm^{-1} is due to the C-Cl bonding under x-polarization and y-polarization, as demonstrated in Figure 3.9 (b). The sensing capacity of the sensor has also been studied with the index changing from n=1.1 to n=1.5 for y-polarization, as presented in Figure 3.9 (c). The sensitivity of MPA is

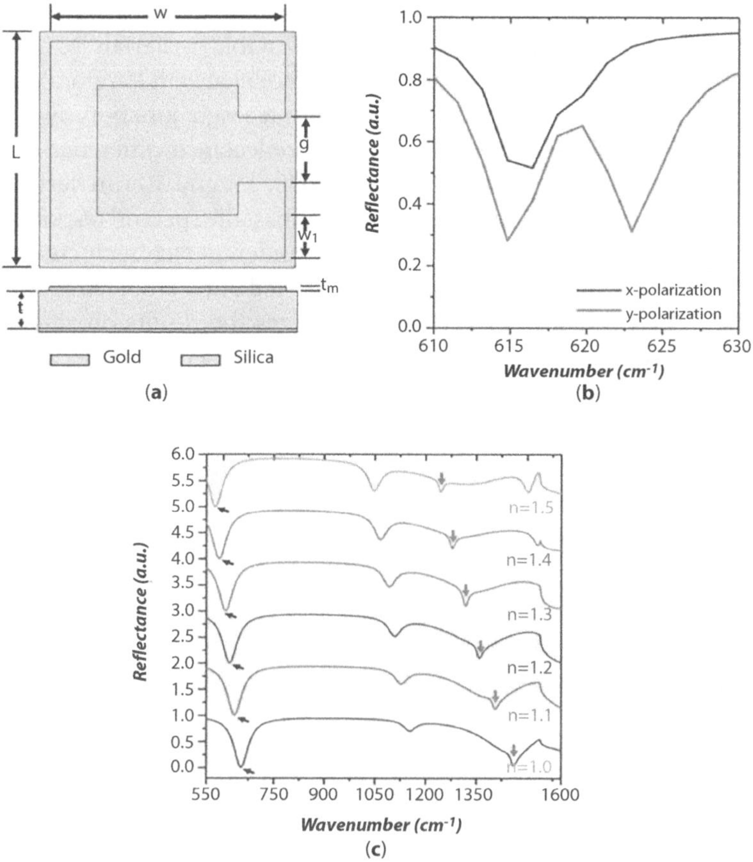

Figure 3.9 a) Scheme of a split-ring-based metamaterial perfect absorber (MPA). b) Simulated reflectance spectrum of the proposed MPA with PVC under two different polarizations. For y polarization, the original reflectance dip was turned into a reflectance peak with two turning points. Meanwhile, there was only a reflectance dip for x-polarization. Such a contrast could corroborate the excitation of Fano-resonance, i.e., the coupling between the plasmonic modes of the MPA and the vibrational mode of PVC. c) Resonance wavelengths of the MPA concerning the refractive index. Sensitivities in the simulation were 4045 and 2361 nm/RIU, respectively. Reproduced with permission from [97].

estimated based on the refractive index for the first and second reflectance peaks as 4045nm/RIU and 2361nm/RIU, respectively.

Although SEIRA spectroscopy is a powerful methodology for detecting molecular vibrations, its detection capacity is limited since a thick insulator layer is needed and in some cases creates a weak field on the surface. In their study, Hwang *et al.* proposed an MA with a vertical nanogap of 10 nm thickness and were able to experimentally detect the single-layer 1-octadecanethiol

(ODT) target molecule with a 36% reflectance difference. The proposed MA structure consists of 3 layers with Gold-SiO_2-Gold materials as shown in Figure 3.10 (a). Vertical surface nanogaps are modeled with temporal coupled-mode theory (TCMT) as paired cavities with a single junction. Figure 3.10 (b) shows the ODT reflection spectra and reflectance difference obtained from the baseline of the MA structure with 30, 15, and 10 nm vertical nanogaps, respectively. In the designed nanoantennas, the spectral response of the SEIRA signal was blueshifted as the depth of undercut etching increased [98].

In recent years, many studies have been carried out on molecular detection with the SEIRA method. Table 3.2 summarizes the results obtained in the studies performed. According to this table, the molecule, molecular structure, antenna structure, and spectral signatures to be determined are shown.

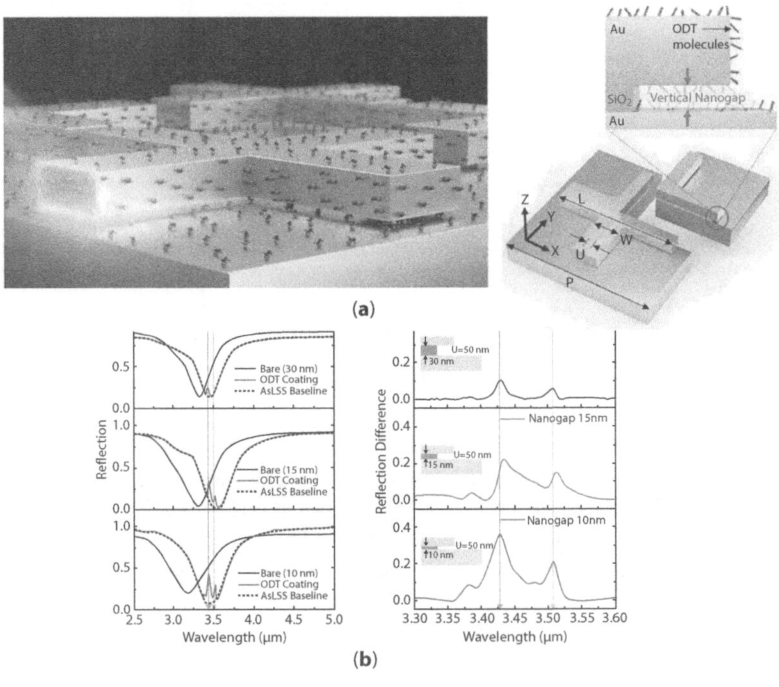

Figure 3.10 a) Schematic images of the MA with a vertical nanogap and its unit cell structure and zoomed-in image of the vertical nanogap edge portion of the MA with ODT monolayer. b) Measured reflection spectra of the MA with 30 nm (top panel), 15 nm (middle panel), and 10 nm (bottom panel) vertical nanogaps before (black) and after (red) ODT coating. The blue-dotted curve indicates the numerically calculated baseline using the AsLSS algorithm, and the blue and orange vertical lines indicate the two vibrational absorption peaks of the ODT molecule. Measured reflection difference SEIRA signal for the MA with 30 nm (top panel), 15 nm (middle panel), and 10 nm (bottom panel) vertical nanogap structures. Reproduced with permission from [98].

Table 3.2 The molecule, molecular structure, antenna structure and vibration bands detected in recent studies.

Detected molecule	Antenna structure	Molecular structure	Vibrational band	Reference
PECA (Poly(ethyl cyanoacrylate))	Si/Au/MgF_2/Au	C=O C-O-C	5724 nm 7987 nm	[95]
PVC (polyvinyl chloride)	Au/Silica/Au	C-Cl	16260 nm	[97]
ODT (1-octadecanethiol)	Au/SiO_2/Au	CH_2	3427 nm 3509 nm	[98]
PMMA (Poly(methyl methacrylate))		C = O	5900 nm	[93]
Protein IgG	Au /MgF_2/Au	Amide-1 Amide-2 Amide-3	6025 nm 6562 nm 6905 nm	
DNT (2,4-dinitrotoluene)	Au/Dielectric Layer/Au	$C_7H_6N_2O_4$	7500 nm	[99]

3.2 Conclusion and Future Work

Since the discovery of metamaterials in 2008, MMPAs with subwavelength structures have gained great interest in many potential application areas over the decade. Here, we have reviewed the theory of MMPAs and one of its applications, biosensing based on perfect absorption. Based on recent studies of MMPAs in the literature, the use of refractive index sensing and SEIRA spectroscopy has become popular in bio- and chemical sensing. The label-free, compact, low-cost, lightweight, real-time measurement non-invasive, and so on properties of MMPA systems make them a unique tool for precise and highly sensitive bio- and chemical detections. The deep learning concept has been used in many studies to gain a comprehensive understanding of massive data and to simplify the design and outcomes. Recently, deep learning and artificial intelligence techniques have been applied in transforming the areas of optical design, integration, and measurements. Combining photonics research with these techniques allows researchers to simplify the design strategies with unique functionalities and optical characterizations, such as real-time detection and transformation of the data, simultaneous measurements, high-speed ultra-high-resolution imaging, high-efficiency energy conversion platforms, and quantum-related applications. The effort of implementing AI and machine learning strategies in all photonic applications ultimately provides efficient high-throughput devices, especially massively multiplex biosensing and super-resolution imaging applications.

References

1. Hoffman, A.J., Alekseyev, L., Howard, S.S., Franz, K.J., Wasserman, D., Podolskiy, V.A., Narimanov, E.E., Sivco, D.L., Gmachl, C., Negative refraction in semiconductor metamaterials. *Nat. Mater.*, **6**, 946–950, 2007.
2. Haxha, S., Malek, F., Ouerghi, F., Charlton, M.D.B., Aggoun, A., Fang, X., Metamaterial superlenses operating at visible wavelength for imaging applications. *Sci. Rep.*, **8**, 16119, 2018.
3. Alitalo, P. and Tretyakov, S., Electromagnetic cloaking with metamaterials. *Mater. Today*, **12**, 22–29, 2009.
4. Rhee, J.Y., Yoo, Y.J., Kim, K.W., Kim, Y.J., Lee, Y.P., Metamaterial-based perfect absorbers. *J. Electromagn. Waves Appl.*, **28**, 1541–1580, 2014.
5. Zhang, H., Zhang, Z., Song, W., Li, Y., He, Y., Wang, Y., Bai, L., Tunable surface plasmon polaritons in the graphene and metamaterials structures. *J. Opt. Soc. Am. B.*, **32**, 1421, 2015.

6. Veselago, V.G., The electrodynamics of substances with simultaneously negative values of ε and μ. *Sov. Phys. Usp.*, **10**, 509, 1968.
7. Pendry, J.B., Holden, A.J., Robbins, D.J., Stewart, W.J., Magnetism from conductors and enhanced nonlinear phenomena. *IEEE Trans. Microwave Theory Tech.*, **47**, 2075–2084, 1999.
8. Smith, D.R. and Kroll, N., Negative refractive index in left-handed materials. *Phys. Rev. Lett.*, **85**, 2933–2936, 2000.
9. Landy, N.I., Sajuyigbe, S., Mock, J.J., Smith, D.R., Padilla, W.J., Perfect metamaterial absorber. *Phys. Rev. Lett.*, **100**, 207402, 2008.
10. Yu, J., Lang, T., Chen, H., All-metal terahertz metamaterial absorber and refractive index sensing performance. *Photonics*, **8**, 164, 2021.
11. Desouky, M., Mahmoud, A.M., Swillam, M.A., Silicon based mid-IR super absorber using hyperbolic metamaterial. *Sci. Rep.*, **8**, 2036, 2018.
12. Ning, J., Dong, S., Luo, X., Chen, K., Zhao, J., Jiang, T., Feng, Y., Ultra-broadband microwave absorption by ultra-thin metamaterial with stepped structure induced multi-resonances. *Results Phys.*, **18**, 103320, 2020.
13. Al-Badri, K.S.L., Design of a perfect metamaterial absorber for microwave applications. *Sci. J. King Faisal Univ.*, **22**, 144–147, 2021.
14. Chen, X.L., Tian, C.H., Che, Z.X., Chen, T.P., Selective metamaterial perfect absorber for infrared and 1.54 μm laser compatible stealth technology. *Optik (Stuttg)*, **172**, 840–846, 2018.
15. Wu, P., Chen, Z., Jile, H., Zhang, C., Xu, D., Lv, L., An infrared perfect absorber based on metal-dielectric-metal multi-layer films with nano circle holes arrays. *Results Phys.*, **16**, 102952, 2020.
16. Schalch, J., Duan, G., Zhao, X., Zhang, X., Averitt, R.D., Terahertz metamaterial perfect absorber with continuously tunable air spacer layer. *Appl. Phys. Lett.*, **113**, 061113, 2018.
17. Duan, G., Schalch, J., Zhao, X., Zhang, J., Averitt, R.D., Zhang, X., An air-spaced terahertz metamaterial perfect absorber. *Sens. Actuators A: Phys.*, **280**, 303–308, 2018.
18. Zhang, B., Li, Z., Hu, Z., Zhang, J., Wang, J., Analysis of a bidirectional metamaterial perfect absorber with band-switchability for multifunctional optical applications. *Results Phys.*, **34**, 105313, 2022.
19. Cao, T., Wei, C.W., Simpson, R.E., Zhang, L., Cryan, M.J., Broadband polarization-independent perfect absorber using a phase-change metamaterial at visible frequencies. *Sci. Rep.*, **4**, 3955, 2014.
20. Rufangura, P. and Sabah, C., Dual-band perfect metamaterial absorber for solar cell applications. *Vacuum*, **120**, 68–74, 2015.
21. Wu, X.X., Lu, C.Y., Huang, T.Y., Creating hot spots within air for better sensitivity through design of oblique-wire-bundle metamaterial perfect absorbers. *Sci. Rep.*, **12**, 3577, 2022.
22. Li, W. and Valentine, J., Metamaterial perfect absorber based hot electron photodetection. *Nano Lett.*, **14**, 3510–3514, 2014.

23. Watts, C.M., Liu, X., Padilla, W.J., Metamaterial electromagnetic wave absorbers. *Adv. Mater.*, **24**, OP181, 2012.
24. Guo, L., da An, Q., Xiao, Z.Y., Zhai, S.R., Cui, L., Li, Z.C., Performance enhanced electromagnetic wave absorber from controllable modification of natural plant fiber. *RSC Adv.*, **9**, 16690–16700, 2019.
25. Liu, X., Li, K., Meng, Z., Zhang, Z., Wei, Z., Hybrid metamaterials perfect absorber and sensitive sensor in optical communication band. *Front. Phys.*, **9**, 637602, 2021.
26. Henry, G.R., Anomalous skin effect in thin films. *J. Appl. Phys.*, **43**, 2996–3001, 1972.
27. Inampudi, S. and Mosallaei, H., Fresnel refraction and diffraction of surface plasmon polaritons in two-dimensional conducting sheets. *ACS Omega*, **1**, 843–853, 2016.
28. Eldlio, M., Che, F., Cada, M., Drude-Lorentz model of semiconductor optical plasmons. *IAENG Trans. Eng. Technol.*, 41–49, 2013.
29. Rakhshani, M.R. and Rashki, M., Metamaterial perfect absorber using elliptical nanoparticles in a multilayer metasurface structure with polarization independence. *Opt. Express*, **30**, 10387, 2022.
30. Wanghuang, T., Chen, W., Huang, Y., Wen, G., Analysis of metamaterial absorber in normal and oblique incidence by using interference theory. *AIP Adv.*, **3**, 102118, 2013.
31. Aydin, K., Ferry, V.E., Briggs, R.M., Atwater, H.A., Broadband polarization-independent resonant light absorption using ultrathin plasmonic super absorbers. *Nat. Commun.*, **2**, 517, 2011.
32. Yu, P., Besteiro, L.V., Wu, J., Huang, Y., Wang, Y., Govorov, A.O., Wang, Z., Metamaterial perfect absorber with unabated size-independent absorption. *Opt. Express*, **26**, 20471, 2018.
33. Rakhshani, M.R. and Rashki, M., Numerical simulations of metamaterial absorbers employing Vanadium Dioxide. *Plasmonics*, 39, 2589–2595, 2022.
34. Huang, L., Chowdhury, D.R., Ramani, S., Reiten, M.T., Luo, S.N., Azad, A.K., Taylor, A.J., Chen, H.T., Impact of resonator geometry and its coupling with ground plane on ultrathin metamaterial perfect absorbers. *Appl. Phys. Lett.*, **101**, 101102, 2012.
35. Zhao, L., Niu, Q., He, Z., Dong, S., Absorptivity enhancement of higher-order electric sextupole plasmonic modes by the outer-square inner-ring coupled resonators. *Opt. Mater. Express*, **8**, 3359, 2018.
36. Xie, T., Chen, D., Yang, H., Xu, Y., Zhang, Z., Yang, J., Tunable broadband terahertz waveband absorbers based on fractal technology of graphene metamaterial. *Nanomaterials*, **11**, 1–12, 2021.

37. Cheng, Y., Chen, H., Zhao, J., Mao, X., Cheng, Z., Chiral metamaterial absorber with high selectivity for terahertz circular polarization waves. *Opt. Mater. Express*, **8**, 1399, 2018.
38. Yang, X., Li, M., Hou, Y., Du, J., Gao, F., Active perfect absorber based on planar anisotropic chiral metamaterials. *Opt. Express*, **27**, 6801, 2019.
39. Chen, T., Li, S., Sun, H., Metamaterials application in sensing. *Sensors*, **12**, 2742–2765, 2012.
40. Brückner, J.-B., le Rouzo, J., Escoubas, L., Berginc, G., Calvo-Perez, O., Vukadinovic, N., Flory, F., Metamaterial filters at optical-infrared frequencies. *Opt. Express*, **21**, 16992, 2013.
41. Cai, W., Chettiar, U.K., Kildishev, A.V., Shalaev, V.M., Optical cloaking with metamaterials. *Nat. Photonics*, **1**, 224–227, 2007.
42. Ghobadi, A., Hajian, H., Gokbayrak, M., Butun, B., Ozbay, E., Bismuth-based metamaterials: From narrowband reflective color filter to extremely broadband near perfect absorber. *Nanophotonics*, **8**, 823–832, 2019.
43. Lu, Y., Chi, B., Liu, D., Gao, S., Gao, P., Huang, Y., Yang, J., Yin, Z., Deng, G., Wideband metamaterial absorbers based on conductive plastic with additive manufacturing technology. *ACS Omega*, **3**, 11144–11150, 2018.
44. Feng, S., Zhao, Y., Liao, Y.L., Dual-band dielectric metamaterial absorber and sensing applications. *Results Phys.*, **18**, 103272, 2020.
45. Luo, Z., Ji, S., Zhao, J., Wu, H., Dai, H., A multiband metamaterial absorber for GHz and THz simultaneously. *Results Phys.*, **30**, 104893, 2021.
46. Davidson, D.B., *Computational electromagnetics for RF and microwave engineering*, Cambridge University Press, Cambridge, 2010.
47. Cao, Y., Ruan, C., Chen, K., Zhang, X., Research on a high-sensitivity asymmetric metamaterial structure and its application as microwave sensor. *Sci. Rep.*, **12**, 1255, 2022.
48. Lu, Y., Li, J., Zhang, S., Sun, J., Yao, J.Q., Polarization-insensitive broadband terahertz metamaterial absorber based on hybrid structures. *Appl. Opt.*, **57**, 6269, 2018.
49. Hoque, A., Islam, M.T., Almutairi, A.F., Alam, T., Singh, M.J., Amin, N., A polarization independent quasi-TEM metamaterial absorber for X and Ku band sensing applications. *Sensors (Switzerland)*, **18**, 4209, 2018.
50. Pendry, J.B., Holden, A.J., Robbins, D.J., Stewart, W.J., Magnetism from conductors and enhanced nonlinear phenomena. *IEEE Trans. Microw. Theory Tech.*, **47**, 2075–2084, 1999.
51. Bagci, F. and Medina, F., Design of a wide-angle, polarization-insensitive, dual-band metamaterial-inspired absorber with the aid of equivalent circuit model. *J. Comput. Electron.*, **16**, 913–921, 2017.
52. Ning, R., Li, D., Yang, T., Chen, Z., Qian, H., Wideband and large angle electromagnetically induced transparency by the equivalent transmission line in a metasurface. *Sci. Rep.*, **9**, 15801, 2019.

53. Liyanage, T., Alharbi, B., Quan, L., Kerscher, A., Slaughter, G., Plasmonic-based biosensor for the early diagnosis of prostate cancer. *ACS Omega*, **7**, 2411–2418, 2022.
54. Federici, J.F., Schulkin, B., Huang, F., Gary, D., Barat, R., Oliveira, F., Zimdars, D., THz imaging and sensing for security applications - explosives, weapons and drugs. *Semicond. Sci. Technol.*, **20**, S266, 2005.
55. Park, J.-H., Cho, Y.-W., Kim, T.-H., Recent advances in surface plasmon resonance sensors for sensitive optical detection of pathogens. *Biosensors (Basel)*, **12**, 180, 2022.
56. Balbinot, S., Srivastav, A.M., Vidic, J., Abdulhalim, I., Manzano, M., Plasmonic biosensors for food control. *Trends Food Sci. Technol.*, **111**, 128–140, 2021.
57. Tantiwanichapan, K. and Durmaz, H., Herbicide/pesticide sensing with metamaterial absorber in THz regime. *Sens. Actuators A: Phys.*, **331**, 112960, 2021.
58. Pechprasarn, S., Ittipornnuson, K., Jungpanich, T., Pensupa, N., Albutt, N., Surface plasmon biosensor platform for food industry. *Appl. Mech. Mater.*, **891**, 103–108, 2019.
59. Usman, F., Dennis, J.O., Aljameel, A.I., Ali, M.K.M., Aldaghri, O., Ibnaouf, K.H., Zango, Z.U., Beygisangchin, M., Alsadig, A., Meriaudeau, F., Plasmonic biosensors for the detection of lung cancer biomarkers: A review. *Chemosensors*, **9**, 326, 2021.
60. Cetin, A.E., Kocer, Z.A., Topkaya, S.N., Yazici, Z.A., Handheld plasmonic biosensor for virus detection in field-settings. *Sens. Actuators B: Chem.*, **344**, 130301, 2021.
61. Manoharan, H., Kalita, P., Gupta, S., Sai, V.V.R., Plasmonic biosensors for bacterial endotoxin detection on biomimetic C-18 supported fiber optic probes. *Biosens. Bioelectron.*, **129**, 79–86, 2019.
62. Bai, J., Guo, H., Li, H., Zhou, C., Tang, H., Flexible microwave biosensor for skin abnormality detection based on spoof surface plasmon polaritons. *Micromachines (Basel)*, **12**, 1550, 2021.
63. Chahkoutahi, A., Emami, F., Rafiee, E., Sensitive hemoglobin concentration sensor based on graphene-plasmonic nano-structures. *Plasmonics*, **17**, 423–431, 2022.
64. Hajshahvaladi, L., Kaatuzian, H., Danaie, M., A high-sensitivity refractive index biosensor based on Si nanorings coupled to plasmonic nanohole arrays for glucose detection in water solution. *Opt. Commun.*, **502**, 127421, 2022.
65. Govindaraman, L.T., Arjunan, A., Baroutaji, A., Robinson, J., Olabi, A.G., Metamaterials for energy harvesting. *Encycl. Smart Mater.*, 2, 522–534, 2022.
66. Piao, R. and Zhang, D., Ultra-broadband perfect absorber based on nano-array of titanium nitride truncated pyramids for solar energy harvesting. *Physica E: Low-Dimens. Syst. Nanostruct.*, **134**, 114829, 2021.

67. Liang, C., Zhang, Y., Yi, Z., Chen, X., Zhou, Z., Yang, H., Yi, Y., Tang, Y., Yao, W., Yi, Y., A broadband and polarization-independent metamaterial perfect absorber with monolayer Cr and Ti elliptical disks array. *Results Phys.*, **15**, 102635, 2019.
68. Liu, B., Wu, P., Li, Y., Zhu, H., Lv, L., Design and photoelectric performance of perfect solar absorber based on GaAs grating. *Front. Mater.*, **8**, 821431, 2022.
69. Argyropoulos, C., Le, K.Q., Mattiucci, N., D'Aguanno, G., Alù, A., Broadband absorbers and selective emitters based on plasmonic Brewster metasurfaces. *Phys. Rev. B*, **87**, 205112, 2013.
70. Diem, M., Koschny, T., Soukoulis, C.M., Wide-angle perfect absorber/thermal emitter in the THz regime. *Phys. Rev. B*, 79, 033101, 2008.
71. Hu, H., Gao, J., Wang, W., Tang, S., Zhou, L., He, Q., Wu, H., Zheng, X., Li, X., Li, X., Rogachev, A.A., Cao, H., Ultra-broadband perfect absorber based on self-organizing multi-scale plasmonic nanostructures. *Appl. Mater. Today*, **26**, 101266, 2022.
72. Deng, H., Li, Z., Stan, L., Rosenmann, D., Czaplewski, D., Gao, J., Yang, X., Broadband perfect absorber based on one ultrathin layer of refractory metal. *Opt. Lett.*, **40**, 2592, 2015.
73. Song, S., Chen, Q., Jin, L., Sun, F., Great light absorption enhancement in a graphene photodetector integrated with a metamaterial perfect absorber. *Nanoscale*, **5**, 9615–9619, 2013.
74. Liu, Z., Zhan, P., Chen, J., Tang, C., Yan, Z., Chen, Z., Wang, Z., Dual broadband near-infrared perfect absorber based on a hybrid plasmonic-photonic microstructure. *Opt. Express*, **21**, 3021–3030, 2013.
75. Park, C.S. and Lee, S.S., Narrowband and flexible perfect absorber based on a thin-film nano-resonator incorporating a dielectric overlay. *Sci. Rep.*, **10**, 17727, 2020.
76. Pan, M., Su, Z., Yu, Z., Wu, P., Jile, H., Yi, Z., Chen, Z., A narrowband perfect absorber with high Q-factor and its application in sensing in the visible region. *Results Phys.*, **19**, 103415, 2020.
77. Korkmaz, S., Turkmen, M., Aksu, S., Mid-infrared narrow band plasmonic perfect absorber for vibrational spectroscopy. *Sens. Actuators A: Phys.*, **301**, 111757, 2020.
78. Lan, G., Jin, Z., Nong, J., Luo, P., Guo, C., Sang, Z., Dong, L., Wei, W., Narrowband perfect absorber based on dielectric-metal metasurface for surface-enhanced infrared sensing. *Appl. Sci.*, **10**, 2295, 2020.
79. Lu, X., Narrowband absorption platform based on graphene and oblique incidence in the infrared range. *J. Lightwave Technol.*, **39**, 1530–1536, 2021.
80. Cheng, Y.C., Chang, Y.J., Chuang, Y.C., Huang, B.Z., Chen, C.C., A plasmonic refractive index sensor with an ultrabroad dynamic sensing range. *Sci. Rep.*, **9**, 5134, 2019.

81. Rodrigo, D., Limaj, O., Janner, D., Etezadi, D., García De Abajo, F.J., Pruneri, V., Altug, H., Mid-infrared plasmonic biosensing with graphene. *Science*, **349**, 165–168, 1979.
82. Hornemann, A., Eichert, D., Flemig, S., Hoehl, A., Ulm, G., Beckhoff, B., Probing biolabels for high throughput biosensing via synchrotron radiation SEIRA technique. *AIP Conf. Proc.*, **1741**, 050007, 2016.
83. Liu, B., Wu, P., Zhu, H., Lv, L., Ultra narrow dual-band perfect absorber based on a dielectric–dielectric–metal three-layer film material. *Micromachines (Basel)*, **12**, 1552, 2021.
84. Sherry, L.J., Chang, S.H., Schatz, G.C., van Duyne, R.P., Wiley, B.J., Xia, Y., Localized surface plasmon resonance spectroscopy of single silver nanocubes. *Nano. Lett.*, **5**, 2034–2038, 2005.
85. Altug, H., Oh, S.H., Maier, S.A., Homola, J., Advances and applications of nanophotonic biosensors. *Nat. Nanotechnol.*, **17**, 5–16, 2022.
86. Farhadi, S., Miri, M., Farmani, A., Plasmon-induced transparency sensor for detection of minuscule refractive index changes in ultra-low index materials. *Sci. Rep.*, **11**, 21692, 2021.
87. W.H.O., Coronavirus disease 2019 (COVID-19): Situation report, 51, 2019.
88. Akib, T.B.A., Mou, S.F., Rahman, M.M., Rana, M.M., Islam, M.R., Mehedi, I.M., Parvez Mahmud, M.A., Kouzani, A.Z., Design and numerical analysis of a graphene-coated SPR biosensor for rapid detection of the novel coronavirus. *Sensors*, **21**, 3491, 2021.
89. Shahamat, Y., Ghaffarinejad, A., Vahedi, M., Plasmon induced transparency and refractive index sensing in two nanocavities and double nanodisk resonators. *Optik (Stuttg)*, **202**, 163618, 2020.
90. Alipour, A., Mir, A., Farmani, A., Ultra high-sensitivity and tunable dual-band perfect absorber as a plasmonic sensor. *Opt. Laser Technol.*, **127**, 106201, 2020.
91. Janneh, M., Surface enhanced infrared absorption spectroscopy using plasmonic nanostructures: Alternative ultrasensitive on-chip biosensor technique. *Results Opt.*, **6**, 100201, 2022.
92. Hartstein, A., Kirtley, J.R., Tsang, J.C., Enhancement of the infrared absorption from molecular monolayers with thin metal overlayers. *Phys. Rev. Lett.*, **45**, 201, 1980.
93. Durmaz, H., Li, Y., Cetin, A.E., A multiple-band perfect absorber for SEIRA applications. *Sens. Actuators B: Chem.*, **275**, 174–179, 2018.
94. Adato, R. and Altug, H., *In-situ* ultra-sensitive infrared absorption spectroscopy of biomolecule interactions in real time with plasmonic nanoantennas. *Nat. Commun.*, **4**, 2154, 2013.
95. Li, D., Zhou, H., Hui, X., He, X., Huang, H., Zhang, J., Mu, X., Lee, C., Yang, Y., Multifunctional chemical sensing platform based on dual-resonant infrared plasmonic perfect absorber for on-chip detection of Poly(ethyl cyanoacrylate). *Adv. Sci.*, **8**, 2101879, 2021.

96. Cheng, F., Yang, X., Gao, J., Ultrasensitive detection and characterization of molecules with infrared plasmonic metamaterials. *Sci. Rep.*, **5**, 14327, 2015.
97. Lin, C.T., Yen, T.J., Huang, T.Y., PVC detection through a hybrid seira substrate and refractive index sensor based on metamaterial perfect absorbers. *Coatings*, **11**, 789, 2021.
98. Hwang, I., Kim, M., Yu, J., Lee, J., Choi, J.H., Park, S.A., Chang, W.S., Lee, J., Jung, J.Y., Ultrasensitive molecule detection based on infrared metamaterial absorber with vertical nanogap. *Small Methods*, **5**, 2100277, 2021.
99. Fabas, A., el Ouazzani, H., Hugonin, J.-P., Dupuis, C., Haidar, R., Greffet, J.-J., Bouchon, P., Dispersion-based intertwined SEIRA and SPR effect detection of 2,4-dinitrotoluene using a plasmonic metasurface. *Opt. Express*, **28**, 39595, 2020.

4

Insights and Applications of Double Positive Medium Metamaterials

Anupras Manwar[1†], Tanmay Bhongade[1†], Prasad Kulkarni[1†], Ajinkya Satdive[2], Saurabh Tayde[2], Bhagwan Toksha[2], Aniruddha Chatterjee[2] and Shravanti Joshi[1*]

[1]Functional Materials Laboratory, Department of Mechanical Engineering, G. S. Mandal's Marathwada Institute of Technology, Aurangabad, Maharashtra, India
[2]Centre for Advanced Materials Research and Technology, Department of Plastic and Polymer Engineering, Maharashtra Institute of Technology, Aurangabad, Maharashtra, India

Abstract
Metamaterials are artificially fabricated composite constructs with tailored features having applications in optoelectronics, sensors, energy regulation and implementation, antennas, super lenses, and even cloaking technology. Among many types, electromagnetic metamaterials are widely documented with special attention on the double negative (DNG) medium metamaterials. However, double positive (DPS) medium metamaterials were rarely discussed in detail. Briefly, DPSs are the subset of metamaterials displaying electric permittivity and magnetic permeability values as positive, that is, greater than zero ($\varepsilon>0$, $\mu>0$). All naturally occurring dielectrics are good examples of such materials, wherein the signals are transmitted in a straight track. DPSs are also called positive index metamaterials (PIM) or right-handed or forward wave's medium. The current review centers on the basic understanding, excellent properties, and novel employment of the DPSs reported to date. First, we have discussed the background in the introduction section followed by an elaboration on insights, fascinating spectacles stemming from extraordinary material properties, and lastly, the accompanying applications.

Keywords: Metamaterials, permittivity, refractive index, electromagnetism, permeability

**Corresponding author*: shravanti.joshi@mit.asia; shravantijoshi@gmail.com
†*Contributed equally to the work.*

Inamuddin and Tariq Altalhi (eds.) Electromagnetic Metamaterials: Properties and Applications, (85–100)

4.1 Introduction

For more than 4000 years, researchers around the globe have been continuously expanding their knowledge of synthesis techniques intending to enhance materials properties for varied applications [1]. The material's property in general is dependent on various parameters, such as the synthesis route, operating temperature, crystal structure, size, shape, dopant concentration, chemical bonding, and so on [2–4]. The last few centuries have seen considerable progress in the fields of materials engineering and nanotechnology resulting in concerted efforts to understand, improve and manipulate materials properties at the nano and atomic scale that were beyond imagination concerning conventional materials [1]. Among the many such novel materials reported in recent years, metamaterials have gained considerable attention [5–7]. The first report was observed toward the end of the 1940s when *Winston E. Kock* affiliated with AT&T Bell Laboratories showcased structures having properties similar to the metamaterials [8]. Thereafter, *David Smith* 2000 demonstrated the metamaterials principle through stacked split-ring resonators in a horizontal manner using thinly structured wires [9].

Briefly, metamaterials are artificially fabricated materials achieving extraordinary features from the formed shapes on the lines of subwavelength scale instead of chemical concentration [10]. The metamaterials showcase optical properties that can be easily reproduced in dimensions lower than that of light wavelength [11]. The research on such materials requires knowledge from interdisciplinary fields involving physics, nanotechnology, composite technology, electrical and computer engineering, microwave engineering, optics and optoelectronics, and material sciences [12]. Metamaterials are widely documented as part of electromagnetism [13–16]. Metamaterials forayed electromagnetic science mainstream owing to applications in the field of optics, human health monitoring, outer space, sensors, detection units in buildings, regulation, and implementation of solar energy, during riots for controlling masses, war equipment, antennas, super lenses, invisible technology, and so on [17–24]. Bulk properties depend on the change in permittivity and permeability, which in turn are affected by the generated electric and magnetic moments due to the interaction with small inhomogeneities in the medium [1]. The materials display rare properties arising because of customized design and shape. Such artificial fabrication provides immense flexibility in terms of morphological alterations and chemical properties of foreign entities to be included in the base matrix, resulting in novel design and development of the metamaterials for cutting-edge performance [20–24]. The composite assemblies usually made up of metallics and polymers are arranged in recurring forms at a measurement lower than the wavelength of the influencing phenomenon [16–20]. It is designed thus allowing the metamaterials to source their features from customized design and shape, rendering their ability to maneuver electromagnetic waves [5, 9, 14]. Engineered metamaterials alter electromagnetic waves by blocking, bending, and absorbing them in a way that is never observed in conventional materials with natural occurrence. Interestingly, from the past few years, metamaterial science has largely been focused on fine-tuning, interconvertible, and intermittent, with excellent detection and monitoring performances [25, 26].

Metamaterials are categorized into different types, namely electromagnetic, frequency-selective surface, elastic, acoustic, structural, thermal, and nonlinear [27]. They are also further classified based on the Hall Effect and frequency bands (such as terahertz, photonic, and plasmonic). Among the aforementioned types, electromagnetic metamaterials are widely documented for various applications and are further classed as single negative (SNG), hyperbolic metamaterials (HMMs), bandgap (BG), double negative (DNG), chiral, Mu-negative (MNG), bi-isotropic and bi-anisotropic, Epsilon negative (ENG), and lastly the double positive (DPS). Recently, *Hassan et al.* reported on a reflector made-up of DNG metamaterial having frequency aligned with 5G antenna technology [28]. *Hakim et al.* studied metamaterial absorbers based on FR-4 transducing platforms for Ku-band monitoring [29]. Moreover, *Shahzad et al.* reported on a DNG metamaterial and its application in estimating the dielectric properties of various coal powders [30]. Despite several such fascinating articles on DNG metamaterials, studies mentioning double-positive medium (DPS) metamaterials were rarely reported in the literature. This book chapter aims to present a deeper understanding limited only to DPS—a subset of electromagnetic metamaterials and review the progress reported in the literature to date acting as a springboard to advance comprehension.

4.2 Insights on the Electromagnetic Metamaterials

Two basic parameters, first, permittivity (ε), and second, permeability (μ), define the overall electromagnetic (EM) medium [11]. In general, ε (μ) characterizes the intent up to which an electric (magnetic) field influences media that in turn is governed by the material's capability to polarize in the presence of a magnetic and an electric field [1]. Figure 4.1 showcases the categorization of the EM metamaterials based on the electric ε and μ for a defined material parameter space under observation. It is one of the ultimate abilities of the metamaterial to showcase a refractive index (RI) < 0, which remains absent in natural materials [31]. Nearly 99% of materials studied in optics and optoelectronics have a refractive index greater than zero. But many noble metals show ε < 0 in the 400 to 800 nm range. However, a material having either values of ε and μ less than zero is considered impervious (that is, opaque) by the EM waves. The non-metamaterial transparent material is defined by the parameters ε, μ, and refractive index (n) depicting ε and μ greater than zero. Nevertheless, selective materials have ε and μ lower than zero. Since the multiplication value of ε*μ is greater than zero, hence n is taken as a real number.

Figure 4.1 illustrates the different types of metamaterials based on electromagnetic parameters, namely ε and μ. The first quadrant (Q-1) details where both the parameters, namely ε and μ, are greater than zero. Such materials are referred to as double positive (DPS), and the most common example is a dielectric material. Here, both the parameters are +ve with signals moving in a forward path and displaying a positive index of refraction metamaterials (PIMs) (Figure 4.2a). The materials are also referred to as "right-handed medium" or "forward-wave medium" or "positive refractive index materials." The second quadrant (Q-2) displays electromagnetic properties, such as permittivity less than zero and permeability greater than zero, hence materials are known as Epsilon negative (ENG). Common examples include metals, ferroelectrics, and defect-induced semiconductors. The third quadrant (Q-3) informs about the material

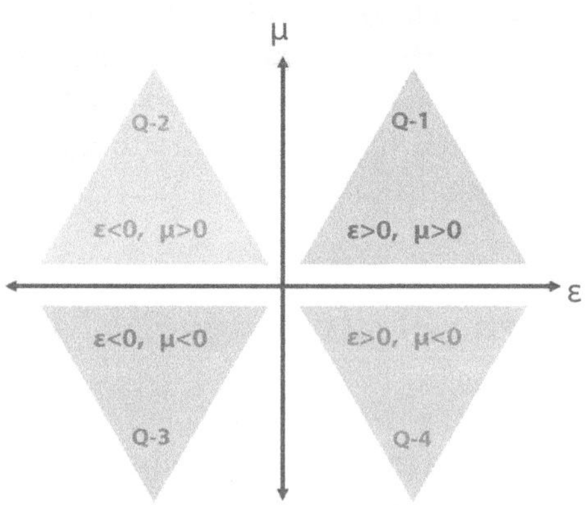

Figure 4.1 Categorization of the electromagnetic metamaterials based on ε and μ for a defined material parameter space under observation.

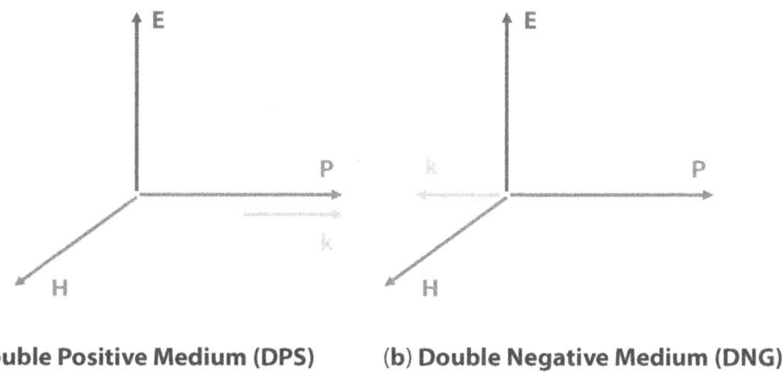

Figure 4.2 Schematic illustration depicting the difference between double positive and double negative medium metamaterials. Here, the P—Poynting vector constitutes the energy motion path, the k—vector constitutes wave transmission, E—electric field, and H—perpendicular magnetic field.

with negative permeability and permittivity. Such artificially synthesized materials are known as double negative (DNG), which have an index of refraction lower than zero and hence referred to as the negative index of refraction metamaterials (NIMs) (Figure 4.2b). Lastly, the fourth quadrant (Q-4) covers the materials with permittivity more than zero and permeability less than zero. These kinds of materials in the literature review are referred to as Mu negative (MNG). Some known examples include gyrotropic and ferrite materials. As mentioned earlier that double negative media metamaterials showcase a negative refractive index, whereas DPS media metamaterials display a positive refractive index for EM waves going through the respective mediums. But in DPS, the wave vector (k) and Poynting vector (P) have the same directional path, while in the case of DNG, the directional path is opposite. Due to this, an incident beam at the DPS-DNG interface will refract to the same side it passed at the start (Figure 4.3).

4.3 Applications of DPS Metamaterials

In 2007, *Ishikawa et al.* detailed the use of magnetic metamaterials with positive permeability reflecting s-polarized instead of p-polarized light by excluding the reflection centered on the Brewster phenomenon [32]. They were able to confirm the single direction of magnetic material with electromagnetism parameters directly proportional to the material's direction using the p and s divergence of Brewster windows. The numerical modeling established diffusion of light takes place without reflection

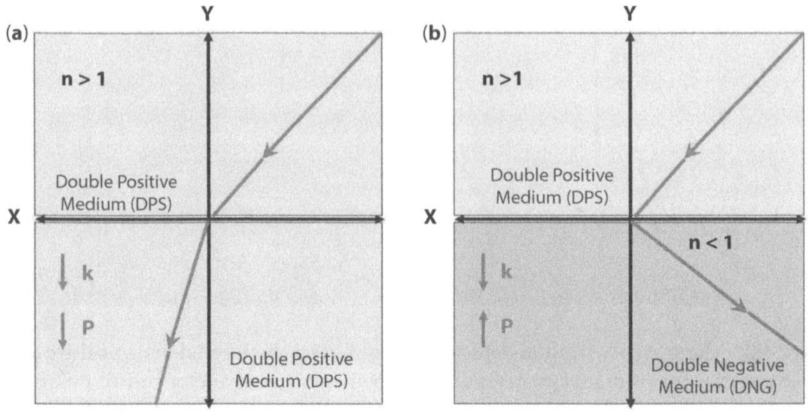

Figure 4.3 Refraction in (a) DPS-DPS and (b) DPS-DNG interfaces. Here, the P—Poynting vector constitutes the energy motion path, and the k—vector constitutes wave transmission.

right through the boundary sandwiched among the materials with the varied index of refraction. They also stated that the 3D metallic nanostructured magnetic metamaterials preparation is cumbersome by the traditional lithography method and hence proposed a new methodology built on the two-photon absorption principle [33, 34]. In 2009, *Aylo et al.* examined the transmission features of electromagnetic waves through a consistent disseminative, periodic/quasiperiodic/randomly altering multi-layered lots of positive index metamaterials (PIMs) or DPS and negative-index metamaterials (NIMs) or DNG employing a transfer matrix method [35]. They observed the wave localization in the arbitrarily disconcerted assembly and concluded that though the length of localization is more than the length of decay in DNG/DPS and DPS/DPS interfaces for equally intermittent and arbitrary assemblies, still they were similar for DNG/DPS in comparison to the DPS/DPS, wherein the observed variance was huge. In 2009, *Serebryannikov et al.* investigated a circular dielectric cylinder showing invisibility at several frequencies upon coating with an isotropic metamaterial integrated ring shell displaying simultaneous large values of positive, as well as negative ε and μ [36]. They deduced that the invisibility stems from using the resonant property in the radial direction. They also state that cylindrical objects can be employed for several variations in the values of diameter/wavelength, in addition to the subwavelength values of the resonant-sized entities. Further, it was found that the existence of shell material frequency scattering largely influences the probability of a multifrequency process. In 2010, *Ghosh et al.* reported on a ray optics methodology situated alongside the central axis for a body transmitting a beam with monochromatic Gaussian nature through a mound of alternated sheets of metamaterials having a refractive index with positive and negative values [37]. The proposed techniques also have applications to achieve representations by modifying the Gaussian beam size or fabricating detection units estimating the features affecting the refractive index or thickness of the altered metamaterials sheet. Additionally, during the same year, *Shi et al.* explored theoretical resonance properties in planar DPS and DNG metamaterials conked out regularity in a highly entrapped situation [38]. The UV-Vis spectroscopic data in the form of reflection and transmission mode for the metamaterial illustrated resonance to the incident EM signal just below the normal incidence value. In addition, the quality factor was observed to be completely reliant on the geometric features of the materials.

In 2011, *Cojocaru* detailed the transverse electric and magnetic of the comparable plate waveguide packed with PIM/PIM or NIM/NIM and NIM/PIM sheets and offered straightforward normalized scattering equations for two different types of modes [39]. The mathematical models were proposed to display scattering curvatures, overall normalized intensity borne by the respective modes during the transmission, and the fields within the comparable plate device guiding the waves. Similarly in 2012, *Popov et al.* discussed superior features of short-pulse second harmonic generation for DPS/DNG materials, appearing in complete contradiction with the equivalents in conventional materials [40]. They fabricated carbon nanotube-based metamaterials supplementing the concurrence of regular elementary and rearward second-harmonic EM signals. Lastly, predictions on unprecedented features of frequency replicating disordered ocular meta mirror were presented. Again in the year 2013, *Korotkevich et al.* reported on the disordered combination and resultant energy surge amid light pulses transmitting in positive and negative refractive index systems contained by a metamaterial showing resonance as a result of doping at atomic level [41]. They revealed the formation of enmeshed, copropagating, disordered pulses transmitting none of the cluster velocities during the mathematical modeling. In rare sequences, the pulse conducts itself in a seesaw manner and promulgates alongside except with slow velocity compared to the initial signal. The emission yielded by the set transmits in the orientation preferred by the linear cluster velocities of its particular colors. They further noticed a pulse showing color matching to DPS converting to DNG region, where emission originated moving rearward with all intents. Concurrently, a hotbed medium of everlasting agitation is created when the pulsation in the DPS zone declines and ultimately comes to an end. The electric field in the DNG zone starts transmitting emissions rearward exactly from the hotbed site. The power accumulated in the hotbed seems to stay still in space upon increasing the period. Simply they successfully envisioned a transparent kind of color switching with convenient experimentation [42, 43]. In 2014, *Pearson et al.* presented a new approach to developing metamaterials for antennas and additional microwave instruments [44]. Earlier reports showed that a decrease in the dimension of antennas centered around innovation in the materials with a greater value of ε_r. Nonetheless, *Pearson et al.* pointed out that by fine-tuning the values of μ and ε, a major decrease in device dimension is achievable while simultaneously perfecting the impedance values for enhancement in the effectiveness and bandwidth when compared to prior modifications that were based only on the value of ε_r. Such reports can result in superior accomplishment for different types of antennas and additional microwave instruments assimilating insulated matter with an optimum value of impedance. The initial stage of the work focused on generating different forms of ferrites and analyzing the parameters of the formed composite. The investigation of parameters, namely μ and ε carried out for preliminary samples based on ferrite-metamaterials revealed that it is possible to optimize the values of wavelength as well as impedance. Following the success, many such ferritic samples were prepared in the form of solid pellets using a binder. The pellets displayed dielectric constant and effective permeability much higher than the free space establishing a novel method for the fabrication of DPS material, giving access to several advantages from the ease of fabrication, miniaturization, and commercialization viewpoints.

Again in 2015, *Pearson et al.* reported on a new procedure to produce a DPS material made up of nickel-zinc ferrite composite [45]. The composite was prepared at many reduced temperatures compared to sintered products. Prior estimation illustrated that the composite owned ε and μ values greater than 4 and less than 6 for frequencies ranging from 0.2 to 2 gigahertz (GHz). The report further elaborated on arithmetical data for dielectric strength further confirming the economic feasibility and efficiencies of the nickel-zinc ferrite composites. In the same year, *O'Connor et al.* reported on magnetodielectric feasibility for application in antenna technology through simultaneous comparison between helical and conical shapes [46]. The conical pattern showed a frequency downshift inversely proportional to the $\sqrt{(\kappa.\varepsilon_r)}$. Magnetodielectricity was a feasible solution upon increasing the value of $(\kappa.\varepsilon_r)$ to 9 owing to optimal frequency downshift in terms of adequate return loss as well as the bandwidth. Owing to signal transmission in materials and ambient air, it was also observed that the conical pattern showed higher variations in the operating bands compared to a helical antenna. The ability to control both features proved to be advantageous in employing metamaterials displaying strong magnetoelectricity. Furthermore, in the year 2016, *Pearson et al.* developed a novel DPS media by combining different compositions of nickel-zinc ferrites in powdered form for applications in pulsed power systems, and similar devices [47]. The measurements showed μ_r and ε_r values greater than 3 and less than 6 for frequencies ranging from 0.02 to 2 gigahertz (GHz). In addition, *Noel et al.* reported on emerging dielectric metamaterial composites with a drastic decrease in device dimensions functioning on impedance values that of free space and sometimes even lower [48]. Before this, they developed a metamaterial synthesis methodology involving a core made up of ferromagnetic material encased by a $BaTiO_3$ matrix that allowed flexibility in achieving a preferred permittivity value. Several composite samples were prepared by increasing the thickness of $BaTiO_3$ sheets and analyzed to estimate electromagnetism parameters concerning frequency. The surface area of the metamaterials was used to evaluate the elemental composition that was further taken as a reference altering the μ and ε values. It was also concluded that an increase in $BaTiO_3$ concentration in contact with a ferromagnetic core even in small quantity enhanced the composite's permittivity in many manifolds but with minimal change in permeability.

Figure 4.4 details a report by *Xia et al.* demonstrating that at the boundary of dielectric and bianisotropic metamaterial with only positive permeability and permittivity, transverse electric magnetic modes can exist [49]. The such artificially designed metamaterial can be successfully used for applications in chemical sensing, super-resolution imaging, and many similar devices. In 2020, Amanollahi *et al.* reported on the transverse magneto-optical Kerr effect (TMOKE) feature by analogizing s and p polarization signals in a bi-gyrotropic matter [50]. They further demonstrated that DPS and DNG are competent enough to accomplish the highest TMOKE however with contrasting signals (Figures 4.5 and 4.6). It was further evinced from Figure 4.5, that for s and p-polarized waveforms, gyration coefficients (β and α) showed a direct correlation with TMOKE. Subsequently, TMOKE is observed in DPS and DNG media irrespective of the parametric value. Lastly, the highest and lowest values supersede at 40° for in the p polarization signal (Figure 4.5 f-h). Interestingly TMOKE is absent for zero gyration (Figure 4.6). By

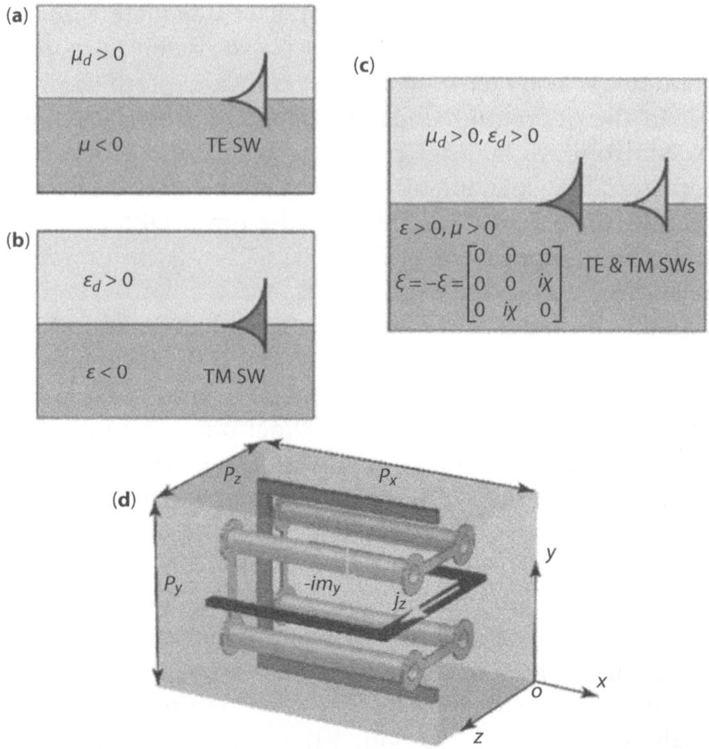

Figure 4.4 Schematic illustration of three types of interface with the corresponding surface wave and one realistic model of a bianisotropic metamaterial. (a) TE surface wave in a ferromagnet/dielectric interface. (b) TM surface wave at the metal/dielectric interface. (c) TE and TM surface waves coexist in a bianisotropic material/dielectric interface. (d) The unit cell of a bianisotropic metamaterial. The copper solid represents copper, and the light gray region is filled with dielectric material (F4BM) with relative permittivity $\varepsilon = 2.2$. The copper structure is identical to two split ring resonators, denoted by red and blue. The bianisotropic response can be understood from the field and current coupling between the two split-ring resonators. The periods in the x-, y-, and z-directions are $P_x = 4.67$ mm, $P_y = 3$ mm, and $P_z = 3$ mm. Reproduced with permission from Ref. [49]. Copyright 2019 De Gruyter.

switching from DPS to DNG, the highest and lowest values are replaced. A similar observation was found in SNG metamaterials giving lower TMOKE when μ' and ε' are swapped. Nevertheless, the report lays the foundation strongly confirming the TMOKE's significance in data storage and imaging techniques.

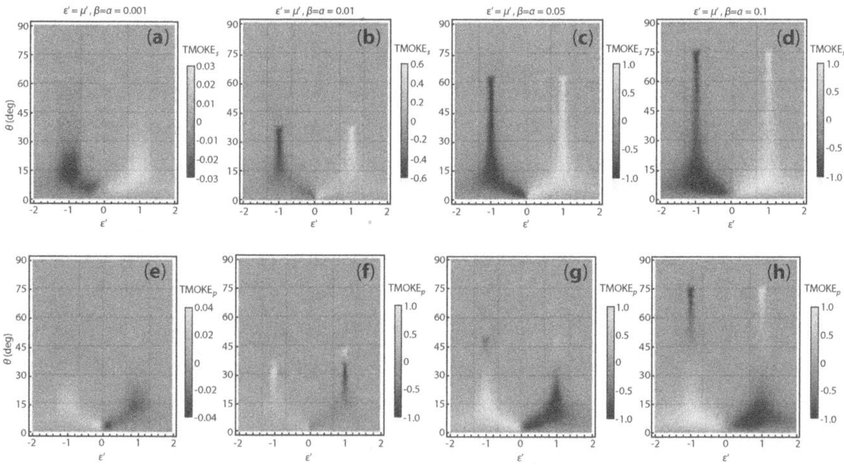

Figure 4.5 Transverse magneto-optical Kerr effect for both s- and p-polarizations as a function of the angle of incidence and ε in double-positive and double-negative metamaterials with μ″ = 0.01, ε″ = 0.04 and different values of β and α. Reproduced with permission from Ref. [50]. Copyright 2020 Elsevier B.V.

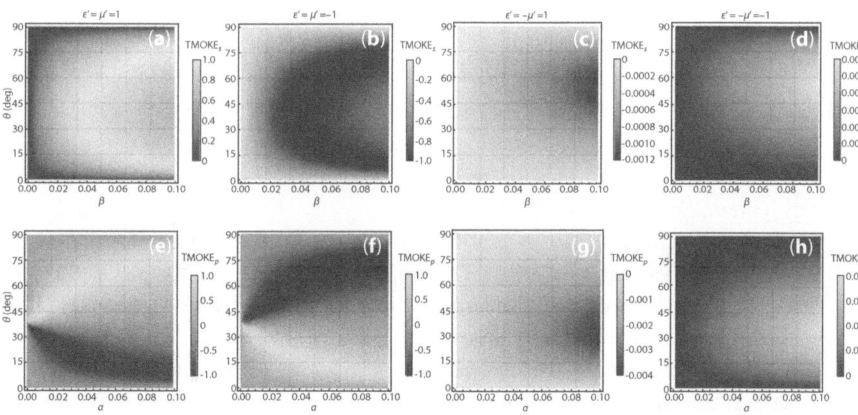

Figure 4.6 Transverse magneto-optical Kerr effect for both s- and p-polarizations as a function of the angle of incidence and gyration factor in (a) and (e) double positive, (b) and (f) double negative and (c), (d), (g) and (h) single-negative metamaterials. In all cases μ″ = 0.01 and ε″ = 0.04. Reproduced with permission from Ref. [50]. Copyright 2020 Elsevier B.V.

4.4 Conclusion

Double-positive medium metamaterials will continue to emerge as a rare yet interesting genre in the fields of materials science and nanotechnology. The extraordinary features shown by such metamaterials are anticipated to garner the extensive interest of physicists, engineers, and scientists from various backgrounds. This will act as a driving force to blend research knowledge around multidisciplinary fields leading to cutting-edge technology and advancements in the form of solutions for societal and environmental concerns. The last few decades showed several notable innovations in the form of chemical sensing, energy harvesting, super-resolution imaging, high-power antennas, optical information processing, integrated photonic devices, super lens, and even cloaking technology. However, man's nature is curious and hopeful of a good future, which is expected to pave way for several other fascinating inventions and applications that are waiting for us to discover.

Acknowledgments

SJ is grateful to accept the project grant approved by the Department of Science and Technology, Government of India.

References

1. Liu, Y. and Zhang, X., Metamaterials: A new frontier of science and technology. *Chem. Soc. Rev.*, 40, 2494, 2011.
2. Joshi, S., Ippolito, S.J., Periasamy, S., Sabri, Y.M., Sunkara, M.V., Efficient heterostructures of Ag@CuO/$BaTiO_3$ for low-temperature CO_2 gas detection: Assessing the role of nanointerfaces during sensing by operando DRIFTS technique. *ACS Appl. Mater. Interfaces*, 9, 27014, 2017.
3. Tonde, S., More, S., Hazra, C., Kundu, D., Joshi, S., Satdive, A., Tayde, S., Bornare, D., Toksha, B., Naik, N., Chatterjee, A., 1D sub 10 nm nanofabrication of ultrahydrophobic Ag@TiO_2 nanowires and their photocatalytic, UV shielding and antibacterial properties. *Adv. Powder Technol.*, 33, 103404, 2022.
4. Kulkarni, A., Satbhai, M., Li, W., Bornare, D., Syed, K., Joshi, S., Oleic acid induced tailored morphological features and structural defects in CuO for multifunctional applications. *Mater. Adv.*, 3, 418, 2022.
5. Smith, D.R., Pendry, J.B., Wiltshire, M.C.K., Metamaterials and negative refractive index. *Science*, 305, 788, 2004.

6. Ramakrishna, S.A., Physics of negative refractive index materials. *Rep. Prog. Phys.*, 68, 449, 2005.
7. Fang, N., Xi, D.J., Xu, J.Y., Ambati, M., Srituravanich, W., Sun, C., Zhang, X., Ultrasonic metamaterials with negative modulus. *Nat. Mater.*, 5, 452, 2006.
8. Zharov, A.A., Zharova, N.A., Noskov, R.E., Shadrivov, I.V., Kivshar, Y.S., Birefringent left-handed metamaterials and perfect lenses for vectorial fields. *New J. Phys.*, 7, 220, 2005.
9. Smith, D.R., Padilla, W.J., Vier, D.C., Nasser, S.C., Schultz, S., Composite medium with simultaneously negative permeability and permittivity. *Phys. Rev. Lett.*, 84, 4184, 2000.
10. Kshetrimayum, R.S., A brief introduction to metamaterials. *IEEE Potentials*, 23, 44, 2004.
11. Ciattoni, A., Rizza, C., Palange, E., Extreme nonlinear electrodynamics in metamaterials with very small linear dielectric permittivity. *Phys. Rev. A*, 81, 043839, 2010.
12. Page, J., Neither solid nor liquid. *Nat. Mater.*, 10, 565, 2011.
13. Pendry, J.B., Holden, A.J., Stewart, W.J., Youngs, I., Extremely low frequency plasmons in metallic mesostructures. *Phys. Rev. Lett.*, 76, 4773, 1996.
14. Shelby, R.A., Smith, D.R., Schultz, S., Experimental verification of a negative index of refraction. *Science*, 292, 77, 2001.
15. Linden, S., Enkrich, C., Wegener, M., Zhou, J., Koschny, T., Soukoulis, C.M., Magnetic response of metamaterials at 100 terahertz. *Science*, 306, 1351, 2004.
16. Soukoulis, C.M., Linden, S., Wegener, M., Negative refractive index at optical wavelengths. *Science*, 315, 47, 2007.
17. Brun, M., Guenneau, S., Movchan, A.B., Achieving control of in-plane elastic waves. *Appl. Phys. Lett.*, 94, 061903, 2009.
18. Rainsford, T.J., Mickan, S.P., Abbott, D., T-ray sensing applications: A review of global developments. *Proc. SPIE Smart Structures, Devices, and Systems II*, vol. 5649, p. 826, 2005.
19. Alici, K.B. and Özbay, E., Radiation properties of a split ring resonator and monopole composite. *Phys. Status Solidi B*, 244, 1192, 2007.
20. Enoch, S., Tayeb, G., Sabouroux, P., Guérin, N., Vincent, P., A metamaterial for directive emission. *Phys. Rev. Lett.*, 89, 213902, 2002.
21. Li, W. and Valentine, J., Metamaterial perfect absorber based hot electron photodetection. *Nano Lett.*, 14, 3510, 2014.
22. Yu, P., Besteiro, L.V., Huang, Y., Wu, J., Fu, L., Tan, H.H., Chennupati, J., Wiederrecht, G.P., Govorov, A.O., Broadband metamaterial absorbers. *Adv. Opt. Mater.*, 7, 1800995, 2018.
23. Pendry, J.B., Negative refraction makes a perfect lens. *Phys. Rev. Lett.*, 85, 3966, 2000.
24. Schurig, D., Mock, J.J., Justice, B.J., Cummer, S.A., Pendry, J.B., Starr, A.F., Smith, D.R., Metamaterial electromagnetic cloak at microwave frequencies. *Science*, 314, 977, 2006.

25. Modi, A.Y., Alyahya, M.A., Balanis, C.A., Birtcher, C.R., Metasurface based method for broadband RCS reduction of dihedral corner reflectors with multiple bounces. *IEEE Trans. Antennas Propag.*, 67, 298, 2019.
26. Modi, A.Y., Balanis, C.A., Birtcher, C.R., Shaman, H.N., Novel design of ultrabroadband radar cross section reduction surfaces using artificial magnetic conductors. *IEEE Trans. Antennas Propag.*, 65, 5406, 2017.
27. Kumar, R., Kumar, M., Chohan, J.S., Kumar, S., Overview on metamaterial: History, types and applications. *Mater. Today: Proc.*, 56, 3016, 2022.
28. Hasan, M.M., Islam, M.T., Moniruzzaman, M., Soliman, M.S., Alshammari, A.S., Sulayman, I.I.M.A., Samsuzzaman, M., Islam, M.S., Symmetric engineered high polarization insensitive double negative metamaterial reflector for gain and directivity enhancement of Sub-6 GHz 5G Antenna. *Mater.*, 15, 5676, 2022.
29. Hakim, M.L., Alam, T., Soliman, M.S., Sahar, N.M., Baharuddin, M.H., Almalki, S.H.A., Islam, M.T., Polarization insensitive symmetrical structured double negative (DNG) metamaterial absorber for Ku-band sensing applications. *Sci. Rep.*, 12, 479, 2022.
30. Shahzad, W., Hu, W., Ali, Q., Raza, H., Abbas, S.M., Ligthart, L.P., A low-cost metamaterial sensor based on DS-CSRR for material characterization applications. *Sensors*, 22, 2000, 2022.
31. Kumar, R. and Kumar, P., Fabry–Pérot cavity resonance based metamaterial absorber for refractive index sensor at infrared frequencies. *Opt. Commun.*, 514, 128142, 2022.
32. Ishikawa, A., Tanaka, T., Kawata, S., Photon tunneling at the material boundary by positive permeability metamaterials. *Proc. SPIE Photonic Metamaterials*, vol. 6638, p. 663808, 2007.
33. Tanaka, T., Ishikawa, A., Kawata, S., Two-photon induced reduction of metal ions for fabricating three dimensional electrically conductive metallic microstructure. *Appl. Phys. Lett.*, 88, 081107, 2006.
34. Ishikawa, A., Tanaka, T., Kawata, S., Improvement in the reduction of silver ions in aqueous solution using two-photon sensitive dye. *Appl. Phys. Lett.*, 89, 113102, 2006.
35. Aylo, R., Banerjee, P.P., Nehmetallah, G., Perturbed multilayered structures of positive and negative index materials. *J. Opt. Soc. Am. B*, 27, 599, 2010.
36. Serebryannikov, A.E. and Ozbay, E., Multifrequency invisibility and masking of cylindrical dielectric objects using double-positive and double-negative metamaterials. *J. Opt. A: Pure Appl. Opt.*, 11, 114020, 2009.
37. Ghosh, A., Verma, P., Banerjee, P.P., Aylo, R., Propagation of a Gaussian beam through a stack of positive and negative refractive index materials. *Proc. SPIE Metamaterials: Fundamentals and Applications III*, vol. 7754, p. 775415, 2010.
38. Shi, J., Zhu, Z., Guan, C., Li, Y., Wang, Z., Highly resonant positive and negative metamaterials. *Advances in Optoelectronics and Micro/Nano-Optics*, vol. 1, 2010.

39. Cojocaru, E., Waveguides filled with bilayers of double-negative (DNG) and double-positive (DPS) metamaterials. *Prog. Electromagn. Res. B*, 32, 75, 2011.
40. Popov, A.K., Myslivets, S.A., Nefedov, I.S., Generation of short contra propagating pulses of second harmonic on frequency double-domain positive/negative index metamaterials. *Photonics Global Conference(PGC)*, 2012.
41. Korotkevich, A.O., Rasmussen, K.E., Kovačič, G., Roytburd, V., Maimistov, A.I., Gabitov, I.R., Optical pulse dynamics in active metamaterials with positive and negative refractive index. *J. Opt. Soc. Am. B*, 30, 1077, 2013.
42. Maimistov, A.I. and Sklyarov, Y.M., Coherent interaction of light pulses with a three-level medium. *Opt. Spectrosc.*, 59, 459, 1985.
43. Byrne, J.A., Gabitov, I.R., Kovačič, G., Polarization switching of light interacting with a degenerate two-level optical medium. *Physica D*, 186, 69, 2003.
44. Pearson, A.M., Curry, R.D., O'Connor, K.A., First results in the development of double-positive metamaterials for high power microwave components. *IEEE International Power Modulator and High Voltage Conference (IPMHVC)*, pp. 13–16, 2014.
45. Pearson, A.M., Curry, R.D., Noel, K.M., Mounter, S.A., O'Connor, K.A., High voltage breakdown analysis of Nickel-Zinc ferrite double-positive metamaterials. *IEEE Pulsed Power Conference (PPC)*, pp. 1–4, 2015.
46. O'Connor, K.A. and Curry, R.D., Simulation of compact high power antenna concepts loaded by double-positive metamaterials. *IEEE Pulsed Power Conference (PPC)*, vol. 1, 2015.
47. Pearson, A.M., Curry, R.D., Noel, K.M., Characterization of Ni–Zn ferrite double-positive metamaterials for pulsed power systems. *IEEE Trans. Plasma Sci.*, 44, 1978, 2016.
48. Noel, K.M. and Curry, R.D., Tailoring permittivity and permeability in double positive metamaterials fabricated with barium titanate. *IEEE International Power Modulator and High Voltage Conference (IPMHVC)*, pp. 225–227, 2016.
49. Xia, L., Yang, B., Guo, Q., Gao, W., Liu, H., Han, J., Zhang, W., Zhang, S., Simultaneous, T.E., and TM designer surface plasmon supported by bianisotropic metamaterials with positive permittivity and permeability. *Nanophotonics*, 8, 1357, 2019.
50. Amanollahi, M. and Zamani, M., Performance of transverse magneto-optical Kerr effect in double-positive, double-negative and single-negative bi-gyrotropic metamaterials. *J. Magn. Magn. Mater.*, 502, 166451, 2020.

5

Study on Application of Photonic Metamaterial

Anupama Rajput[1]* and Amrinder kaur[2]

[1]Department of Applied Sciences, Faculty of Engineering & Technology, Manav Rachna International Institute of Research & Studies, Faridabad, Haryana, India
[2]Department of Electrical &Electronic Engineering, Faculty of Engineering & Technology, Manav Rachna International Institute of Research & Studies, Faridabad, Haryana, India

Abstract
A meta material is a man-made material having special properties that are not present in naturally occurring materials. These properties can be altered by changing the composition of the materials. The composition of these materials consists of metals and plastic also. These are arranged in repeated patterns. This pattern results in properties that are quite different from base materials. A study of meta-materials goes back to 1898 till date discovery of chiral material. Negative Index Meta Material, terahertz metamaterial, and plasmonic metamaterial are nowadays popular and used in lots of applications. This chapter deals with the basic structure and application of meta-materials. Change in properties of materials with change in composition is discussed.

Keywords: Photonic-metamaterials, terahertz, plasmonics, index, structure

5.1 Introduction

A man-made material that has the property which is not present in naturally befalling materials is known as metamaterial. These have been prepared by using different elements shaped from compound materials like

**Corresponding author*: anupama.fet@mriu.edu.in

Inamuddin and Tariq Altalhi (eds.) Electromagnetic Metamaterials: Properties and Applications, (101–118) © 2023 Scrivener Publishing LLC

metals and plastics. Metamaterials are generally organized in patterns that are repeated; at measures, those are lesser as compared to the wavelengths of the phenomena of their inspiration. Metamaterials develop their characteristics from the structural design of newly designed materials. Their specific profile, geometry, dimension, direction, and organization give these materials the properties by using which these can block, absorb, enhance, or manipulate electromagnetic waves. By having these specific properties, metamaterials can have advantages that are more than conventional materials.

By suitably designing metamaterials these can be used to alter/change waves of electromagnetic radiation or sound differently as compared to bulk materials. Some specific metamaterials having a negative index of refraction for particular wavelengths have generated a large amount of research interest nowadays [1]. So these materials are classified as negative-index metamaterials. Because metamaterials could be created by implanting man-made additions in a specific base medium, it is possible to design a huge group of self-governing parameters, e.g., Host material properties, magnitude, profile, and composition of additions. For achieving the results, design parameters could prove an important factor. Thus, basically, the form of additions/insertion offers a new way of processing metamaterials

5.2 Types of Metamaterials

In general, whenever any system is observed for its response toward the presence of electromagnetic fields, that response depends on the properties of the materials that are part of that system. There are macroscopic parameters like permittivity ε and permeability μ which are used to define the properties of these materials. Based on permittivity ε and the permeability μ, the types of metamaterials can be graphically represented as presented in Figure 5.1.

Thus, metamaterials are defined as a constant medium having active parameters, known as effective dielectric constant and effective magnetic permeability. These parameters can be altered, i.e., can have large, small, and even negative values by the proper choice of material used and the geometric arrangement of cells. There exist subclasses of metamaterials based on the value of permittivity and permeability. A medium having permittivity and permeability both greater than zero ($\varepsilon > 0$, $\mu > 0$) is known as a double positive medium (DPS). Large number of materials like dielectrics come under this category. An epsilon negative medium

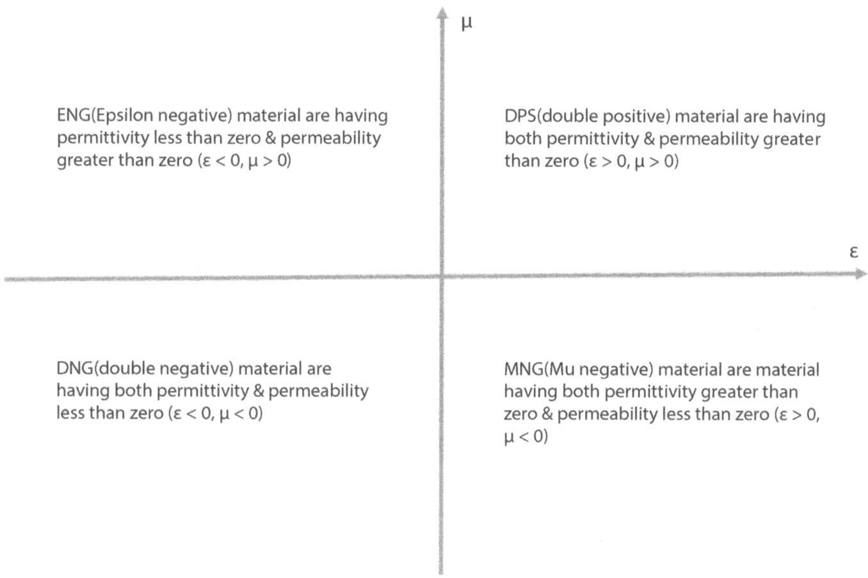

Figure 5.1 Types of metamaterials.

(ENG) is a medium with a permittivity that is less than zero and a permeability greater than zero ($\varepsilon < 0$, $\mu > 0$). For certain frequency ranges, there are many plasmas those are having these properties. Further, a medium with a permittivity larger than zero and permeability below zero ($\varepsilon > 0$, $\mu < 0$) is called a Mu negative medium (MNG). Few gyro tropic materials unveil this property in specific frequency ranges. A double negative medium (DNG) is having both permittivity and permeability below zero ($\varepsilon < 0$, $\mu < 0$). This material class has so far only been proven in artificial constructions. Further, for other materials, the subclass of metamaterials that has effective parameters, i.e., permittivity and negative permeability, i.e., ($\mu < 0> 0$) and permeability as negative, i.e., ($\mu < 0> 0$) in a certain frequency range; they are called negative epsilon (ENG) materials. The design of metamaterials is based on the values of permittivity and permeability. As with changing these two effective parameters, the effective electromagnetic behavior and response of the metamaterial vary, resulting in the different categories of metamaterials, such as electromagnetic metamaterials, chiral metamaterials (arrays of gammadion dielectrics, or planar metal on a substrate. A planar metal will become elliptically polarized by interaction with polarized light. There are also photonic metamaterials (operating at optical frequencies), terahertz metamaterials (in the range of 0.110 THz), etc.

5.3 Negative Index Metamaterial

Left-handed metamaterials (LHM) were first recognized historically as having a negative index value of their refractive index. When functions that respond towards electromagnetic fields, namely permittivity ε and permeability μ are preciously altered to have negative/less than zero values of their real part then the resultant material is left-handed metamaterials. To distinguish these from chiral materials, these are also known as negative refractive index materials (NRM), as chiral are also Kyrgyzstani monetary units sometimes referred to as "left-handed" [2–4]. Sometimes to distinguish two different structures made of the same material but different permeability or permittivity, few pieces of literature refer to another name which is double negative materials (DNG). DNG is different from single negative materials (SNG).

For the optical variety of frequencies, there has been no naturally existing negative index material discovered. Thus, it is essential to show that artificial materials are having a composition resulting in an averaged (effective) index of refraction less than zero, i.e., $N_{eff}' < 0$. Once a cloth was designed, it will show such properties of photonic crystals (PC), but in that case, the internal arrangement of the fabric is not sub-wavelength. Thus, PCs failed to prove a complete range of attainable advantages of left-handed materials. There are a few examples, such as super-resolution, that was foreseen by Pen dry, but it is not possible with photonic band gap materials for the reason that their cyclist that is within the range of λ. For optical wavelengths, it is required that metamaterials should so be nanocrafted. It is possible to attain a negative refractive index by designing a cloth such that wherever the (isotropic) permittivity $\varepsilon = \varepsilon' + i\varepsilon''$ and then the (isotropic) porousness $\mu = \mu' + i\mu''$ are having values that satisfy the equation as under

$$\varepsilon'\mu + \mu'\varepsilon < zero \tag{5.1}$$

This results in a negative real that means a certain portion of the index of refraction.

$n = \varepsilon\mu$ is having a negative value. Equation 5.1 is fulfilled, if $\varepsilon' < 0$ and $\mu' < 0$ But still, there is a tendency to observe that it can be not an obligatory condition. It is possible that there could be a magnetically active media having $\mu \neq 1$ with a positive real part μ' such that Equation (5.1) is consummated, and this proves a negative value n′. Till now, there is consideration of only isotropic media, wherein ε and μ are complicated scalar quantiles. The use of anisotropic media is a totally promising technique in

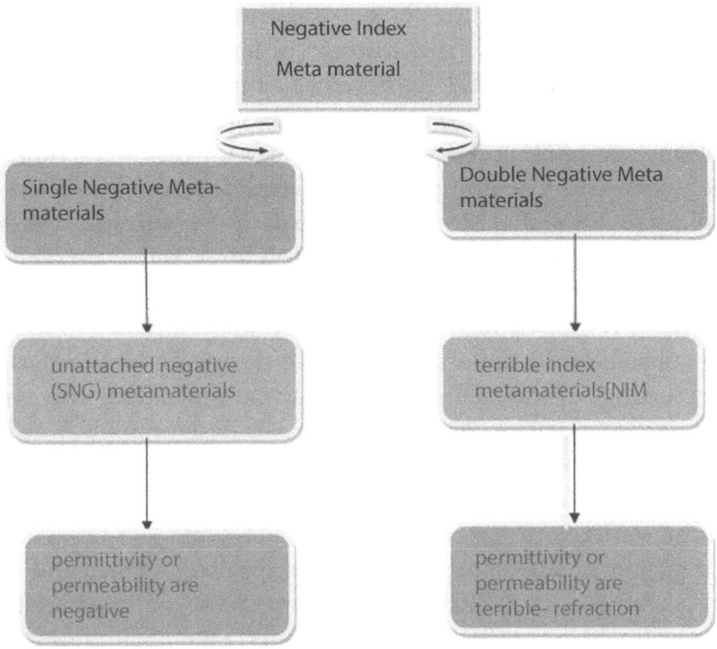

Figure 5.2 Different kinds of metamaterials.

metamaterials; still, this part of metamaterial is neglected. This is especially due to the fact for optical frequencies, a poor index has been achieved to date by using the technique of magnetically active media [5–7]. Different Classes of Negative Index Metamaterials are shown in Figure 5.2.

5.4 Terahertz Metamaterials

By using conventional planar-patterning tactics, it is possible to design Terahertz (THz) and optical metamaterials, even though a large number of sensible programs need metamaterials having 3-D resonators. It is a problem to create arrays of unique 3-D microresonators and nanoresonators [8]. By using a special technique to design metamaterials having 3-D resonators that are rolled up from stressed films, it is possible to have novel THz metamaterials/structures. These designs/structures show large polarization movements through numerous chiral metamaterials/structures. There is the presence of ultra-sharp quasi-periodic peaks in the polarization spectra of chiral metamaterials when these are studied on semiconductor substrates. With the help of 3-D printing, it is possible to allow the assembling of extra complicated structures those consists of the bi anisotropic gadget with ultimate microhelices. This further confirmed a severe polarization azimuth rotation of around 85° with a drop through 150° while having a frequency shift of 0.4%. Quasi-periodic peaks are referred to inside of the polarization spectra of metamaterial structures for the interaction of various echoes, consisting of strange chiral waveguide resonance. There is currently no other availability of technology to design. For destiny generations of industrial metamaterials, gadgets, and structures, rolled-up meta-atoms could be considered as the best constructing blocks [9].

By using the rolling up of 3-D resonators from the strained films and also due to a substantial increase in the field of designing metamaterials, the primary blessings of the rolling-up approach are invented. This approach results in precise precision and freedom in 3-D designing, the flexibility of substances, and also change in 3-d factors which have a significant decrease in micrometers to 3 nm. It is possible to manufacture THz metamaterials because this variety is actively evolved nowadays and lacks critical optical factors, e.g., wave plates. This is due to the absence of basic and artificial substances with practical THz houses. A comparison of metamaterials and metamaterials with THz frequency, i.e., terahertz metamaterials is given in Table 5.1. Nowadays, metamaterials with helical resonators are the center of study. In helical resonators, a helix reveals a green coupling each with magnetic and electric-powered fields alongside the axis. In this case, both the oscillations, i.e., along the magnetic dipole and the electric dipole are inseparable. This result in magnetoelectric polarizability.

It is possible to design a particular custom-made electric-powered, magnetic, and chiral responses to design a helix which is a unique hobby having few conventional 3-d designs of metamaterial factors. There is demand for Helical resonators due to chiral, magnetic, negative-index, "chiral nihility" sixty-four, and different metamaterials. All this is favorable for transformation involving polarization, imaging beneath the diversion limit, invisibility wrapping, non-reflectivity, and different programs. There are also rolled-up helices that additionally have the potential for sensible programs except for metamaterials. There is improvement in smoothness and precision by using the rolling-up approach for making helices. This results in different appraise of the manufacturing of helical structures and opinions on rolled-up systems, such as helical ones. Metal-semiconductor microhelices show a large polarization azimuth rotation when the measurements in terahertz variety are done with a Fourier-rework spectrometer. When 3-D printing is applied to find metamaterial primarily based structures with similarly greater functionality, there is a change in the finished product due to the interaction of the half-wave resonance of helices, waveguide, and

Table 5.1 Metamaterials vs terahertz metamaterials.

Component	Metamaterials	Terahertz metamaterials
Losses	Significant losses	Not significant losses
Structure	Need 3D structures	Single layer can be used
Band	Narrow band	Broadband suitable for imaging
Working	Require active control having negative index value in general	Have monolithic integration leading to unit cell control method

Fabric-Perot resonances. In polarization rotation spectra, these interactions lead to the development of ultra-sharp quasi-periodic peaks. This can be understood as a frequency shift of THz radiation of 0.4% consequences with inside the alternation of the azimuth of transmitted radiation through a 150 degree. There is a different setting up of the 3-D printing era and the rolling-up era but their incorporation creates a foundation that results in the introduction of novel structures for the handling of radiation, having possibilities for mass application.

5.5 Plasmonic Materials

There is engrossed interest in plasmonic materials due to their surfaces that have the property to localize incident light and also spread the optical path length. DSSC performance is reinforced by the optical path of plasmon materials [10]. Silica-coated Au nanotubes, Au and atomic number 47 core/shells, Zn, and Cu are a few examples of plasmonic materials. When a material uses surface Plasmon to attain optical properties it is known as a plasmonic metamaterial. When sunshine interacts with metal-dielectric materials at that time plasmons are formed. Subject to certain precise circumstances, the incident light combines with the surface plasmon's to make self-sustaining, propagating magnetism waves acknowledged as SPP, i.e., surface plasmon polar tons. On the metal-dielectric interface, the flow of surface plasmon polar tons also begins along with SPP. The SPPs will have a largely shorter wavelength when these are compared with the incident light. The distinctive structure of the results of the metal-dielectric composite in properties stems from the options of having wavelengths that are smaller than the wavelength of sunlight [11]. Some metamaterials are reworked into surface plasmon polar tons, those having wavelengths shorter than the incident light. Plasmons include the interface between light and free charges when in resonant modes. Examples of plasmas are ionosphere and metals. When a media contain freely mobile chargers, is known as plasma [12]. Longitudinal plasma oscillations are oscillations having a frequency that crates conditions for resonance, and these upsurges are due to the restoring force exerted on the mobile charges by altered charge distribution when there is a disturbance in equilibrium(may be due to passage of an electron nearby) can be sustained by the majority of plasmas (plasmons). The plasmon frequency, x_p, for bulk plasma is given by

$$x_p^2 = \frac{ne^2}{\varepsilon_0 m} \tag{5.2}$$

where

n =Mobile charge carrier's Density Number
e = Charge of mobile charge carrier
m = Mass of mobile charge carriers
ε_0 = Relative permittivity of free space.

When electromagnetic radiation (light) at a frequency less than plasma frequency incident on the plasma, there will be motion in the charge carriers. This motion results in screening out the incident field which causes the reflection of incident waves. At a frequency more than plasma frequency, waves are transmitted because now there is no swift response from charges to screen out the incident field [13]. Here, ionosphere will be working as a plasma, and with the use of equation (5.2), it has been found that the plasma frequency values lie in the MHz range. This shows the significance of the ionosphere working as a reflector for long-wavelength radio communications. When the study is carried out for conduction electrons that cause the formation of a plasma in silver, it shows that it is a frequency lying range of UV. This is also the reason that metals reflect light in the visible spectra of light and one of the prominent reasons that these are used as mirrors in history. There are still a few types of carbon also those are having high charge carrier densities required to generate plasmons when using UV.

During the movement of the charge carriers in the material, these experience dissipative processes. This motion is represented by an imaginary component of the relative permittivity, which is a complex quantity [14]. The dissipation is included in the concept of plasma by adding plasma frequency having a damping term. This concept is recognized by accepting a

Table 5.2 Plasma frequency of important metals.

Metal	Plasma frequency
Aluminum	3570THz
Silver	2175THz
Gold	2175THz

model known as the Drude model, in which the frequency is dependent on the relative permittivity of the metal. When this relative permittivity of metal is negative, an interesting phenomenon of resonance occurs and different values of resonance frequency are available. The plasma Frequency range for a few common metals is given in Table 5.2.

It is not possible to excite a large number of plasmons using light because the wavering charges present in the plasmon have longitudinal characteristics as well as oblique characteristics of the electric field of light. To overcome this, a different technique is used by bombarding the metal with electrons to excite plasmon. This already established method is a dominant way to examine the three-dimensional distribution of the electromagnetic fields related to plasmons. By rearranging the structure of metal, it is possible to change the plasmonic response. This will result in the coupling of light to the associated plasmon mode. In general, structured metals are the combination of two or a lot of element materials. These materials possess properties/characteristics different from the basic properties of the element materials on their own. These characteristics count on the pure mathematics of the selected structure. These materials units are called metamaterials. A metamaterial is a model that is well matched with plasmonic. Nevertheless, a setup of metal-like wires could be used for manufacturing non-natural, synthetic materials with ample lower plasma frequencies as compared to metals. If the designed structures show a resonance at a few frequencies, then there is a possibility that negative permeableness exists with a condition that the reflag is tough enough. When negative permeableness (l) is joined with a negative permittivity, (e) then it is possible to attain a negative ratio. This will be in opposition to the standard of Snell's law for refraction. Shelby accomplished properties related to this set. There is the production of arrays of split-ring resonators by the United Nations agency. These arrays offer a magnetic resonance and validated the microwave regime with a negative ratio. To discuss the reliability of this approach, it needs to be assured that satisfactory results are obtained at microwave frequencies as metals absorb very little at low frequencies. Further, when the frequency is increased, so absorption becomes essential. Because of all these reasons, it becomes difficult to have information about artificial resonances for structures like split-ring resonators. To find out other shapes having better results as compared to split-ring resonators, incontestable shaped resonators were anticipated by Zhou *et al.*, these are also lying within the microwave regime. For other shapes even a combination of rods may provide a resonance, here the capacitance between the ends of the rods helps to complete conducting path of the rods. There is still a chiral structures area unit that is not tested. Recently, maltreatment gold double posts and cut-wire structures having magnetic resonances are found within the visible spectral vary. Further, when a weak field penetrates metal, there is poor captivity of the sphere along the rest face which is the mode that may not be a Plasmon but a grazing gauge boson. However, by punching the urban center face with a secondary group of other wavelength holes, it is promising to create a coat on the surface that would effectually allow the sphere to pierce, through temporary fields associated with lower cut-off modes of the holes. Hibbin *et al.* confirmed this theoretical prediction through experimentation.

Moreover, semiconductors provide a novel and interesting opportunity for optical plasmon control which is excitation using the photons having electricity large than the semiconductor band gap that must regulate the conduction through orders of magnitude [15]. By using this method, it is possible to change the modes of houses of plasmon on a semiconductor floor and the same can be switched through a mild supply inclusive of a laser. By using laser pulse width to switch the instances, an interesting opportunity for speedy photo-switching can be provided by Ultra-short (sub-picoseconds) laser pulses. Furthermore, the lifestyles of excessive high-satisfactory substances and handling skills for semiconductors are already in existence; that is a place where this could be assumed to have captivating developments.

5.6 Applications

5.6.1 In Optical Field

Metamaterials are preciously conspired erections having negative refractive index as exclusive electromagnetic properties, that are imaginary in usual materials [16–25]. The properties of metamaterial have been improved with the quiet measured ascents in sponginess to form advanced lenses [26, 27]. One of the most important practical applications of metamaterial has been just taken out by using spectroscopic ellipsometry [28–30] and hiddenness screens [31–35]. Based on nonlinear metamaterials [36-40] and the marvelous application, the maximum work done is in the microwave domain [41–44]. To design novel optical absorbers, photonic metamaterials can be used and this proves to be a major application of metamaterials. Electromagnetic metamaterials are used to create a high electromagnetic absorber [45]. The metamaterial can be used in antennae to increase the performance of the antenna system. Zero refractive index material is mostly used for making high directivity antennas [46]. Star-shaped dielectric resonator antenna (SDRA) expansion, a project by Kumar *et al.* [47], is recently started under this application. Further, there are super lenses made up of used metamaterials to increase diffraction limit. Ramakrishna (2005) proved that superlenses made of metamaterials have more resolution capabilities than ordinary microscopes having conventional optical materials as a super lens [48].

An ultrathin nonlinear chiral metamirror has been validated by Kang *et al.* [49]. An array of SRRs composed of amorphous silicon (α-Si) is used to design the proposed metamirror. It has been testified that the met mirror shows reckless optical polarization switching. There are also hyperbolic metamaterials that could be incorporated into several applications, such as sub-diffraction imaging [50], sub-wavelength modes [51], thermal emission engineering (TEE), [52–54] high-sensitivity sensors, etc. Recently, microfluidic sensing techniques have been using designs inspired by electromagnetic metamaterials [55, 56]. High sensitivity and powerful interaction between the electric field, and the analysts are key properties of this type of sensing. Ho *et al.* [57] deliberate applications of microwave metamaterials for biomedical sensing.

5.6.2 In Medical Devices

Metamaterials are also used in medical devices as super lenses [58] and RF microelectromechanical systems (MEMS) strain sensors [59]. The monitoring of the whole recovery process of fractured bones is done by SRR-based wireless strain sensors. Metamaterial resonators and microwave devices [60] can be used for cancer detection. Additionally, they can assist in the protection of medical personnel from harmful X-ray radiation [61]. Metamaterial sensors are used in agriculture, biomedical, etc. Some sensors are based on resonant material. These use SRR to gain better sensitivity and are hence used in agriculture. Goren kit *et al.* [2012] showed in their research studies that metamaterial could be used in wireless sensors to enhance the sensitivity. It is also used in designing sensors for doors [62]. Photo sonic material, such as CNT is used in ultrafast sensors [63–65] along with microwave array applicators that are used for the effective treatment of large tumors utilizing a left-handed metamaterial lens [66].

5.6.3 In Aerospace

Photonic metamaterials are used to reduce jet noise, aircraft noise and diminish community noise (one of the major uses) [67]. It is counted as the most important application in noise and sound trapping [68] (Figure 5.3).

5.6.4 In Solar Power Management

Microporous photonic can be adopted in the metamaterial design method [69, 70]. Wang *et al.* reported in their studies about the absorption efficiency of the metamaterial through gradient index effects and also explained solar energy harvesting [71] (Figure 5.4). Elastic Met material (EMM) worked as band gaps, zero refractive indexes, as well as negative mass density. They act as elastic waves for leading co finishing and isolation [72].

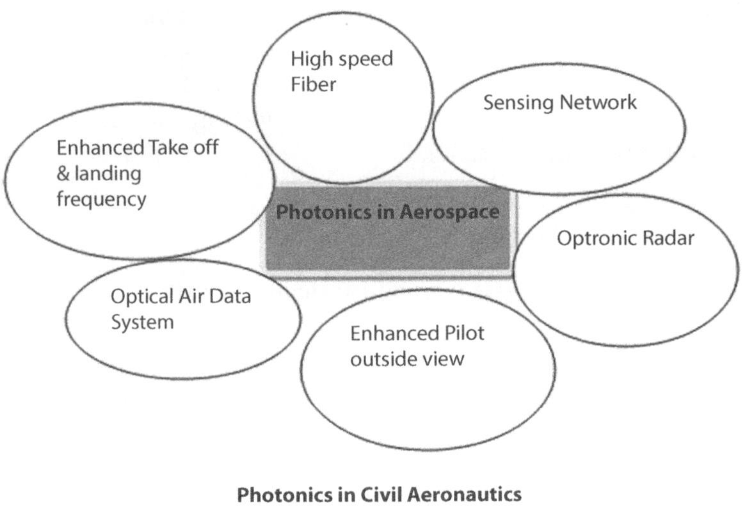

Figure 5.3 Photonic metamaterial in aerospace.

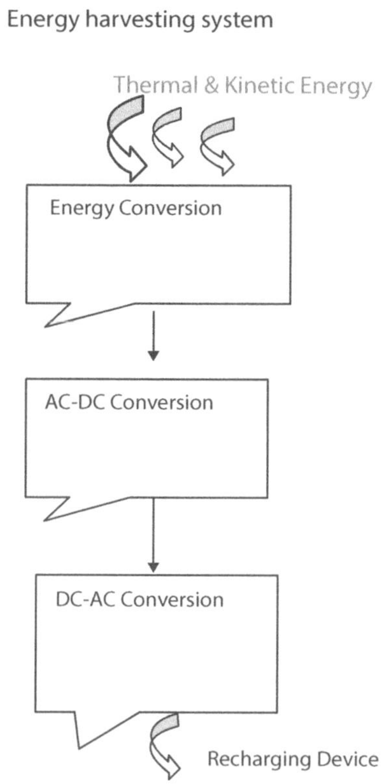

Figure 5.4 Energy harvesting system.

5.7 Conclusion

During the last decade, a revolution has started in the field of industrial applications of photonic metamaterials. After the discovery of terahertz metamaterials, Negative index metamaterial, Now scientists are in search to obtain aerospace metamaterial. The healthcare industry needs skilled engineers and physicists for the development of suitable equipment. Researchers seem to be keenly interested in metamaterials in the past decade which carries a high scope of application in the field of healthcare. Thus, the use of metamaterials has extensively increased. The study of mechanical metamaterials and their respective applications in orthopedics would work for a great cause. As stated metamaterials concerning their biocompatibility, will open up doors for research and manufacturing of permanent prosthetics on practical grounds.

References

1. Liu, A., Zhu, W., Tsai, D., Zheludev, N., Micro machined tunable metamaterials: A review. *J. Opt.*, 14, 114009, 2012.
2. Atre, A.C., Etxarri, A., Alaeian, H., Dionne, J.A.A., Broadband negative index metamaterial at optical frequencies. *Adv. Opt. Mater.*, 1, 327–333, 2013.
3. Vesseur, E.J.R., Coenen, T., Caglayan, H., Engheta, N., Polman, A., Experimental structures for visible light. *Phys. Rev. Lett.*, 110, 013902, 2013.
4. Park, I.-Y., Kim, S., Choi, J., Lee, D.H., Kim, Y.J., Kling, M.F., Stockman, M.I., Kim, S.W., Plasmonic generation of ultra short extreme-ultraviolet light pulses. *Nat. Photonics*, 5, 677–681, 2011.
5. Zhang, X. and Iu, Y., Metamaterials a new frontier of science and technology. *Chem. Soc. Rev.*, 40, 2494–2507, 2011.
6. Soukoulis, C.M. and Wegener, M., Past achievements and future challenges in the development of three-dimensional photonic metamaterials. *Nat. Photonics*, 5, 523–530, 2011.
7. Xu, T., Agrawal, A., Abashin, M., Chau, K.J., Lezec, H.J., All-angle negative refraction and active flat Lansing of ultraviolet light. *Nature*, *497*, 470–474, 2013.
8. Dulkeith, E., Morteani, A.C., Niedereichholz, T., Klar, T.A., Feldmann, J., Levi, S.A., van, F.C.J.M., Reinhoudt, D.N., Möller, M., Gittins, D.I., Fluorescence quenching of dye 11 molecules near gold nanoparticles: Radiative and non-radiative effects. *Phys. Rev. Lett.*, 89, 203002, 2002.
9. Lawandy, N.M., Localized surface plasmon singularities in amplifying media. *Appl. Phys. Lett.*, 85, 21, 5040–5042, 2004.
10. Avrutsky, I., Surface plasmon at nanoscale relief gratings between a metal and a dielectric medium with optical gain. *Phys. Rev. B*, 70, 155416, 2004.
11. Seidel, J., Grafström, S., Eng, L., Stimulated emission of surface plasmon at the interface between a silver film and an optically pumped dye solution. *Phys. Rev. Lett.*, 94, 177401, 2005.
12. Bergman, D.J. and Stockman, M.I., Surface plasmon amplification by stimulated emission of radiation quantum generation of coherent surface plasmon in nano systems. *Phys. Rev. Lett.*, 90, 027402, 2003.
13. Ou, J.Y., Plum, E., Zhang, J., Zheludev, N.I., An electromechanically reconfigurable plasmonic meta material operating in the near-infrared. *Nat. Nanotechnol.*, 8, 252–255, 2013.
14. Ramakrishna, S.A. and Pendry, J.B., Removal of absorption and increase in resolution in a near-field lens via optical gain. *Phys. Rev. B*, 67, 201101, 2003.
15. Urzhumov, Y., Lee, J.S., Tyler, T., Dhar, S., Nguyen, V., Jokerst, N.M., Smith, D.R., Electronically reconfigurable metal on silicon meta material. *Phys. Rev. B*, 86, 075112, 2012.
16. Landy, N.I., Sajuyigbe, S., Mock, J.J., Smith, R., Padilla, J., Perfect metamaterial absorber. *Phys. Rev. Lett.*, 100, 207402, 2008.

17. Chen, H., Padilla, W.J., Zide, J.M.O., Gossard, A.C., Taylor, A.J., Averitt, R.D., Active terahertz metamaterial devices. *Nature*, 444, 597–600, 2006.
18. Alù, A. and Engheta, N., Achieving transparency with plasmonic and metamaterial coatings. *Phys. Rev. E*, 72, 016623, 2005.
19. Gansel, J.K., Thiel, M., Rill, M.S., Decker, M., Bade, K., Saile, V., von Freymann, G., Gold helix photonic metamaterial as broadband circular polarizer. *Science*, 325, 1513–1515, 2009.
20. Enoch, S., Tayeb, G., Sabouroux, P., Vincent, A., Metamaterial for directive emission. *Phys. Rev. Lett.*, 89, 213902, 2002.
21. Shelby, A., Smith, D.R., Nasser, S.C., Schultz, S., Microwave transmission through a two-dimensional isotropic and left-handed metamaterial. *Appl. Phys. Lett.*, 78, 489–491, 2001.
22. Dolling, G., Enkrich, C., Wegener, M., Soukoulis, C.M., Linden, S., Simultaneous negative phase and group velocity of light in a metamaterial. *Science*, 312, 5775, 2006.
23. Liu, X., Starr, A.F., Padilla, W.J., Infrared spatial and frequency selective metamaterial with near-unity absorbance. *Phys. Rev. Lett.*, 104, 207403, 2010.
24. Dolling, G., Wegener, M., Soukoulis, C.M., Linden, S., Negative-index metamaterial at 780 nm wavelength. *Opt. Lett.*, 32, 53–55, 2007.
25. Papasimakis, N., Fedotov, V.A., Zheludev, N.I., Prosvirnin, S.L., Metamaterial analog of electro magnetically induced transparency. *Phys. Rev. Lett.*, 101, 253903, 2008.
26. Kundtz, N. and Smith, R., Extreme-angle broadband metamaterial lens. *Nat. Mater.*, 9, 129–132, 2010.
27. Cao, W., Dong, B., Reissman, T., Zhang, W., Sun, C., Additive manufacturing of a 3d terahertz gradient-refractive index lens. *Adv. Opt. Mater.*, 4, 1034–1040, 2016.
28. Zhang, C., Hong, N., Zhu, C.J.W., Chen, X., Agrawal, A., Zhang, Z., Tiwald, T.E., Schoeche, S., Hilfiker, J.N., Guo, L.J., Lezec, H.J., Robust extraction of hyperbolic metamaterial permittivity using total internal reflection ellipsometry. *ACS Photonics*, 5, 2234–2242, 2018.
29. Yermakov, O.Y., Permyakov, D.V., Porubaev, F.V., Dmitriev, P.A., Samusev, A.K., Malureanu, I.V.R., Avrinenko, A.V.L., Bogdanov, A.A., Effective surface conductivity of optical hyperbolic meta surfaces: From far-field characterization to surface wave analysis. *Sci. Rep.*, 8, 14135, 2018.
30. Dilts., J., Hong, C., Siahmakoun, A., Syed, M., Alisafaee, H., Low-MSE extraction of permittivity in optical hyperbolic meta materials. *Opt. Lett.*, 44, 4303–4306, 2019.
31. Li., J. and Pendry, J.B., Hiding under the carpet: A new strategy for cloaking. *Phys. Rev. Lett.*, 101, 203901, 2008.
32. Landy, N. and Smith, D.R., A full-parameter unidirectional metamaterial cloak for microwaves. *Nat. Mater.*, 12, 25–28, 2013.

33. Schurig, D., Mock, J.J., Justice, B.J., Cummer, S.A., Pendry, J.B., Starr, A.F., Smith, D.R., Metamaterial electromagnetic cloak at microwave frequencies. *Science*, 314, 977–980, 2006.
34. Ergin, T., Stenger, N., Brenner, P., Pendry, J.B., Wegener, M., Three-dimensional invisibility cloak at optical wavelengths. *Science*, 328, 5976, 2010.
35. Zhou, F., Bao, Y., Cao, W., Stuart, C.T., Zhang, J., Gu, W., Sun, C., Hiding a realistic object using a broadband terahertz invisibility cloak. *Sci. Rep.*, 1, 78, 2011.
36. Zharov, A.A., Shadrivov, I.V., Kivshar, Y.S., Nonlinear properties of left-handed metamaterials. *Phys. Rev. Lett.*, 91, 037401, 2003.
37. Gorkunov, M. and Lapine, M., Tuning of a nonlinear metamaterial band gap by an external magnetic field. *Phys. Rev. B*, 70, 235109, 2004.
38. Brien, S.O., McPeake, D., Ramakrishna, S.A., Pendry, J.B., Near-infrared photonic band gaps and nonlinear effects in negative magnetic metamaterials. *Phys. Rev. B*, 69, 241101, 2004.
39. Popov, A.K. and Shalaev, V.M., Unidirectional Doppler-free gain and generation in optically pumped lasers. *Appl. Phys. B*, 27, 63–67, 1982.
40. Feng, S. and Halterman, K., Parametrically shielding electromagnetic fields by nonlinear metamaterials. *Phys. Rev. Lett.*, 100, 063901, 2008.
41. Gil, I., Garcia, J., Bonache, J., Martin, F., Sorolla, M., Marques, R., Varactor-loaded split ring resonators for tunable notch filters at microwave frequencies. *Electron. Lett.*, 40, 21, 2004.
42. Shadrivov, I., Morrison, S., Kivshar, Y., Tunable split-ring resonators for nonlinear negative-index metamaterials. *Opt. Express*, 14, 9344–9349, 2006.
43. Degiron, A., Mock, J.J., Smith, D.R., Modulating and tuning the response of metamaterials at the unit cell level. *Opt. Express*, 15, 1115–1127, 2007.
44. Powell, D.A., Shadrivov, I.V., Kivshar, Y.S., Nonlinear electric metamaterials. *Appl. Phys. Lett.*, 95, 084102, 2009.
45. Adnan, N., *Metamaterial electromagnetic absorbers and plasmonic structures*, pp. 1–185, University of Manchester, Manchester, UK, 2010.
46. Khushnuma, P., Metamaterials types, applications, development, and future scope. *IJARIIT*, 4, 3, 2018.
47. Kumar, A., Kapoor, P., Kumar, P., Kumar, J., Kumar, A., Metamaterial loaded aperture coupled biodegradable star shaped dielectric resonator antenna for WLAN and broadband applications. *Microw. Opt. Technol. Lett.*, 62, 264–277, 2019.
48. Wenshan, C. and Shalaev, V., Optical metamaterials: Fundamentals and applications. *Phys. Today*, 63, 57, 2010.
49. Kang, L., Wang, C.-Y., Guo, X., Ni, X., Liu, Z., Werner, D.H., Nonlinear chiral meta-mirrors: Enabling technology for ultrafast switching of light polarization. *Nano Lett.*, 20, 2047–2055, 2020.
50. Liu, Z., Lee, H., Xiong, Y., Sun, C., Zhang, C., Far-field optical hyperlens magnifying sub-diffraction-limited objects. *Science*, 315, 1686, 2007.

51. Iorsh, I., Poddubny, A., Orlov, A., Belov, P., Kivshar, Y.S., Spontaneous emission enhancement in metal–dielectric metamaterials. *Phys. Lett. A*, 376, 185–187, 2012.
52. Guo, Y., Cortes, C.L., Molesky, S., Jacob, S., Broadband super-planckian thermal emission from hyperbolic metamaterials. *Appl. Phys. Lett.*, 101, 131106, 2012.
53. Biehs, S.A., Tschikin, M., Abdallah, P., Hyperbolic metamaterials as an analog of a blackbody in the near field. *Phys. Rev. Lett.*, 109, 104301, 2012.
54. Nefedov, I.S. and Simovski, C.R., Giant radiation heat transfer through micron gaps. *Phys. Rev. B*, 84, 195459, 2011.
55. Gordon, J.A., Holloway, C.L., Booth, J., Kim, S., Wang, Y., Jarvis, J., Novotny, D.R., Fluid interactions with meta films/meta surfaces for tuning, sensing, and microwave-assisted chemical processes. *Phys. Rev. B*, 83, 205130, 2011.
56. Awang, R.A., Lopez, F.J., Baum, T., Sriram, S., Rowe, W.S., Meta- atom micro fluidic sensor for measurement of dielectric properties of liquids. *J. Appl. Phys.*, 121, 094506, 2017.
57. Ho, J.S. and Li, Z., Microwave metamaterials for biomedical sensing, in: *Reference Module in Biomedical Sciences*, Elsevier, Amsterdam, Poland, 2021.
58. Fang, N., Lee, H., Sun, C., Zhang, X., Sub-diffraction-limited optical imaging with silver superlens. *Science*, 308, 534–537, 2005.
59. Melik, R., Unal, E., Perkgoz, N.K., Puttlitz, C., Demir, H.V., Wireless metamaterial RF-MEMS strain sensors. *IEEE Sens.*, 95, 1, 2173–2176, 2009.
60. Spada, L., Bilotti, F., Vegni, L., Metamaterial biosensor for cancer detection. *IEEE. Sens.*, 627–630, 2011.
61. Zhang, S., Genov, D.A., Sun, C., Zhang, X., Cloaking of matter waves. *Phys. Rev. Lett.*, 100, 123002, 2011.
62. Kiti, G., Radoni, V., Bengin, V., Soil moisture sensors based on metamaterials. *Songklanakarin J. Sci. Technol.*, 34, 689–693, 2012.
63. Set, S.Y., Yaguchi, H., Tanaka, Y., Jablonski, M., Laser mode locking using a saturable absorber incorporating carbon nanotubes. *J. Lightwave Technol.*, 22, 51, 2004.
64. Nicholson, J.W., Windeler, R.S., DiGiovanni, D.J., Optically driven deposition of single-walled carbon-nanotube securable absorbers on optical fiber end-faces. *Opt. Express*, 15, 9176–9183, 2007.
65. Schibli, T.R., Minoshima, K., Kataura, H., Itoga, E., Minami, N., Kazaoui, S., Miyashita, K., Tokumoto, M., Sakakibara, Y., Ultrashort pulse-generation by saturable absorber mirrors based on polymer-embedded carbon nanotubes. *Opt. Express*, 13, 8025–8031, 2005.
66. Tao, Y. and Gang, W., Hyperthermia of large superficial tumor with a flat LHM lens, IEEE/MTT-S. *International Microwave Symposium Digest.*, pp. 1–3, 2012.
67. Lambert, J., Champelovier, P., Blanchet, R., Lavandier, C., Terroir, J., Márki, F., Griefahn, B., Iemma, U., Janssens, K., Bisping, R., Human response to

simulated airport noise scenarios in home-like environments. *Appl. Acoust.*, 90, 116–125, 2015.

68. Jiang, X., Liang, B., Zou, X.Y., Yang, J., Yin, L.L., Yang, J., Cheng, J.C., Acoustic one-way metasurfaces asymmetric phase modulation of sound by subwavelength layer. *Sci. Rep.*, 6, 28023, 2016.
69. Roca, D., Valls, O., Cante, J., Oliver, J., A computational multiscale homogenization framework accounting for inertial effects: Application to acoustic metamaterials modelling. *Comput. Methods Appl. Mech. Eng.*, 330, 415–446, 2018.
70. Sridhar, A., Kouznetsova, V.G., Geers, M.G.D., Homogenization of locally resonant acoustic metamaterials towards an emergent enriched continuum. *Comput. Mech.*, 57, 423–435, 2016.
71. Wang, Z. and Cheng, P., Enhancements of absorption and photo thermal conversion of solar energy enabled by surface plasmon resonances in nanoparticles and metamaterials. *Int. J. Heat Mass Transf.*, 140, 45, 2019.
72. Ma, T.X., Fan, Q.S., Li, Z.Y., Zhang, C., Wang, Y.S., Flexural wave energy harvesting by multi-mode elastic metamaterial cavities. *Extreme Mech. Lett.*, 41, 101073, 2020.

6

Theoretical Models of Metamaterial

Hira Munir[1]* and Areeba Kashaf[2]

[1]*Department of Biochemistry, Government College Women University Faisalabad, Faisalabad, Pakistan*
[2]*Department of Biochemistry and Biotechnology, University of Gujrat, Gujrat, Pakistan*

Abstract
Metamaterials have growing attention for the past few decades in scientific societies. Although metamaterials can be acknowledged as left-handed substances otherwise materials having a negative value of the refractive index. To get the required response to electromagnetic metamaterials by way of sub-wavelength resonant structures can be artificially engineered. On the other hand, features of natural materials remain inadequate as a result of their chemical configurations and constructions. Unit cells known as meta-atoms are used to control the optical activity of metamaterials. In metamaterials, many unusual advances have been proposed, including negative value, the refractive index, cloaking, perfect absorber, etc. To comprehend the fundamental physics of metamaterial (MTMs), several theoretical models or theories have been introduced. Each model is abstractly distinct and yields characteristic equations and assumptions. This chapter reviews the metamaterials' theoretical models, including lumped equivalent circuit model, theory of effective medium, transmission-line theory, coupled-mode concept, and interference model in addition Casimir-Lifshitz theory. The theoretical methods offer an understanding of MTMs from dissimilar points of opinion. Models have their separate benefits and drawbacks.

Keywords: Metamaterials, unit-cell, left-handed materials, coupled mode theory, impedance

**Corresponding author*: dr.hiramunir@gcwuf.edu.pk

Inamuddin and Tariq Altalhi (eds.) Electromagnetic Metamaterials: Properties and Applications, (119–136)

6.1 Introduction

According to Wikipedia, "a substance which achieves its characteristics through its structure instead straight from its composition" is known as a metamaterial. The word metamaterial got familiarized in nineteen ninety-nine by scientist Rodger M. Walser. The metamaterials are macroscopic complexes containing an artificial, 3-D, interrupted cellular structure aimed to form an adjusted combination that is unavailable in the environment and shows multiple reactions to particular excitation. The mentioned characterizations reveal the definite natures of the metamaterial. Mainly, in actuality a MTMs stands for a macroscopic complex having the structure of periodic otherwise non-periodic nature, equally the engineering of cells, as well as the chemical alignment, plays a role in its functioning. If metamaterials act as the effective medium the cellular size should be reduced or else equal to the sub-wavelength. The principal advantage of metamaterials is their great flexibility due to cellular design, with the help of metamaterial structure novel substances that are not available then be able to recognize in use in surroundings [1].

Permittivity and permeability of metamaterial composites can take not only much large or small values in clouding zero but also can be negative in certain free quency band. The latter case makes it possible to implement a medium with a negative refractive index. Thus, we can directly affect the electrophysical properties of the material by modifying material constants. Permittivity and permeability of metamaterial composites can take not only many large or small values including zero but also can be negative in certain frequency bands. The latter case makes it possible to implement a medium with a negative refractive index. Thus, we can directly affect the electrophysical properties of the material by modifying material constants.

Metamaterials can form a variety of complex structures containing synthetic inclusions known as unit cells; which have definite forms and are entrenched into the base medium, characteristically dielectric substrate. Exceptional properties of metamaterials are hard to realize technically and in natural materials cannot be detectable. Certain unit cell parameters and the base substrate help attain the properties. The unit cells have a large wavelength than their size. Thus, ε plus μ be there for the characterization of the particular material. The ε, as well as μ values of metamaterials, can be 0 and large or else small. It can be negative in a particular frequency band, which gives a negative refractive index to a medium. Hence, the alteration of material constants impacts the electrophysical characteristics of the material [2]. Metamaterials have greatly more qualities further than these characteristics and are broader than left-handed materials (LHM). Subject to diverse necessities, metamaterials can be styled as inadequately and extremely anisotropic. The flexibility and optical transformation make metamaterial likely to regulate electromagnetic waves [1].

The materials have a negative refractive index, thus, when light enters metamaterials from a vacuum it bends in erroneous means. The light bends away from normal in metamaterials contrary to the normal bending of light. The alteration of a sign of permittivity and permeability in the four Maxwell's equations immediately gives negative values of them in metamaterials that result in an LH triplet of E, electrical inductance also k (phase trajectory). Although it is an RH triplet for typical constituents, for metamaterials, this infers the phase vector in addition to the group velocity in the opposed path. So, the wave movement is in a backward manner (whereas the energy circulates beside the incident in a forward track). The k specifies the route of the phase rate, while the Poynting vector/group velocity designates the energy's route.

The metamaterial shows the reverse Doppler effect. It makes the pitch of sound high and low depending on the requirement. The phenomenon has applications in mobile wireless communication. Similarly, metamaterials have the reverse Cerenkov effect: that is emission of light from a cone behind the particle instead of from the front. Cerenkov radiation arises due to the faster speed of the charged particle in a material [3].

6.2 Background of Metamaterials

Metamaterial was initially considered a Left-handed material. The idea was theoretically offered by the Russian physicist, Victor Veselago in nineteen sixty-eight, while the -ve refraction, as well as the backward-wave movement was introduced greatly before metamaterials. He explored the propagation of electromagnetic wave activities and their properties in materials with negative ε and μ [4]. Then, in 1996, revolt on Left Handed Material occurred. John B. Pendry proposed the artificially electric plasma of the copper wire having negative ε. After a few years, he with his co-worker introduced the artificial magnetic plasma via an interrupted array of split-ring resonators (SRRs) whose μ has a negative value [5, 6]. The metamaterials were created most simply by constructing an intermittent structure consisting of two elements; one of the element arrays produces negative permeability and the other element results in (-) μ.

The first artificial metamaterial was constructed in April 2001, by David Smith *et al.* by combining copper lines and SRRs. During this experimentation, researchers used a 2-D array of recurring copper unit-cells bands, as well as SRR, arranged at the intertwining bands of typical circuit panel constituents. The scattering angle made-up of these metamaterials showed a negative refraction phenomenon [7]. These metamaterials gain popularity in both theoretical and practical investigation, in addition to sensational innovation of super lens and perfect lens. After that in 2002, three groups proposed an alternative to Left-handed material using the transmission-line method [1]. Commonly known traditional TL contains dispersed series inductance and shunt capacitance, and can characterize 1D right-handed materials while new transmission-line have series capacitance and shunt inductance that can characterize Left-handed materials. An overall model of a RLH TL was projected to signify Right-handed and left-handed material. Many microwave constituents and antennas have been introduced based on CRLH concept [8].

In 2005, second revolt on metamaterial originated, after the gradient refractive index medium showed bending of EM waves, in addition, the optical transformation was introduced in 2006. The optical transformation is used for the creation of imperceptible cloaks plus involves the controlling of propagation of EM waves through the metamaterials [9–11]. After that, the metamaterial is considered a broader term, and does not need the negative values of ε and/or μ thus opening new areas of research.

6.3 Theoretical Models of Metamaterials

To design the tunable meta-devices or metamaterials to understand the foundation models of MTMs is necessary. Thus, the existing theoretical approaches are lumped element circuit idea, temporal coupled-mode concept, interference, transmission line theory, etc. Every model has its compensations and shortcomings. All approaches are favored for diverse purposes. Let us say the lumped-element model permits originally scheming the structure geometries, bandwidth, as well as elimination of the substances plus resonance frequency. On the other hand, the computation procedure of this model is tiresome and imprecise because of the unregular field charge in addition to the current supply in MTMs. The temporal coupled-mode model streamlines the difficulty into theory having a small number of overall factors, such as coupling power and exterior and interior damping degrees. This method is beneficial for joined systems using numerous resonators. However, the mentioned constraints usually are difficult to calculate but need to be taken out from simulations or experimentations. In addition, MTMs having planar structure is explainable over optics with the customary interference model in a condition once the near-field connection is insignificant in the system. Lastly, regardless of understanding as well as appreciated strategy directions accessible through theoretical models, the precise adding up stands quite greatly reliant on statistical tools sensitive to the FDTD scheme, finite-element technique, etc. [12]. Following are theoretical models of the metamaterials:

6.3.1 Lumped Equivalent Circuit Model

There is wide use of lumped element models in integrated circuit technology, automated systems, high-temperature transmission, sound quality, etc. This model provides the abridged consideration of spatially dispersed systems. Particularly, lumped RLC circuits are used in microelectronic devices. The circuit has 3 essentials this includes the capacitor (C), inductor (L), and resistor (R).

- In MTMs, the movement of charges is due to the E-field factor of the incidence of ray, this forms an electret and is thus known as a micromeasure or nanomeasure device (C).
- As we know that there is fluctuation in the outer field, and *I* use to pass through MTMs, inside the MTMs inductance must be measured.
- Last of all, the incident energy dissipates due to loss of energy in materials, or else soakage in the adjacent channel. This is what the character of R is within the electrical track.

Consequently, MTMs be able to certainly demonstrated using the model of RLC model [13]. Consider the simplest circuit in which there is the association of R, L, and C in series. The resonance *f*, bandwidth, and quality factor indicated as

$$f = \frac{1}{2\pi\sqrt{LC}} f = \frac{1}{2\pi\sqrt{LC}} \tag{6.1}$$

$$\Delta f = \frac{1}{2\pi} \cdot \frac{R}{L} \Delta f = \frac{1}{2\pi} \cdot \frac{R}{L} \tag{6.2}$$

$$Q = \frac{1}{R} \cdot \sqrt{\frac{L}{C}} \; Q = \frac{1}{R} \cdot \sqrt{\frac{L}{C}} \tag{6.3}$$

The reflection can be determined via the traditional TL model underneath the illumination from free space:

$$R = \left| \frac{Z - Z_0}{Z - Z_0} \right|^2 R = \left| \frac{Z - Z_0}{Z - Z_0} \right|^2 \tag{6.4}$$

In this,

Z_0 = 377 ohm is the free space's Z
Z = impedance of the equivalent circuit

Generally, elements can be linked to difficulty. Hence, the effective impedance should be measured. For instance, Chen and co-workers described tunable THz MTMs using SRR [14]. They utilized the equivalent circuit. In that design, the adaptable *R* is presented in the circuit via monitoring the decrease in the flow of charges of substratum by outer preference. The added *R* impacts Z_{eff} of the connection circuit. This primes a great variation of terahertz transmission through fifty percent. This lumped ECM is extensively functional to analyze the Microelectromechanical systems motivated MTMs, specifically the pre-stressed cantilever-based MTMs. This kind of MTM has the unit cell of a bimorph. Here, the unrestricted point curves up while the other end-point is attached to the substratum. According to equation (one), the *f* is controllable due to the effective L plus C. Bending down of the cantilever due to exterior stimulus causes variation of the effective C and L because of the alteration within the disconnected network in addition length of the projection of the biomorph, founding adjustable portions of the inductor-capacitor circuit.

There are some drawbacks at this point, such that it is not easy to calculate the values of these elements. Numerous properties must be completely monitored such as irregular *q* or *I* delivery also dissemination in metals. Subsequently, there is an unavailability of the correct closed-form formulas are to calculate the resistor, inductor, and capacitor. As an alternative, effective dimensions must be assessed, e.g., addition of an empirical computing parameter. Lastly, communal joining among adjacent components must remain in consideration for similar results of simulated or measured ones [12].

6.3.2 Effective Medium Theory

It can be defined as modeling a natural or artificial material (one or the other) by way of an assembly of dipoles. This theory has had a long history from Clausius, Ottaviano-Fabrizio Mossotti, Lorentz, and Lorenz in the late 19th century. MTMs can be defined in EMT, as a collection of intensely sub-wavelength substances or else elements, which perform cooperatively as a homogeneous material in particular conditions. These elements can be periodic or non-periodic. The objects that have MTMs are expected to remain sufficiently small-headed for labeling using point electric dipoles [15]. Thus the effective homogeneous structure has an average cell size greatly lesser concerning the guided λ. Consequently, the average p ought to be lesser than a quarter of λ, i.e., $p < \lambda_g /4$. This condition can be known as the effective-homogeneity limit or condition. This certifies that refractive-index phenomena would be dominating over the scattering/diffraction process once a wave transmits through the metamaterial medium. Due to the satisfaction of the effective-homogeneity condition, the structure acts as the real material. It means that EW is biased toward the lattice and merely probes the effective, macroscopic, and well-distinct parameters. These parameters are determined by the behavior of the unit cell p, thus the configuration has even electromagnetic behavior alongside the path of transmission. The fundamental factors include the electric permittivity also the magnetic permeability, linked to the n as a result of

$$n = \pm\sqrt{\varepsilon_r \mu_r} \tag{6.5}$$

where

ε_r = relative permittivity
μ_r = relative permeability

(associated to the free area ε also the μ through $\varepsilon_o = \varepsilon/\varepsilon_r$ = 8 points 854 as well as $\mu_0 = \mu/\mu_r$ = four π separately)

The mark ± is for the double-valued square root task and is known for generality [16].

Lorentz in 1878 presented for the first time in his thesis that point sources must be used to describe the continuous material's body in a vacuity that contains the microscopic Maxwell's or Lorentz calculations [15]. Conferring to Einstein, Lorentz therefore well discussed the separation of matter and ether. This provides the main context through which a microscopic structure of point dipoles may result in the derivation of the effective substantial factors that are utilized in the macroscopic Maxwell's equations [17, 18]. Planck, Madelung, Hoek, and De Groot upgraded Lorentz's primary [18–21]. Clausius-Mossotti's equation well described homogenization and proposed the formula for the ε or μ of a substance. It is likewise worth stating that Lewin presented the scheming of the effective ε and μ of spherical elements loaded channel [22] then Holloway *et al.* utilize this approach in the metamaterial perspective [23]. Effective medium models of metamaterials are simple and effectual, however, the rationality, as well as accuracy, are inadequate by the long-wavelength approx., necessitating a smaller unit cell dimension than the λ in the EM [24].

EMT delivers a direct knowledge of metamaterial absorbers based on material. It features the disappearing reflection and transmission to impedance similarity as well as greater loss attained instantaneously, which leads to impeccable absorption. This gives an instinctive tactic to obtain quality absorption [25]. The EMT theory of elastic metamaterials results in the accurate reproduction of the band structure, even at frequencies adjacent to the built-in resonances. Additionally, the theory shows the beneficial direction to searching MTMs with anticipated properties by changing the scattering features of the scatterers. In the study of super-lensing, quite suitable to define a piece of metamaterial concerning the effective medium and use its ε_{eff} and μ_{eff} in the Maxwell equations [26].

6.3.3 Transmission Line Theory

Metamaterials can be fundamentally demonstrated using 1D transmission lines because of having efficiently homogeneous structures. The 1D TL's propagation direction signifies all (any) directions in the material [16]. Filling a host TL medium with sensitive elements is another technique to understand the LHMs. For instance, to synthesize an LHM in 2D, series C also shunts L and is loaded periodically within a moderator micro-strip line network. Here the basic concept is that the (-) ε and a shunt inductor: (-) μ and a series capacitor are corresponding. Through this artificial MTMs with (-) ε and (-) μ, then henceforth a (-) refractive index. The array can be observed as a homogeneous, EM when the unit cell dimension is lesser than the impinging and guided λ. It can be described by effective constitutive factors ε_{eff} and μ_{eff} which are the results of a laborious periodic exploration [27].

Transmission line theory is functional for essential features concerning transmission lines. The decodable case is considered. The complex propagation constant (γ), propagation constant (β), specific impedance (Z_c), phase velocity (v_p), also the group velocity (v_g) of the transmission line is represented as

$$\gamma = \gamma = -j\frac{1}{\omega\sqrt{L_L'C_L'}}\frac{1}{\omega\sqrt{L_L'C_L'}} \tag{6.6}$$

$$\beta = -\frac{1}{\omega\sqrt{L_L'C_L'}}\ \beta = -\frac{1}{\omega\sqrt{L_L'C_L'}} < 0 \tag{6.7}$$

$$Zc = +\sqrt{\frac{L_L'}{C_L'}} > 0\ \ Zc = +\sqrt{\frac{L_L'}{C_L'}} > 0 \tag{6.8}$$

$$v_p = -\omega^2\sqrt{L_L'C_L'} < 0\ \ v_p = -\omega^2\sqrt{L_L'C_L'} < 0 \tag{6.9}$$

$$v_g = +\omega^2\sqrt{L_L'C_L'} > 0\ \ v_g = +\omega^2\sqrt{L_L'C_L'} > 0 \tag{6.10}$$

The last two equations of v_p and v_g in a transmission line are parallelly opposite, v_p -|| v_g. The v_p related to the path of β has a –ve value, while the group velocity, linked to the way of the flow of power, has a positive value. Hence, the TL is LH. Studies shows that the left-handed transmission line medium would be fixed up structurally in the circumstance in which normal dimensions of the cell should be greatly lesser as compared to group velocity (v_g). This helps to form a non-resonant type, perfectly realizable Left-handed medium. The microstrip is the first applied two-dimensional version of the LH transmission line.

Because of non-resonant quality, transmission line metamaterials can be proposed to have low-loss and broad bandwidth. The low loss will be due to the stable design of the structure and good similarity, while broad bandwidth is controlled by LC parameters of the transmission line, this helps to find the corner frequency of the structure. Secondly, transmission-line metamaterials are advantageous structures because of engineering in planar configurations. It can be well matched with modern microwave-integrated circuits. The metamaterials became benefited from TL theory for the effective strategy of microwave applications. There are more than 25 new devices that have unique properties and performance [16].

6.3.4 Coupled-Mode Theory

Concerning electromagnetics, coupled mode concept might be outlined back to the early 1950s, then by 1954, Pierce used the CMT to analyze the microwave TWT [28]. Miller presented the CMT to design microwave waveguides in addition to submissive tools. The development of CMT stood to couple the resonant circuit to 1 otherwise additional port, or else to further the resonant circuit [29]. Structures using lumped factors, for example, mode amplitudes, and degree of deterioration, in addition to constants of pairing, are described by this theory. All parameters link to a precise physical connotation concerning the entire arrangement as an alternative to separate deposits. Through brief substantial ideas besides the least necessities proceeding to algebra, CMT is extensively active to research waveguides [30], optical resonators [31], photonic crystals [32], MTMs [33, 34], among others.

MTMs TL assemblies have lead to innovative coupled-line couplers having unparalleled features which include arbitrary coupling level and the wide *B* of conservative coupled-line couplers. There are two categories of couplers taking on MTM TL recorded. The one includes the two alike CRLH TL functioned in their left-handed *f* series, and the second category comprises a conservative right-handed and the other one composite right-left-handed. The thoroughly examined CRLH–CRLH couplers manufactured via the even or odd mode decomposition method while RH–CRLH couplers have been considered from a speculative view [35].

The temporal coupled-mode theory is a significant technique broadly applied in MTMs. The temporal CMT makes simpler the systems into multiple resonant modes using mutual connections. Meanwhile, as a substitute for elements, this theory uses mode. The method gives flawless and instinctive physical intuitions that are particularly valuable to explore the novel occurrences. Preliminary from the most naive one, the case has a single resonator having one port.

There are essential equations:

$$\frac{d_a}{d_t} = i\omega_\circ a\left(\frac{1}{\tau_\circ} + \frac{1}{\tau_e}\right)a + \sqrt{\frac{2}{\tau_e}}S + \frac{d_a}{d_t} = i\omega_\circ a\left(\frac{1}{\tau_\circ} + \frac{1}{\tau_e}\right)a + \sqrt{\frac{2}{\tau_e}}S + \tag{6.11}$$

$$S_- = S_+ + \sqrt{\frac{2}{\tau_e}}a \quad S_- = S_+ + \sqrt{\frac{2}{\tau_e}}a \tag{6.12}$$

where

a = amplitude
ω_o = angular f_o
$1/\tau_o$ = decay rate (because of the fundamental substantial damage)
$1/\tau_e$ = power loss from the resonator,
S_+ = largeness of incident
S_- = reflected wave

Taking place the right hand of Eq. 6.10, there is an indication of E interchange in the resonator by former value, while second term presents energy expended inside the resonant device. The third value signifies absconding energy; the final term represents the energy of incidence joined to the resonant device. The total of these terms represents the variation of energy in the resonant device. The second equation is about the overall reflected wave and represents the summation of the reflected measure of the wave of incidence and the wave of radiation after the *r*. Thus, via uniting eq 6.11 and 6.12, the reflection coefficient can be considered as

$$r = \frac{S_-}{S_+} = \frac{\left(\frac{1}{\tau_e}\right) - (1/\tau_\circ) - \imath(\omega - \omega_\circ)}{\left(\frac{1}{\tau_e}\right) - (1/\tau_\circ) + \imath(\omega - \omega_\circ)} \quad r = \frac{S_-}{S_+} = \frac{\left(\frac{1}{\tau_e}\right) - (1/\tau_\circ) - \imath(\omega - \omega_\circ)}{\left(\frac{1}{\tau_e}\right) - (1/\tau_\circ) + \imath(\omega - \omega_\circ)} \tag{6.13}$$

According to the equation, a seamless absorber at the resonance *f* is attained by a precarious coupling state (one/τ_o = one/τ_e). Also, modulated phase $\varphi = \arg(r)$ can be obtained [36]. One port particular mode resonator will be made by employing the MIM cavity although the MTM at the upper metal sheet is bimorph. This electrostatically activates biomorph, operates the waste of radiations energetically also inherent absorption damage remains secure. Due to adjusting association b/w 2 damages, an active phase alteration can be established [37]. Here to define the system, two parameters are represented. Precisely, this is the symbolize as Q_a and absorptive quality factor (Q_r), which can be well-defined by way of $Q_a = \omega_0\tau_0/2$ and $Q_r = \omega_0\tau_e/2$. Qa or Qr is the proportion of energy deposited inside to the energy absconding from *r*. Absorption can be depicted as a purpose of Qa and Qr on the absorptive peak *f* demonstrating an impeccable incorporation after the absorptive quality parameter is equal to radiative quality parameter.

CMT designates MTMs absorber by three factors, i.e., ω_o, Q_a, and Q_r commencing based on energy. The expression is sophisticated and simple. Furthermore, for MTM absorbers by powerfully coupled resonant modes, there is a need for additional factors to be deliberated, to embrace offerings of other resonant modes [25]. A disadvantage of the CMT includes the appropriate parameters comprising resonance frequency, and inner and outer damping rates, which should be taken out from the mathematical model or intended by the lumped-element concept. Yet, CMT is appreciated for physical insight. It can escort the strategy procedure, which is particularly valuable in coupled systems, e.g., electromagnetically induced transparency is explained by the CMT. Additionally, Fano resonance and plasmonically enhanced molecule sensors are well explained by coupled mode theory [12].

6.3.5 Interference Theory

From an appropriate tool, light is distributed into two rays which are at that point superimposed. The strength within the area of superposition varies from point to point among maximum values (beyond the summation of the intensities) in the stream of light, and minimum values (possibly 0). This occurrence stands as interference. It can be due to more than two or multiple beams, known as multiple-beam interference [38]. Metamaterials most of the time involve various beams which may share a contact by the distant field intrusion as well as nearby field coupling. MTMs can be suitably demonstrated through the conventional interference theory during insignificant adjacent field coupling [39, 40]. The technique is particularly appropriate for constituents having multiple-layer constructions.

Using the interference model's primary phase, Fresnel equations or simulations extraction is used to determine the composite transference plus reflection constants of individual interaction. Then complete distant field reaction is present interference of total light fields subsequently seeing several inner reflections. Astonishingly, in MTM's perfect absorber the magnetic mode in addition to near-field coupling plays a role. This marvel can be explained by the interference model. The intricate transmission also reflection constants can be represented as $\widetilde{rr}_{12}$, $\widetilde{rr}_{21}$, $\widetilde{tt}_{21}$, and $\widetilde{tt}_{12}$. The complexity of the coefficients specifies a phase incoherence by the side of the crossing point; controlled by the resonant plasmons. The propagation phase remains indicated by β. Here the point to be noted is that the value of the r of the reflector put at the back of the cell presents as minus one. The reflection's amplitude coefficient may be indicated as:

$$\begin{aligned} r &= \widetilde{r_{12}} + \widetilde{t_{12}}\widetilde{t_{21}} \sum_{n=1}^{\infty} (\exp(j.2\beta))^n (-1)^n (\widetilde{r_{21}})^{n-1} \\ &= \widetilde{r_{12}} - \frac{\widetilde{t_{12}}\widetilde{t_{21}} \exp(j.2\beta)}{1-\widetilde{r_{21}} \exp(j.2\beta)} = \widetilde{r_{12}} - \frac{\widetilde{t_{12}}\widetilde{t_{21}} \exp(j.2\beta)}{1-\widetilde{r_{21}} \exp(j.2\beta)} \end{aligned} \tag{6.14}$$

The parameters take out after modeling by eliminating the reflector at the back of the cell. The accounts show a worthy equivalency b/w interference concept and computer-generated consequences. The benefit of this method is to comprehend the reaction of MTMs in a customary context. Still, in the lack of adjacent-field coupling, the model is only effective. Interference theory helps to understand MTM absorbers after an optical basis by using modest calculations.

As well, the proper guiding principle for structure design or materials assortment is not provided by this concept. Additionally, complexes of t, as well as r coefficients at the particular edge, are not able to calculate forthrightly however there is a need for computer-based results. Consequently, though this model has inherent weaknesses although it gives a significant lookout at MTMs. It has difficulty in practical applications as well [12].

6.3.6 Casimir-Lifshitz Theory

Lifshitz and colleagues comprehensively described the theory of Casimir, for the real materials. The innovative origins of Lifshitz were mainly done for the structures having planar geometry; however, it can be used for a plate and a sphere. It has a geometry suitable and extensively used in the experimentations. Boyer proposed in 1974, that there is repulsion because of the quantum fluctuations between two parallel plates; one of the plates is effortlessly conducting, and the other has infinite magnetic permeability. According to the Lifshitz theory, the magnetic media displays that the repulsive forces can act for the materials having high μ. However, this result was not approved meanwhile natural materials magnetic permeability (ω) does not show satisfaction with the condition [41]. As the metamaterials are progressing, known as the artificial assemblies with intended EM properties. Metamaterials are composite materials with inclusions of small size.

The complex on the other hand exceedingly effective Lifshitz model of Casimir forces undertakes isotopes in addition to non-magnetic media.

Based on the deviation of its generalization to magnetodielectric media Casimir forces that involve MTMs might have compelling things, e.g., LH MTMs can show repulsive Casimir forces. Based on supposition, the Lifshitz theory states that the media can be approached as continua labeled through ε and μ. Thus, there is a need that the field wavelengths taking part meaningfully in the direction of F should greater than dimension measures describing the MTMs assemblies in addition to partings. Characteristically, MTMs have narrow bands in magnetic field reactions and are inhomogeneous consequently, for magnetic anisotropic media, the generalized Lifshitz theory should be used.

$$\frac{F(d)}{A} = 2\hbar\int_0 \infty\frac{d\xi}{2\pi}\int\frac{d^2k_{\|}}{(2\pi)^2}K^3\ Tr\frac{R_1.R_{2e^{-2K_3d}}}{1-R_1.R_{2e^{-2K_3d}}}$$
$$\frac{F(d)}{A} = 2\hbar\int_0 \infty\frac{d\xi}{2\pi}\int\frac{d^2k_{\|}}{(2\pi)^2}K^3\ Tr\frac{R_1.R_{2e^{-2K_3d}}}{1-R_1.R_{2e^{-2K_3d}}} \tag{6.15}$$

where,

$R_{1,2}$ = 2*2 reflection matrices

$k_{\|}k_{\|}$ = transverse vector

Force's plus(+) or minus(-) charge resembles attractive and repulsive forces. Based on renowned critical features concerning ε also μ, reflection matrice at this point, remain assessed depending on the imaginary frequencies $\omega = \iota\xi$. To describe the overall anisotropic medium, the reflection matrices can be presented as:

$$R_j = \begin{bmatrix} r_j^{SS}(\iota\xi,k_{\|}) & r_j^{sp}(\iota\xi,k_{\|}) \\ r_j^{ps}(\iota\xi,k_{\|}) & r_j^{pp}(\iota\xi,k_{\|}) \end{bmatrix} R_j = \begin{bmatrix} r_j^{SS}(\iota\xi,k_{\|}) & r_j^{sp}(\iota\xi,k_{\|}) \\ r_j^{ps}(\iota\xi,k_{\|}) & r_j^{pp}(\iota\xi,k_{\|}) \end{bmatrix} (j=1,2) \tag{6.16}$$

where

$r_j^{ab}r_j^{ab}$ = ratio of a **R** field through b-polarization at odds by an inbound field using a-polarization

s = perpendicular polarizations

p = parallel polarizations

(concerning the incident point).

While, in isotropous media, the components that are not comparable disappear, Fresnel eq. are used to give diagonal elements representing permittivity $\epsilon\epsilon(\iota\iota\xi)$ in addition μ(iξ), due to which the equation one moves toward the typical Lifshitz formulation aimed at **F** [42].

There is another method to use MTMs to achieve repulsive forces is as transformation media between two interrelating matters. Deliberate a flawless lens, a slab of MTMs with ε =μ = −1 of thickness b squeeze in b/w 2 conducting metal plates, at a distance and in the middle of them. A theoretical method established in [43], the Casimir force formula rests the same as in the case of conducting plates, yet the space is altered by the MTMs:

$$x' = \begin{matrix} x, & for\ x < 0 \\ -x & for\ 0 \le x \le b \\ x-2b & for\ x > b \end{matrix} \quad x' = \begin{matrix} x, & for\ x < 0 \\ -x & for\ 0 \le x \le b \\ x-2b & for\ x > b \end{matrix} \tag{6.17}$$

Equation three shows that the force which is attractive in the x-space (normal Casimir force in the middle of 2 plates without an MTM within them) is repulsive in x′-space (having the perfect lens).

6.4 Conclusion

Theoretical models offered overhead to clarify the procedure of the MTM having altered viewpoint using different sets of expression. To understand the same device, each theory delivers particular opinions. Based on material, EMT abstracts, comparable ε as well as μ concerning MTM constructed by reflection plus transmission factors. Transmission line modeling deals with the MTMs from the circuit-centered perception. Coupled mode theory describes the metamaterials energy viewpoint while the interference model deals with metamaterials based on the optical concept. It clarifies the reflection is the superimposition of multiple mirror images between the binary crossing points. A metamaterial is a diverse topic, researchers are working to design metamaterials to use for various purposes.

References

1. Cui, T.J., Smith, D.R., Liu, R., *Metamaterials*, Springer, London, 2010.

2. Buriak, I.A., Zhurba, V., Vorobjov, G., Kulizhko, V., Kononov, O., Rybalko, O., Metamaterials: Theory, classification and application strategies. *J. Nano Electron. Phys.*, *8*, 04088, 2016.
3. Kshetrimayum, R.S., A brief intro to metamaterials. *IEEE Potentials*, 23, 44–46, 2004.
4. Veselago, V.G., The electrodynamics of substances with simultaneously negative values of ε and μ. *Phys.-Usp.*, 10, 509–514, 1968.
5. Pendry, J.B., Holden, A.J., Robbins, D.J., Stewart, W., Magnetism from conductors and enhanced nonlinear phenomena. *IEEE Trans. Microw. Theory Tech.*, 47, 2075–2084, 1999.
6. Pendry, J.B., Holden, A., Stewart, W., Youngs, I., Extremely low frequency plasmons in metallic mesostructures. *Phys. Rev. Lett.*, 76, 4773, 1996.
7. Shelby, R.A., Smith, D.R., Schultz, S., Experimental verification of a negative index of refraction. *Science*, 292, 77–79, 2001.
8. Caloz, C. and Itoh, T., *Electromagnetic metamaterials: Transmission line theory and microwave applications*, Wiley-IEEE Press, Hoboken, NJ, USA, 2006.
9. Pendry, J.B., Schurig, D., Smith, D.R., Controlling electromagnetic fields. *Science*, 312, 1780–1782, 2006.
10. Leonhardt, U., Optical conformal mapping. *Science*, 312, 1777–1780, 2006.
11. Smith, D.R., Mock, J.J., Starr, A., Schurig, D., Gradient index metamaterials. *Phys. Rev. E*, 71, 036609, 2005.
12. Chang, Y., Wei, J., Lee, C., Metamaterials–from fundamentals and MEMS tuning mechanisms to applications. *Nanophotonics*, 9, 3049–3070, 2020.
13. Costa, F., Genovesi, S., Monorchio, A., Manara, G., A circuit-based model for the interpretation of perfect metamaterial absorbers. *IEEE Trans. Antennas Propag.*, 61, 1201–1209, 2012.
14. Chen, H., Padilla, W., Zide, J., Gossard, A., Taylor, A., Averitt, R., Active metamaterial terahertz devices. *Nature*, 444, 597–600, 2006.
15. Bowen, P.T., *Metamaterials analysis, modeling, and design in the point dipole approximation*, Duke University, Department of Electrical and Computer Engineering, Durham, 2017.
16. Itoh, T. and Caloz, C., *Electromagnetic metamaterials: Transmission line theory and microwave applications*, John Wiley & Sons, London, 2005.
17. Lorentz, G.L. and Auer, J.A.C.F., *HA Lorentz; Impressions of his life and work*, North-Holland Publishing Company, Amsterdam, 1957.
18. Kranendonk, J. and Sipe, J., *V foundations of the macroscopic electromagnetic theory of dielectric media*, pp. 245–350, Elsevier, North-Holland Publishing, Amsterdam, 1977.
19. Madelung, E., Das. elektrische Feld in Systemen von regelmäßig angeordneten Punktladungen. *Phys. Z.*, 19, 32, 1918.
20. Hoek, H., *Algemeene theorie der optische activiteit van isotrope media*, Luctor et emergo, Leiden, Netherland, 1939.
21. Groot, S., *The maxwell equations, volume iv of studies in statistical mechanics*, North-Holland, Amsterdam, 1969.

22. Lewin, L., The electrical constants of a material loaded with spherical particles. *J. Inst. Electr. Eng.-Part III: Radio Commun. Eng.*, 94, 65–68, 1947.
23. Holloway, C.L., Kuester, E.F., Jarvis, J., Kabos, P., A double negative (DNG) composite medium composed of magnetodielectric spherical particles embedded in a matrix. *IEEE Trans. Antennas Propag.*, 51, 2596–2603, 2003.
24. Kuester, E.F., Memic, N., Shen, S., Scher, A.D., Kim, S., Kumley, K., Loui, H., A negative refractive index metamaterial based on a cubic array of layered non-magnetic spherical particles. *Prog. Electromagn. Res. B.*, 33, 175–202, 2011.
25. Duan, G., Schalch, J., Zhao, X., Li, A., Chen, C., Averitt, R.D., Zhang, X., A survey of theoretical models for terahertz electromagnetic metamaterial absorbers. *Sens. Actuator A Phys.*, 287, 21–28, 2019.
26. Wu, Y., Lai, Y., Zhang, Z.-Q., Effective medium theory for elastic metamaterials in two dimensions. *Phys. Rev. B*, 76, 205313, 2007.
27. Eleftheriades, G.V., EM transmission-line metamaterials. *Mater. Today*, 12, 30–41, 2009.
28. Pierce, J.R., Coupling of modes of propagation. *J. Appl. Phys.*, 25, 179–183, 1954.
29. Haus, H.A. and Huang, W., Coupled-mode theory. *Proc. IEEE*, 79, 1505–1518, 1991.
30. Miller, S., Coupled wave theory and waveguide applications. *Bell Syst. Tech. J.*, 33, 661–719, 1954.
31. Suh, W., Wang, Z., Fan, S., Temporal coupled-mode theory and the presence of non-orthogonal modes in lossless multimode cavities. *IEEE J. Quantum Electron.*, 40, 1511–1518, 2004.
32. Fan, S., Suh, W., Joannopoulos, J.D., Temporal coupled-mode theory for the Fano resonance in optical resonators. *JOSA A*, 20, 569–572, 2003.
33. Cong, L., Pitchappa, P., Wu, Y., Ke, L., Lee, C., Singh, N., Yang, H., Singh, R., Active multifunctional microelectromechanical system metadevices: Applications in polarization control, wavefront deflection, and holograms. *Adv. Opt. Mater.*, 5, 1600716, 2017.
34. Sukhorukov, A.A., Solntsev, A.S., Kruk, S.S., Neshev, D.N., Kivshar, Y.S., Nonlinear coupled-mode theory for periodic plasmonic waveguides and metamaterials with loss and gain. *Opt. Lett.*, 39, 462–465, 2014.
35. Nguyen, H. and Caloz, C., Generalized coupled-mode approach of metamaterial coupled-line couplers: Coupling theory, phenomenological explanation, and experimental demonstration. *IEEE Trans. Microw. Theory Tech.*, 55, 1029–1039, 2007.
36. Haus, H., *Waves and fields in optoelectronics*, Prentice-Hall, Inc., Englewood Cliffs, NJ 07632, USA, 1984.
37. Cong, L., Pitchappa, P., Lee, C., Singh, R., Active phase transition via loss engineering in a terahertz MEMS metamaterial. *Adv. Mater.*, 29, 1700733, 2017.
38. Born, M. and Wolf, E., *Principles of optics*, vol. 461, p. 93, Press Syndicate of the University of Cambridge, United Kingdom, 1999.

39. Saleh, B.E. and Teich, M.C., *Fundamentals of photonics*, John Wiley & Sons, NJ, USA, 2019.
40. Chen, H.-T., Interference theory of metamaterial perfect absorbers. *Opt. Express*, 20, 7165–7172, 2012.
41. Aulin, Y., *Casimir-Lifshitz forces*, pp. 1–21, University of Groningen, Netherland, 2009.
42. Rosa, F., Dalvit, D., Milonni, P., Casimir-Lifshitz theory and metamaterials. *Phys. Rev. Lett.*, 100, 183602, 2008.
43. Leonhardt, U. and Philbin, T.G., Quantum levitation by left-handed metamaterials. *New J. Phys.*, 9, 254, 2007.

7

Frequency Bands Metamaterials

D. Vasanth Kumar[1], N. Srinivasan[2], A. Saravanakumar[3], M. Ramesh[4*] and L. Rajeshkumar[5]

[1]Department of Mechanical Engineering, Dr. NGP Institute of Technology, Coimbatore, Tamil Nadu, India
[2]Department of Mechanical Engineering, Jansons Institute of Technology, Coimbatore, Tamil Nadu, India
[3]Department of Mechanical Engineering, Dhanalakshmi Srinivasan College of Engineering, Coimbatore, Tamil Nadu, India
[4]Department of Mechanical Engineering, KIT-Kalaignarkarunanidhi Institute of Technology, Coimbatore, Tamil Nadu, India
[5]Department of Mechanical Engineering, KPR Institute of Engineering and Technology, Coimbatore, Tamil Nadu, India

Abstract

Unlike most traditional materials, metamaterials' have effective qualities of elasticity and are engineered by humans to be exceptional. In this context, "meta" refers to artificial materials possessing extraordinary characteristics. Due to its limitless potential in triggering a wide variety of applications, including but not limited to band gaps, cloaking devices, electromagnetic, and transformation elastodynamics, metamaterials have prompted an upsurge in interest. Due to its peculiar characteristics, the term metamaterial was nearly always used interchangeably with left-handed material (LHM) in the early years of the uptick in interest surrounding metamaterials. Research into metamaterials has advanced significantly beyond the study of materials with negative refraction in recent years. Considering the recent proliferation of metamaterials and the growing interest in the associated research, three important milestones need to be discussed in this chapter.

Keywords: Frequency bands, metamaterials, electromagnetic, left-handed material, Penta metamaterials

**Corresponding author*: mramesh97@gmail.com

Inamuddin and Tariq Altalhi (eds.) Electromagnetic Metamaterials: Properties and Applications, (137–164) © 2023 Scrivener Publishing LLC

7.1 Introduction

Structures that have been artificially created and that display physical qualities that do not occur naturally are known as metamaterials [1]. The prefix "meta" imp"beyond" in this context. Metamaterials are made up of specially designed periodic patterns that exhibit tunable mechanical, acoustic, optical, or electromagnetic (EM) properties [2]. These properties are beneficial for a variety of functions, such as controlling noise, isolating vibrations, manipulating EM waves, cloaking, and energy harvesting, among other things [3–5]. For instance, EM metamaterials have created sub-wavelength components that enable them to manipulate EM waves. The performance of the metamaterial is determined by the dimensions, geometry, and arrangement of these sub-wavelength units. They have been shown to have higher gain and bandwidth when used as reflectors, filters, polarizers, or antennas [6, 7]. Traditional EM metamaterials, on the other hand, can only display set behaviors during operation since their geometries and materials cannot be altered after fabrication [8]. Alternately, reconfigurable metamaterials, also known as active metamaterials, possess actively modifiable patterns and/or physical properties, which enable them to be used in a manner that is both more adaptable and programmable [9].

The reduction of low-frequency vibrations has been a significant obstacle for a very long time. Architected materials, often known as acoustic or elastic metamaterials, are widely recognized as a potentially fruitful research topic. They may have a phononic band gap, which refers to a frequency range in which no vibration may be transmitted through the material [10]. Although several recent research sought to demonstrate low-frequency band gaps, there is no consensus on whether frequency ranges should be considered "low" or "ultra-low" [11]. Many recent studies attempted to demonstrate low-frequency band gaps. The precise definition of low frequency can range from a fraction of one Hz up to several thousand kHz. The definition of "low" is relative and is determined by the particular circumstances of each application [12].

7.2 Frequency Bands Metamaterials

The term "metamaterial" refers to man-made materials that have qualities that cannot be found in naturally occurring substances [13]. Because of the physical field they exhibit, metamaterials can be divided into several categories, including electromagnetic, optical, acoustic, and elastic metamaterials [14]. Important subfields of metamaterials, such as acoustic metamaterials and elastic metamaterials, do not have distinct borders between them [15]. In recent years, penta mode metamaterials and band gaps have both become popular topics, and for good reason: both have a significant amount of untapped potential in a variety of applications [16].

7.2.1 EM Metamaterials

EM metamaterials are engineered materials with structures whose EM properties are designed to provide a response spectrum that would be extremely challenging, to accomplish with the aid of natural materials or biocomposites [17, 18]. Negative index of refraction, precise lensing, and invisibility cloaks of EM are just a few of the mind-blowing uses for metamaterials [19]. Magnetic permeability medium comprised of conducting non-magnetic components is a notable property of a certain class of metamaterials. Because of this property, artificially magnetic materials can be manufactured at high frequencies, even though no such materials exist in nature [20]. In the Terahertz range, the resonant response of the medium, known as left-handed (LH) behavior, is established (far-infrared), shown below in Figure 7.1.

7.2.2 Metamaterial Response Tuning

The special electromagnetic properties of metamaterials are due to their resonant response but at the cost of built-in restrictions towards bandwidth [21]. One effective strategy for dealing with bandwidth constraints is dynamic tuning, or the capacity to dynamically alter resonance characteristics as shown in Figure 7.2. There are a few different ways to go about this, from incorporating metamaterials into FET electronic devices to creating hybrids and naturally occurring materials with desirable features.

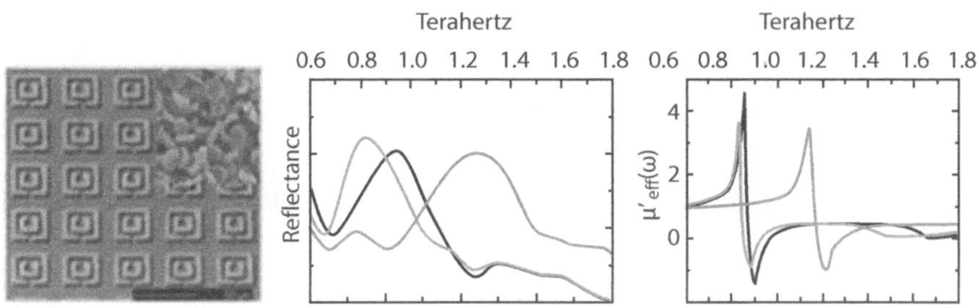

Figure 7.1 Terahertz range metamaterials (*Padilla et. al, Science 2004*).

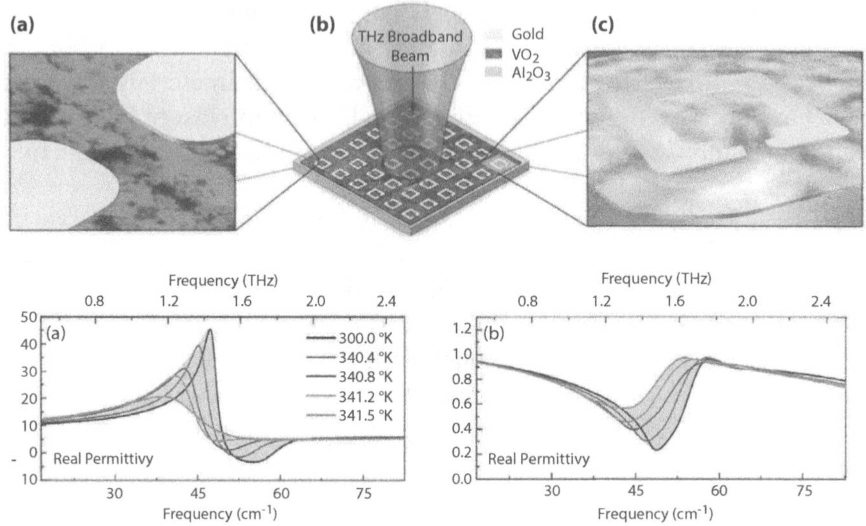

Figure 7.2 Hybrid metamaterials (*Driscoll et al., Applied Physics Letters, 2008*).

The addition of a vanadium dioxide layer to a metamaterial makes it possible to fine-tune its properties by more than 20% using only temperature changes [22].

7.2.2.1 Persistent Tuning

While it has been shown in Figure 7.3, that metamaterials can be tuned to a variety of stimuli (including light, temperature, mechanical stress, and electricity), in almost all situations too far, the stimulus is required to keep

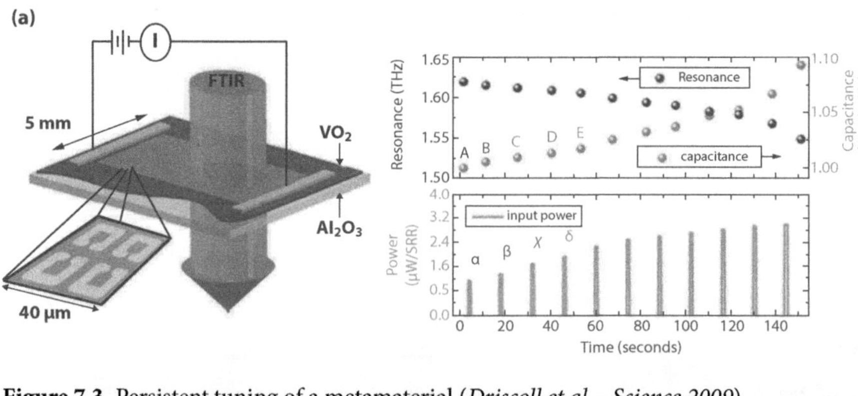

Figure 7.3 Persistent tuning of a metamaterial (*Driscoll et al. - Science 2009*).

the material in tune [23]. If a large number of metamaterial unit cells are to be tuned to different levels, this can be a drawback. The use of VO_2 proved an example of tuning with electrical persistence [24]. Here the metamaterial tuning occurs as a response to the electrical voltage applied and is retained until the voltage is turned off. This configuration's apparent memory can be explained by the phenomenon of memory capacitance, which is connected to memristance [25].

7.2.3 Spectroscopic Investigation

Metamaterial reactions can be spontaneously understood from their responses to magnetic permeability and electric permittivity. The problem is that it is difficult to explicitly test for these characteristics. Important for both design and comprehension, a description of the electric and magnetic response of a metamaterial can be derived from measurements of its reflection and transmission. A voltage pulse applied across the connections accomplishes this, leading to a long-lasting shift in the metamaterial's optical characteristics thanks to the incorporation of VO_2. Because current densities vary with distance from the contacts, the quantity of joule heating varies across the device, causing the observed gradient. Finer control over this gradient may be attainable through better design of contact geometry [26–28].

7.2.4 Optical Metamaterials

The structural units of a metamaterial, are significantly lesser than the evaluated wavelength and the average distance between meta-atoms in the vicinity should be on a gauge that is sub-wavelength. Because the inhomogeneities in a metamaterial are on a sub-wavelength scale, the material is macroscopically uniform. Because of this property, a metamaterial is more accurately described as a "material" than a "device" [29, 30]. Metamaterials can be differentiated from many other electromagnetic media due to the extent of the inhomogeneities that they include [31]. Both traditional materials and man-made metamaterials share the property that the wavelength of the relevant light wave is significantly shorter than the lattice constants of both of these types of materials [32]. These phenomena may be found almost everywhere across the EM spectrum. The reaction of structures is typically characterized by making use of geometric optics and ray tracing when larger scales of homogeneity are dealt with [33].

A forward step was taken in the transition from the theoretical prediction to the experimental validation of the theory during The Veselago media's first experimentation. The work in which an attempt is made to close the gap between innovative metamaterials and intriguing applications on a flawless lens [34, 35]. There are a variety of intentionally created metamaterials that are currently displaying EM properties that have never before been seen in biomaterials and it is not possible to obtain it using approaches that are considered to be the industry standard for synthesis [36, 37]. To accomplish specific capabilities, manipulations can be carried out in the structure of the metamaterial specifically in their composition, and size. shape, and the inclusions which could be constructed by customization to produce the desired results [38, 39].

The field of optics has moved swiftly to embrace the concept of metamaterials. The study of optical metamaterials, which refers to materials that display tailored EM responses at light frequencies, is by far the most exciting and difficult of all the subfields of metamaterial research that are being conducted. Light is the most efficient method for transmitting data from and to the materials' internal structure because it encapsulates the information in a zero-mass signal which moves at an unrivaled speed. The proliferation of research efforts about optical metamaterials is a direct consequence of the combining of a plethora of nanofabrication techniques with developments in nano-scale imaging as well as computational EM design and simulations [40, 41].

The development of photonic metamaterials has radically shifted the manner that researchers and engineers conceive and create practical optical systems. Common optical design practices assume that the media to be utilized are homogeneous and isotropic, making device design primarily an interfacial engineering problem between the various media [42]. In the case of a lens manufacturing sector, many forms of aberrations can be reduced by employing cascaded lenses made from different materials and with precisely controlled curvatures. The development of photonic metamaterials has opened the door to the modification of optical space and the generation of novel responses that are impossible in the underlying materials. With metamaterials, designers must rethink how they approach the creation of optical devices, as the required functionality is delivered not just via the interfacial configuration but through the virtual control of each point in the optical space [4, 43, 44].

7.2.5 Optical Materials and Electronic Structures

By far the most common kind of materials found in optical parts and gadgets are dielectrics. Conventional optical systems typically employ crystalline and glassy materials for nearly all functioning elements [45].

A medium must be at least somewhat dielectric to allow light to be controlled efficiently across it. Maxwell's equations and the two governing equations given below can be used to investigate the physical interaction of dielectric with light [46–49]:

$$D = \varepsilon_0 E + P = \varepsilon_0(1+\chi_e)E = \varepsilon_0\varepsilon_r E$$
$$B = \mu_0(H+M) = \mu_0(1+\chi_m)H = \mu_0\mu_r H$$

Relationships among electric displacement (D), electric field (E), polarization density (P), electric susceptibility (e), vacuum permittivity (ε_0), permeability (μ0), and their magnetic analogs are defined by the above equations.

7.2.6 Optical Properties of Metals

Metals are traditionally used in mirror making and thin optical sheets for optically functional devices. However, most of the concepts now under investigation for optical metamaterials have metals embedded in the metamaterials' atomic structure. Two of the most striking characteristics of metals at optical wavelengths are readily apparent to the naked eye. To start, metals are opaque, so you cannot see through them unless they are extremely thin (on the order of tens of nanometers or less). To add to this, metals have excellent reflectivity. When light hits a shiny metal surface, it is largely reflected into space. They are both rooted in the way electrons behave in metals [50–52].

In metals, the Fermi level has no separation between unoccupied and occupied levels of energy [53]. Therefore, an electron can be excited to a higher energy level by the photon energy of any EM radiation. Light of any frequency that manages to penetrate a metal (not an easy task) can be absorbed throughout its short propagation length (often less than 100 nm) due to the constant availability of unoccupied electron states [54–56]. Metals, such as silver and aluminum, reflect a large portion of the sun's rays and appear silvery when lit from above. This means that the light we detect from gold and copper, for example, is primarily composed of yellow to red wavelengths, which are reflected poorly [57, 58].

7.2.7 Metal-Dielectric Composites

The meta-atoms of photonic metamaterials are often composed of compositional units that are fragile, sub-wavelength structures made up of both dielectric and metallic components. Since the metamaterial response to the EM field is highly dependent on meta-atoms architecture, a simple approach for the property analysis of structural unit metamaterial becomes interesting [59]. However, there are broad mathematical methodologies that allow us to determine the typical EM response of the composites, even when constructing them from metamaterials that lack well-structured building blocks. Since the metal and dielectric components are randomly distributed in such media, the overall optical characteristics of the composite might vary greatly from those of the individual constituents. Multiple fields have found a use for random metal-dielectric metamaterials, from nonlinear optics to biological spectroscopy. In addition, random metal-dielectric composites constitute the basis for the design and implementation of numerous metamaterial devices [60, 61]. The boundary conditions of a composite, with its random distribution of metal and dielectric components, make it extremely difficult, if not impossible, to calculate the composite's EM response by solving Maxwell's equations. Fortunately, things can be made much easier under certain conditions [62, 63]. Transmission electron microscopy (TEM) images of two samples are shown in Figure 7.4.

The inclusions particles lodged in the host material (shown as white) in the first composite sample are quite small and have sharply delineated spherical forms. Maxwell-Garnett geometry is a common term for this. Figure 7.4(b) illustrates the difficulty in distinguishing between the host and the inclusion when the two constituent elements intermix and play

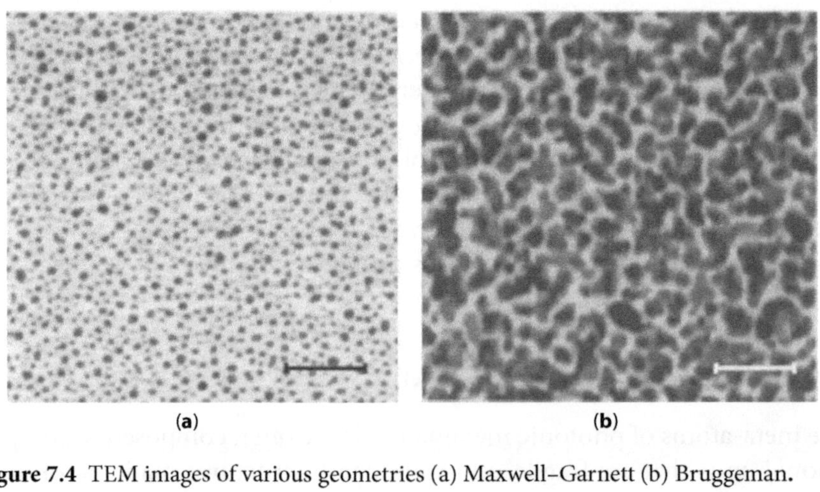

Figure 7.4 TEM images of various geometries (a) Maxwell–Garnett (b) Bruggeman.

symmetric roles. A frequent name for this branch of topology is "Bruges geometry."

7.2.8 Acoustic Metamaterials

For the better part of two decades, researchers have been curious about how acoustic and elastic waves travel through heterogeneous material. Important characteristics of so-called phononic crystals include the presence of frequency band gaps [64–66], mass density, and regulated elastic moduli. They significantly attenuate the transmission spectra of sound and acoustic vibrations in the frequency band gap ranges, and this attenuation occurs in both directions of propagation. Different material combinations using solid and fluid phononic crystals were proposed to increase the width of these acoustic band gaps [67–69]. The propagation of waves could be controlled by phononic crystals by employing the band gap principle. In other words, they act as ideal reflectors of elastic or acoustic waves that fall within the band gap's frequency range. It is so clear that phononic crystals are of fundamental interest for elastic energy regulation and that their prospective applications are well established. Initial research into lamb, bulk, and surface waves has laid the groundwork for understanding the underlying physics behind capturing, directing, and de-multiplexing acoustic waves via single and linear defects [70, 71].

To put it simply, the periodicity of the structure can cause Bragg reflections to serve as the seed for band gaps [72, 73]. Early on, it became clear that this theory had limitations when applied to acoustic applications in low frequency [74, 75]. Earthquakes have been widely recognized as a harmful kind of pollution to the environment and a potentially catastrophic natural hazard [76]. Using an acoustic metamaterial is another way to realize acoustic band gaps, which are significantly smaller than the wavelength of acoustic bands. Generally speaking, these materials are characterized as an artificial structural arrangement with a scale of inhomogeneity smaller than the acoustic wavelength, which is intended to achieve favorable and uncommon acoustic qualities [77]. In terms of standard material properties, their acoustic response can be written.

Different synthetically manufactured metamaterials are currently displaying acoustic properties that have never been seen before in nature. Some examples are acoustic cloaking, negative refraction, and super-prism resolution. Metamaterials' structural units can be artificially modified in terms of shape, size, composition, and morphology; inclusions can be constructed and positioned in a controlled fashion, and a wide range of desired capabilities can be realized [78]. Various literature works reported the construction of individual unit cells of sonic crystals with local resonance along with acoustic resonators. A sonic band gap with lattice constants in second order is lesser than the wavelength of an acoustic band and can adapt to a wide frequency range (10 Hz to 10 kHz) [79, 80]. As an additional benefit, materials with a sound velocity in two folds higher than the actual matrix, like silicon rubber, can be used to fine-tune the local resonance. Band gaps in frequency were found for such composite [81, 82] with the polarization state of the wave affecting the width of the gap, and for lamb waves, with both partial and complete band gaps being reported.

7.2.9 Elastic Metamaterials

Reducing low-frequency tremors has been a difficult problem to solve for a long time. Architected materials, also known as acoustic or elastic metamaterials [83] are a promising area of study. There may be a sonic dead zone, or phononic band gap, in these materials [84, 85]. There is no agreement on which frequency ranges should be labeled low or ultra-low, even though numerous recent research sought to demonstrate such gaps. The precise definition of low frequency ranges from a few hertz to several kilohertz, and anywhere in between. The term “low” is contextual and can be interpreted in a variety of ways. Various strategies have been presented to produce acoustic and elastic LHM, demonstrating the recent application of the metamaterial principle to acoustic and elastic media. In contrast to electromagnetic and acoustic media, an ordinary elastic medium can allow transverse and longitudinal wave patterns to travel through it. These waves are characterized by three different material parameters in an isotropic solid: the bulk modulus, the shear modulus, and the mass density. Similar to EM metamaterials, an elastic metamaterial can achieve negative effective parameters by including resonant structures into its constituent parts [86, 87].

Thin membranes and rubber-coated-lead structures exhibited dipolar resonances that have been demonstrated to cause these devices to attain negative mass density [88]. The bulk modulus can be made negative due to the monopolar resonances of air bubbles present in water [83]. Also, quadrupolar resonance has been linked to a negative shear modulus. It is noted that any resonant structures have been reported to achieve the negative shear modulus over a significant frequency range [78]. There are exciting practical implications of discovering the negative effective shear modulus in addition to the scholarly interest in doing so. Negative refraction may arise, for instance, if the density and shear modulus of an elastic LHM are both negative in each frequency range. However, the elastic LHM can function as a polarizer if the bulk modulus of the material is positive and the longitudinal wave disappears. Later it will be shown that the metamaterial that is in contact with an ordinary solid can likewise accomplish the entire mode conversion from shear waves to longitudinal waves. None of the natural solids can be joined in this way. Multiple scattering theory (MST) band structure studies indicate the presence of a sizable negative band. Using a wedge-shaped sample, MST simulations prove the negative refraction [89–91].

7.3 Penta Metamaterials

The composites are made up of periodic rather than random configurations of several individual lattices fashioned from ordinary base materials like metals or polymers [92, 93]. So, the unique features of metamaterials come from their carefully crafted geometrical structures rather than their chemical makeup. Negative bulk modulus and/or mass density acoustic metamaterials, auxetics with negative Poisson's ratio (NPR), negative permeability, negative thermal expansion coefficient, cosserat and micropolar metamaterials, and the recently emerged penta metamaterials (PMMs) transforming thermodynamics are all commonplace in the metamaterials research domain [94–96].

After 17 years of the first appearance in design, PMMs were manufactured in 2012 using the direct-laser writing (DLW) optical lithography technique. The penta mode properties were achieved through the highly localized deformation by the fused point-like tiny tips of the rigid-body double-cones as shown in Figure 7.5. PMMs are composites of microstructures containing a sufficiently rigid isotropic phase and a sufficiently compliant isotropic phase, in which 5 out of 6 components (eigenvalues) of a diagonalized elasticity tensor are close to 0. In design, PMMs use one

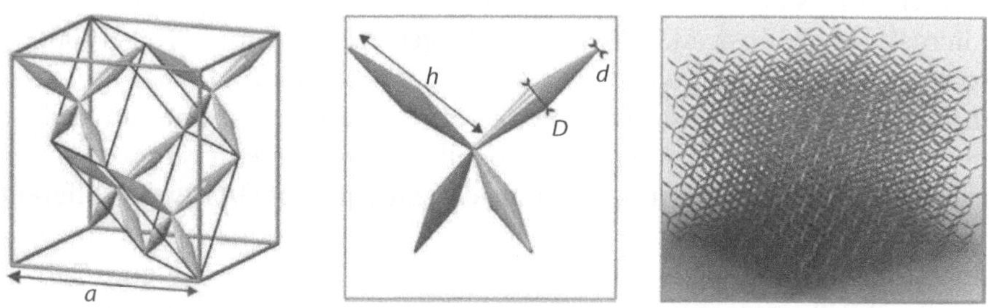

Figure 7.5 Concept of conventional rigid-body double-cone penta mode materials.

isotropic material with extremely large elastic moduli and another with extremely small elastic moduli, to build composites with effective elasticity tensors, to implement 3D transformation elastodynamics analogous to transformation optics in EM metamaterials [97, 98].

Because of this, the shear modulus of a perfect isotropic penta-mode material is zero, and its Poisson ratio is consequently half. According to the research, the liquid behavior of penta-mode materials can also be approximated by rationally designed 3D solid microstructures, which will not 'flow away' like conceptually perfect homogeneous penta-mode materials [99, 100]. This is because the liquid behavior of penta-mode materials can be thought of as a mixture of a liquid and a solid. It is possible to design the microstructures so that they have a substantially higher bulk modulus (B) in comparison to their shear modulus (G). This unique characteristic opens new opportunities. For instance, the ideas that have been proposed for three-dimensional transformation acoustics [101], such as inaudibility cloaks, phononic band gaps, acoustic prisms, or novel concepts for loudspeakers, might one day become a reality. It also regulates local elastic wave propagation and controls transformation elastodynamics designs to create acoustic cloaks that are invisible to sound and do not have the sensation of being touched. Because Poisson's ratio of naturally occurring elastic solids is typically somewhere around 0.3, the value of B/G2 implies that the magnitudes of B and G are comparable to one another.

Penta modes, on the other hand, are solids, but they differ from other solids in that their B dimension is significantly larger than their G dimension. The Poisson's ratio of an isotropic penta mode material has the formula v=(3B2G)/(6B+2G)0.5 when B is perfectly finite and G is equal to 0 [97, 102]. As a result, the PMMs are also referred to as "meta-fluids." In actuality, however, when solid microstructures are used to approximate the penta mode behavior of fluids, it is difficult to make the 3D solid microstructures have an infinitely large B while simultaneously having a nearly zero G. This is because the G is dependent on the orientation of the solid microstructure. Instead, it is feasible to get a B/G ratio of 100–1000 for solid microstructures, which is a high value. In this particular scenario, the transformation elastodynamics is not quite relevant but can still be used in a rough sense [101]. Traditionally, PMMs were developed empirically by continually testing different dependences of B/G upon different sets of parameters (h, D, d) of the rigid-body double-cone lattice. This was done to determine the optimal values for these parameters (Figure 7.5). A smaller d will result in a greater ratio; however, d cannot be endlessly small because such a tiny rigid-body connection would make the manufacturing

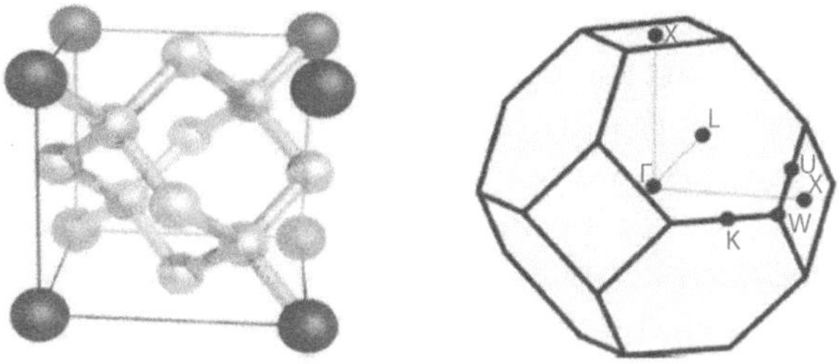

Figure 7.6 Concept of diamond-type of lattice (FCC) for penta-mode materials.

process extremely tough and would readily fall apart upon the least possible mechanical distortion [103, 104].

As a result, the majority of today's ideas for PMMs may be only significant in theory, not in practice. To attain a high ratio of B/G, the use of the elastic deformation of the entire microstructures, as opposed to the highly concentrated deformation that results from double-cone tip connections occurring inside of a diamond structure (FCC lattice) (Figure 7.6). The numerical topology optimization method will be used to design the PMMs that will be used in this research, and additive manufacturing (AM) will be used to prototype them, when applicable [105]. The originality of PMMs is highly dependent on both their design and production. The use of topology optimization as a potent computational design tool has been growing quickly in recent years. This has opened up boundless prospects for the development of novel materials and structures that are both more robust and lightweight. It is, in essence, a numerical process to redistribute materials within a fixed reference domain until the best material distribution and the corresponding topology are achieved, based on the optimization of a specific objective function subject to design constraints. This process continues until the best material distribution and the corresponding topology are achieved. It is a computational method that has gained widespread recognition for its use in discovering novel geometries to enable design innovation in the areas of structures, materials, and processes. There have been many different approaches to topology optimization developed over the past three decades. Some of these approaches include the solid isotropic material with penalization (SIMP), the evolutionary structural optimization (ESO), and the level set method (LSM) [106, 107].

For instance, one well-liked topology optimization method for a variety of structural and material designs is known as the SIMP model. The recent rapid development of technology for AM has made it possible to install structural and mechanical parts that have complicated geometries [108]. AM eliminates the constraints that traditional manufacturing processes impose on the removal of materials and significantly expands the freedom that can be exercised in topological design. This makes it possible to create shapes and geometries that are more intricate and complicated, which frequently leads to the discovery of newly engineered artificial materials and composites with novel properties. Microstructures of bi-mode metamaterials (2D) have been designed with the help of Topology Optimization, as have 2D and 3D elastic metamaterial micro-architectures with crystal symmetries [109, 110]. The purpose is to construct 3D solid elastic microstructures with the features of a penta mode, to obtain many of the effective B above G that is both realistically practicable and attainable. The microstructure is generated by using topology optimization with a single piece of isotropic material [111, 112]. This is done by the overall elastic deformation of the microstructure, and the numerical homogenization approach is used to determine the equivalent property of the microstructure.

7.4 Reconfigurable Metamaterials for Different Geometrics

A structural reconfiguration of the metamaterial has been simulated on a cylindrical-shaped substrate with a radius of 76 mm, as portrayed in Figure 7.7(a) has been created to evaluate the conformability and adaptability of the metamaterial on various geometries. The dimensions of the metamaterial were 90 mm x 90 mm x 2 mm. The results of the structural analysis show that the material adopts smoothly to the surface of the metamaterial in deployed states when subjected to gravity-assisted magnetic actuation [113]. The experimental verification is illustrated in Figure 7.7(b). The conformability of the metamaterial is preserved even after it transforms from its deployed state into its folded state on the cylindrical substrate when subjected to a magnetic field of 60 mT. For facilitating a rapid metamaterial deployment after it has been recovered, the magnetic field of 15 mT in the reverse direction is applied. The magnetic actuation makes it possible to achieve quick (in the subsecond range), reversible, and untethered local and global deformation. The structural analysis in Figure 7.7(c) demonstrates that the metamaterials' structure may also conform to spherical contours towards the application of magnetic actuation. The substrate has a radius of 75 mm. Figure 7.7(d) of the experimental verification shows reliable folding and deploying when subjected to a magnetic field of 60 mT in one direction and a magnetic field of magnitude 15 mT in another direction. This demonstrates compatibility existing amongst the simulated

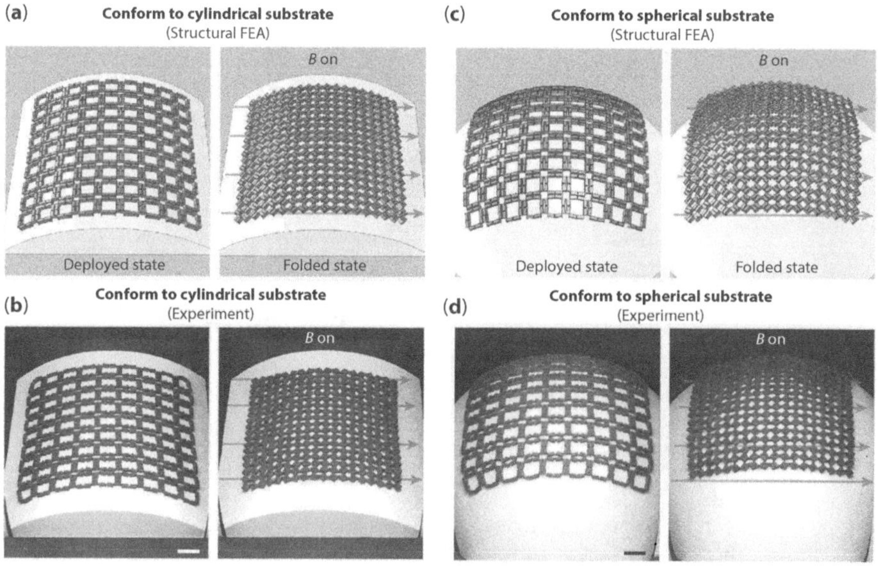

Figure 7.7 Conformation to (a, b) cylindrical substrate, (c, d) spherical substrate.

deformations. The metamaterial demonstrates the capacity to acclimatize to different geometries of varying sizes. Reconfigurability and conformability working together will make it possible for EM metamaterials to have a wide range of property tenability and programmability [114–116].

7.4.1 3D Freestanding Reconfigurable Metamaterial

As depicted in Figure 7.8, the metamaterial, in addition to being able to adapt to a variety of geometries, also possesses the ability to metamorphose into curved forms. This results in a reliable design that achieves reconfigurable metamaterials with an increased number of degrees of freedom by combining the capabilities of shape morphing both globally and locally. Under magnetic actuation, magnetic units are built in this instance [117, 118]. As can be seen in Figure 7.8(a), for the 7×7 array that was presented, each magnetic unit locally contains the alternating in-plane magnetization along both Y-directions. The out-of-plane magnetizations are programmed with 30°, 45°, and 60° angles towards left to the Z negative direction, whereas at right the angles of 60°, 45°, and 30° to the positive Z-direction were used for programming, as shown in Figure 7.8(b). The units that make up the center column only have magnetization in one plane. When a magnetic field is applied in the X-direction, the in-plane

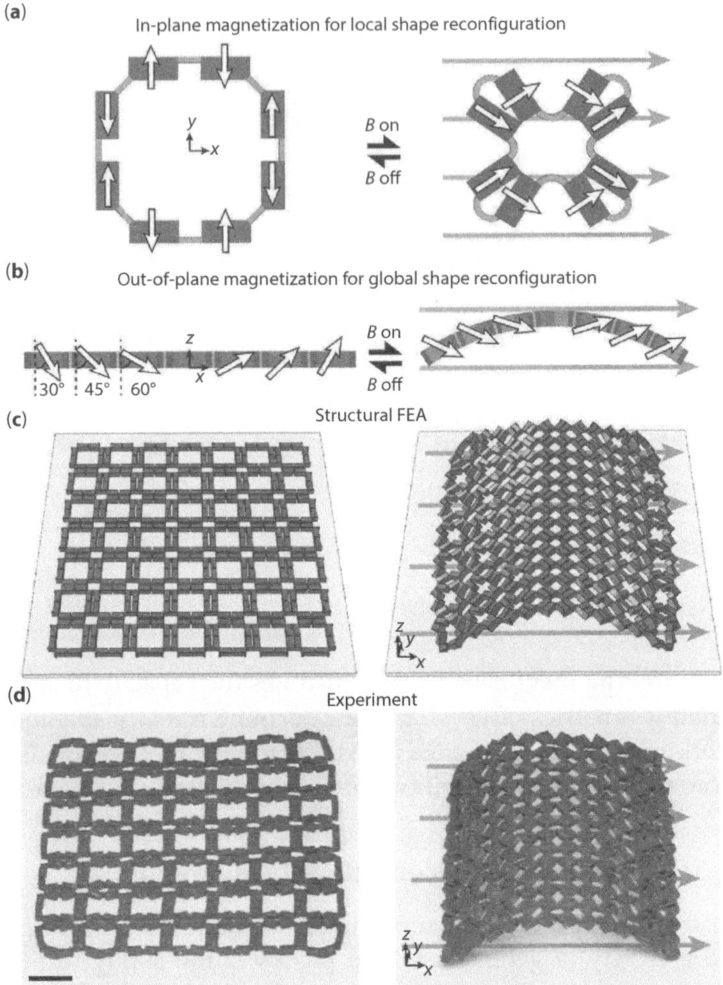

Figure 7.8 Reconfigurable metamaterial magnetization (a) In-plane; (b) out-of-plane; (c) structural FEA and (d) experimentation.

magnetization of the planar metamaterial causes the local folding of every magnetic cell, and the out-of-plane magnetization of the metamaterial causes them to flex upward, creating a freestanding curved structure [119–121]. As can be seen in Figure 7.8(c), the structural finite element analysis (FEA) that has been applied to forecast the magnetic actuation has allowed for the metamaterial shape morphing, which has gone from the flat deployment to a freestanding curved deployment. In addition, experimental validation is carried out in an increasingly homogenous magnetic field up to 110 mT along the X-direction for the metamaterials, as shown in Figure 7.8(d). This verification is carried out following the structural FEA prediction very well. To achieve the required curvature distribution of the metamaterial, the magnetization component in an out-of-plane direction can be tailored to a variety of alternative 3D form reconfigurations with the use of structural FEA [122].

7.4.2 Reconfigurable EM Metamaterials

The metamaterial that is magnetically activated and has effective structural reconfigurability and conformability as a result of coupled shape morphing. Local shape morphing lays the groundwork for the tuning of properties for reconfigurable metamaterials [123]. This allows for the metamaterials' properties to be tuned in a wide range of ways.

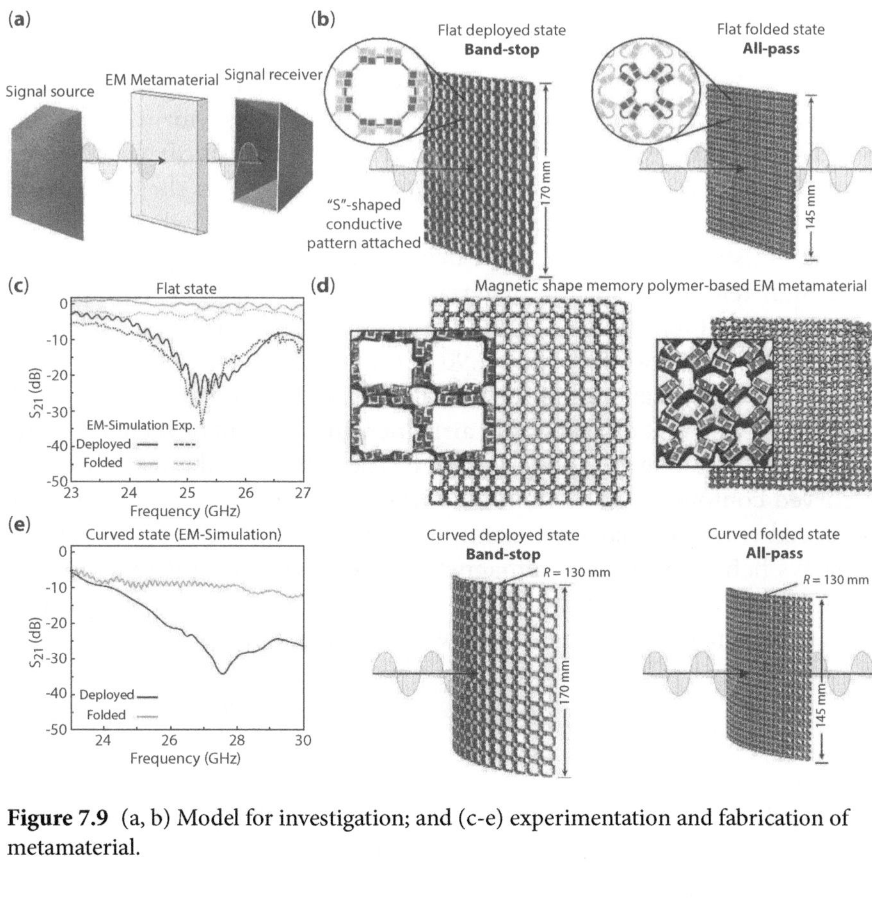

Figure 7.9 (a, b) Model for investigation; and (c-e) experimentation and fabrication of metamaterial.

During the electromagnetic simulation, the metamaterial is in the space between two horn antennas (Figure 7.9a). The first antenna serves as the signal source by producing an incident wave, while the second antenna is responsible for receiving the wave as it travels through the metamaterial. As shown in Figure 7.9(b), the structure transforms from a band-stop to an all-pass filter when the flat metamaterial is folded [23, 24, 124]. This transformation can be seen in the structure. Within the frequency range of 23 to 27 GHz, scattering parameter S21 is shown to be displayed in Figure 7.9(c) using the metamaterial. According to the solid black curve that was acquired from EM simulation, the developed metamaterial demonstrates significant filtering during deployment at approximately 26 GHz. There is more than 6 dB of signal transmission through the band when the structure is in the folded condition. This is represented by the continuous curve (blue) in Figure 7.9(c). This is accomplished by structurally reconfiguring the material via magnetic actuation [53, 54, 57]. Figure 7.9(d) shows the method of construction of shape memory polymer-based magnetic metamaterial with periodical "S"-shaped patterns and it is how the experimental verification is carried out. It could be observed that the fabricated structure exhibits filtering at approximately 26 GHz during deployment (represented by the black dotted lines) and all-pass behavior when in the folded state (represented by the blue curve with dotted lines) as shown in Figure 7.9(c). It is important to point out that the magnetic shape memory polymer that was utilized in this instance is a composite whose stiffness may be adjusted to a certain degree depending on the temperature.

At high temperatures, the material is malleable and pliable, but as it cools to normal temperature, it hardens to the point that it can no longer be deformed as shown in Figure 7.9d [114, 117, 125]. Because of its conformability, the metamaterial is particularly beneficial for usage in curved structural arrangements, such as in airplane wings. As can be seen in Figure 7.9(e), the lattice-like metamaterial that has been presented can easily adapt to curved contours. Furthermore, the EM performance of the material in its curved condition has been examined using EM modeling. The curved 13 arrays behave as an electromagnetic wave filter during deployment at 28.5 GHz. However, when it is in the folded state, it loses its filtering capability, depicting the tunable EM behavior alike a flat metamaterial.

7.5 Conclusion

In the opinion of several of the most prominent authorities in the area, the definition of a metamaterial should include phrases such as "properties unlike any naturally existing substance" or "not observed in nature." These definitions are helpful because they highlight the primary goal of metamaterial research, which is to achieve EM properties that are not seen in normal materials. Nevertheless, it might be too simplistic to rule out any naturally existing feature when discussing the subject of the metamaterial. It is humbling in the face of mother nature, and it would require an excessive amount of effort and might not even be possible at all to investigate naturally occurring materials in sufficient detail and establish the non-existence of property in those materials. As an illustration, even though substances having a negative index of refraction are, unquestionably, negative refraction has been bafflingly seen. The situation is extremely comparable to what occurred in the field of photonic crystals, which are periodic optical structures meant to influence the motion of photons. Even though photonic crystals represent an outstanding success for the efforts of a large number of researchers, it has been known for millions of years that opals and the wings of Morpho butterflies include structures and phenomena that are strikingly comparable.

References

1. Xiao, S., Ma, G., Li, Y., Yang, Z., Sheng, P., Active control of membrane-type acoustic metamaterial by electric field. *Appl. Phys. Lett.*, 106, 91904, 2015.
2. Stavenga, D.G., Invertebrate superposition eyes-structures that behave like metamaterial with negative refractive index. *J. Eur. Opt. Soc. Publ.*, 1, 06010, 2006.
3. Duan, G., Zhao, X., Anderrson, S.W., Zhang, X., Boosting magnetic resonance imaging signal-to-noise ratio using magnetic metamaterials. *Commun. Phys.*, 2, 1–8, 2019.
4. Pendry, J.B. and Smith, D.R., The quest for the superlens. *Sci. Am.*, 295, 60–67, 2006.
5. Lakshminarasimhan, R., Solar-driven water treatment: The path forward for the energy–water nexus, in: *Solar-Driven Water Treatment*, O. Mahian, J. Wei, R.A. Taylor, S. Wongwises, (Eds.), pp. 337–362, Elsevier, USA, 2022.
6. Li, Y., Baker, E., Reissman, T., Sun, C., Liu, W.K., Design of mechanical metamaterials for simultaneous vibration isolation energy harvesting. *Appl. Phys. Lett.*, 111, 251903, 2017.
7. Ji, J.C., Luo, Q., Ye, K., Vibration control based metamaterials origami structures: A state-of-the-art review. *Mech. Syst. Signal Process.*, 161, 107945, 2021.
8. Masana, R., Khazaaleh, S., Alhussein, H., Crespo, R.S., Daqaq, M.F., An origami-inspired dynamically actuated binary switch. *Appl. Phys. Lett.*, 117, 81901, 2020.

9. Jiang, J., Chen, M., Fan, J.A., Deep neural networks for the evaluation design of photonic devices. *Nat. Rev. Mater.*, 6, 679–700, 2021.
10. Bhuvaneswari, V., Priyadharshini, M., Deepa, C., Balaji, D., Rajeshkumar, L., Ramesh, M., Deep learning for material synthesis manufacturing systems: A review. *Mater. Today Proc.*, 46, 3263–3269, 2021.
11. Li, X., Zhou, Y., Yang, Z.-Z., Zou, X.-Y., Cheng, J.-C., Tunable acoustic metasurface based on PVDF/polyimide unimorph sheets. *Appl. Phys. Express*, 15, 14001, 2021.
12. Xu, X., Wu, Q., Pang, Y., Cao, Y., Fang, Y., Huang, G., Cao, C., Multifunctional metamaterials for energy harvesting vibration control. *Adv. Funct. Mater.*, 32, 2107896, 2022.
13. Fawzi, T. and Al-Talib, A.A.M., Composite metamaterials: Classification, design, laws, future applications, in: *21st Century Nanostructured Materials - Physics, Chemistry, Classification, and Emerging Applications in Industry, Biomedicine, and Agriculture*, P. Pham (Ed.), IntechOpen, London, United Kingdom 2022.
14. Veselago, V.G., The electrodynamics of substances with simultaneously negative values of ϵ and μ. *Sov. Phys. Usp.*, 10, 509, 1968.
15. Smith, D.R. and Schurig, D., Electromagnetic wave propagation in media with indefinite permittivity permeability tensors. *Phys. Rev. Lett.*, 90, 77405, 2003.
16. Deepa, C., Rajeshkumar, L., Ramesh, M., Preparation, synthesis, properties characterization of graphene-based 2D nano-materials for biosensors bioelectronics. *J. Mater. Res. Technol.*, 19, 2657–2694, 2022.
17. Dai, S., Ma, Q., Liu, M.K., Andersen, T., Fei, Z., Goldflam, M.D., Wagner, M., Watanabe, K., Taniguchi, T., Thiemens, M., Graphene on hexagonal boron nitride as a tunable hyperbolic metamaterial. *Nat. Nanotechnol.*, 10, 682–686, 2015.
18. Deepa, C., Balaji, D., Bhuvaneswari, V., Rajeshkumar, L., Ramesh, M., Priyadharshini, M., Deep learning for the selection of multiple analogs, in: *Drug Design Using Machine Learning*, Inamuddin, T. Altalhi, J.N. Cruz, M.S.D. Refat (Eds.), pp. 117–142, Wiley, Germany, 2022.
19. Basov, D.N., Fogler, M.M., Abajo, F.J., Polaritons in van der Waals materials. *Science*, 354, aag1992, 2016.
20. Liu, Y. and Zhang, X., Metamaterials: A new frontier of science technology. *Chem. Soc. Rev.*, 40, 2494–2507, 2011.
21. Kats, M.A., Sharma, D., Lin, J., Genevet, P., Blanchard, R., Yang, Z., Qazilbash, M.M., Basov, D.N., Ramanathan, S., Capasso, F., Ultra-thin perfect absorber employing a tunable phase change material. *Appl. Phys. Lett.*, 101, 221101, 2012.
22. Driscoll, T., Kim, H.-T., Chae, B.-G., Kim, B.-J., Lee, Y.-W., Jokerst, N.M., Palit, S., Smith, D.R., Di Ventra, M., Basov, D.N., Memory metamaterials. *Science*, 325, 1518–1521, 2009.

23. Driscoll, T., Palit, S., Qazilbash, M.M., Brehm, M., Keilmann, F., Chae, B.-G., Yun, S.-J., Kim, H.-T., Cho, S.Y., Jokerst, N.M., Dynamic tuning of an infrared hybrid-metamaterial resonance using Vanadium dioxide. *Appl. Phys. Lett.*, 93, 24101, 2008.
24. Driscoll, T., Reev, G.O., Basov, D.N., Palit, S., Cho, S.Y., Jokerst, N.M., Smith, D.R., Tuned permeability in terahertz split-ring resonators for device sensors. *Appl. Phys. Lett.*, 91, 62511, 2007.
25. Driscoll, T., Reev, G.O., Basov, D.N., Palit, S., Ren, T., Mock, J., Cho, S.-Y., Jokerst, N.M., Smith, D.R., Quantitative investigation of a terahertz artificial magnetic resonance using oblique angle spectroscopy. *Appl. Phys. Lett.*, 90, 92508, 2007.
26. Yen, T.-J., Padilla, W.J., Fang, N., Vier, D.C., Smith, D.R., Pendry, J.B., Basov, D.N., Zhang, X., Terahertz magnetic response from artificial materials. *Science*, 303, 1494–1496, 2004.
27. Ramesh, M., Bhuvaneswari, V., Balaji, D., Rajeshkumar, L., Self-healable conductive and polymeric composite materials, in: *Aerospace Polymeric Materials*, Inamuddin, T. Altalhi, S.M. Adnan (Eds.), pp. 231–258, Wiley, Germany, 2022.
28. Ramesh, M., Deepa, C., Rajeshkumar, L., Selvan, M., Balaji, D., Influence of fiber surface treatment on the tribological properties of Calotropis gigantea plant fiber reinforced polymer composites. *Polym. Compos.*, 42, 4308–4317, 2021.
29. Sathishkumar, T.P., Gil, J., García, R., Rajeshkumar, L., Rajeshkumar, G., Rangappa, S., Siengchin, S., Alosaimi, A.M., Hussein, M.A., Redeemable environmental damage by recycling of industrial discarded virgin glass fiber mats in hybrid composites—An exploratory investigation. *Polym. Compos.*, 44, 318–329, 2022.
30. Ramesh, M., Rajeshkumar, L., Balaji, D., Mechanical dynamic properties of ramie fiber-reinforced composites, in: *Mechanical Dynamic Properties of Biocomposites*, S. Krishnasamy, R. Nagarajan, S.M.U. Thiagamani, S. Siengchin (Eds.), pp. 275–291, Wiley, Germany, 2021.
31. Smith, D.R., Padilla, W.J., Vier, D.C., Nasser, S.C., Schultz, S., Composite medium with simultaneously negative permeability permittivity. *Phys. Rev. Lett.*, 84, 4184, 2000.
32. Hong, J.-S. and Lancaster, M.J., Couplings of microstrip square open-loop resonators for cross-coupled planar microwave filters. *IEEE Trans. Microw. Theory Tech.*, 44, 2099–2109, 1996.
33. Ramesh, M., Selvan, M., Rajeshkumar, L., Deepa, C., Ahmad, A., Influence of Vachellia nilotica Subsp. indica tree trunk bark nano-powder on properties of milkweed plant fiber reinforced epoxy composites. *J. Nat. Fibers*, 19, 1–14, 2022.
34. Cai, W., Chettiar, U.K., Yuan, H.-K., de Silva, V.C., Kildishev, A.V., Drachev, V.P., Shalaev, V.M., Metamagnetics with rainbow colors. *Opt. Express*, 15, 3333–3341, 2007.

35. Ramesh, M., Rajeshkumar, L., Bhoopathi, R., Carbon substrates: A review on fabrication, properties, applications. *Carbon Lett.*, 31, 557–580, 2021.
36. Shalaev, V.M., Cai, W., Chettiar, U.K., Yuan, H.-K., Sarychev, A.K., Drachev, V.P., Kildishev, A.V., Negative index of refraction in optical metamaterials. *Opt. Lett.*, 30, 3356–3358, 2005.
37. Ramesh, M., Rajeshkumar, L., Saravanakumar, R., Mechanically-induced self-healable materials, in: *Self-Healing Smart Materials*, Inamuddin, M.I. Ahamed, R. Boddula, T. Altalhi (Eds.), pp. 379–404, Wiley, United Kingdom, 2021.
38. Pendry, J.B., Negative refraction makes a perfect lens. *Phys. Rev. Lett.*, 85, 3966, 2000.
39. Dolling, G., Wegener, M., Soukoulis, C.M., Linden, S., Negative-index metamaterial at 780 nm wavelength. *Opt. Lett.*, 32, 53–55, 2007.
40. Cai, W. and Shalaev, V.M., *Optical metamaterials*, Springer, Singapore, 2010.
41. Devarajan, B., Narasimhan, R., Venkateswaran, B., Rangappa, S., Siengchin, S., Additive manufacturing of jute fiber reinforced polymer composites: A concise review of material forms methods. *Polym. Compos.*, 43, 6735–6748, 2022.
42. Hardy, W.N. and Whitehead, L.A., Split-ring resonator for use in magnetic resonance from 200–2000 MHz. *Rev. Sci. Instrum.*, 52, 213–216, 1981.
43. Lakhtakia, A. and Geddes, J.B., Thin-film metamaterials called sculptured thin films, in: *Trends in Nanophysics*, pp. 59–71, 2010.
44. Ahamed, M.I., Boddula, R., Altalhi, T.A., Self-healable conductive materials, in: *Self-Healing Smart Materials*, Inamuddin, M.I. Ahamed, R. Boddula, T. Altalhi (Eds.), pp. 297–320, Wiley, United Kingdom, 2021.
45. Jackson, J.D., *Classical electrodynamics*, Wiley, India, 1999.
46. Palik, E.D., *Handbook of optical constants of solids*, Academic Press, US, 1998.
47. Ibach, H., Lüth, H., Kittel, C., Kroemer, H., Fine, M.E., Hoyt, J.J., *Phase transformations*, McMaster Innovation Press. Porter, D.A., Easterling, K.E., Sherif, M.Y., *Phase transformations in metals alloys*, CRC Press, Boca Raton, FL, 2009.
48. Batsanov, S.S., Ruchkin, E.D., Poroshina, I.A., *Refractive indices of solids*, Springer, Singapore, 2016.
49. Trager, F., *Handbook of laser and optics*, Springer, Berlin, Heidelberg, 2007.
50. Sihvola, A.H. and Kong, J.A., Effective permittivity of dielectric mixtures. *IEEE Trans. Geosci. Remote Sens.*, 26, 420–429, 1988.
51. Ramesh, M. and Rajeshkumar, L., Technological advances in analyzing of soil chemistry, in: *Applied Soil Chemistry*, Inamuddin, M.I. Ahamed, R. Boddula, T. Altalhi (Eds.), pp. 61–78, Wiley, Germany, 2021.
52. Ross, N.S., Srinivasan, N., Amutha, P., Gupta, M.K., Korkmaz, M.E., Thermophysical, tribological machining characteristics of Hastelloy C276 under sustainable cooling/lubrication conditions. *J. Manuf. Process.*, 80, 397–413, 2022.
53. Monecke, J., Bergman spectral representation of a simple expression for the dielectric response of a symmetric two-component composite. *J. Phys. Condens. Matter.*, 6, 907, 1994.

54. Shalaev, V.M., *Nonlinear optics of rom media: Fractal composites metal-dielectric films*, Springer Science & Business Media, 1999.
55. Rajenthirakumar, D., Srinivasan, N., Sridhar, R., Development of a micro-forming system for micro-extrusion process of micro-pin in AZ80 alloy, in: *Machines, Mechanism, Robotics*, Kumar, R., Chauhan, V.S., Talha, M., Pathak, H. (Eds.), pp. 1227–1229, 2022.
56. Srinivasan, N., Rajenthirakumar, D., Sridhar, R., Niranjan, G., Combined influence of size effect temperature in microscale deformation of 6063 Aluminum gear. *J. Braz. Soc. Mech. Sci. Eng.*, 42, 317, 2020.
57. Black, M.R., Lin, Y.-M., Cronin, S.B., Rabin, O., Dresselhaus, M.S., Infrared absorption in bismuth nanowires resulting from quantum confinement. *Phys. Rev. B*, 65, 195417, 2002.
58. Srinivasan, N., Rajenthirakumar, D., Sridhar, R., Micro-scaled plastic deformation behavior of biodegradable AZ80 magnesium alloy: Experimental numerical investigation. *Int. J. Adv. Manuf. Technol.*, 102, 3531–3541, 2019.
59. Bergman, D.J., Exactly solvable microscopic geometries rigorous bounds for the complex dielectric constant of a two-component composite material. *Phys. Rev. Lett.*, 44, 1285, 1980.
60. Milton, G.W., *The theory of composites*, Cambridge University Press, Cambridge, UK, 2002.
61. Ramesh, M., Case-studies on green corrosion inhibitors, in: *Sustainable Corrosion Inhibitors*, Inamuddin, M.I. Ahamed, M. Luqman, T. Altalhi (Eds.), pp. 204–221, Materials Research Forum, Millersville, USA, 2021.
62. Looyenga, H., Dielectric constants of heterogeneous mixtures. *Physica*, 31, 401–406, 1965.
63. Garnett, J.C., Colours in metal glasses in metallic films. *Phil. Trans. R. Soc. Lond. A*, 203, 385–420, 1904.
64. Deepa, C., Rajeshkumar, L., Ramesh, M., Thermal properties of Kenaf fiber-based hybrid composites, in: *Natural Fiber-Reinforced Composites*, S. Krishnasamy, S.M.K. Thiagamani, C. Muthukumar, R. Nagarajan, S.S.S. Krishnasamy, S.M.U. Thiagamani, C. Muthukumar, S.S. Nagarajan (Eds.), pp. 167–182, Wiley, Germany, 2022.
65. Goffaux, C., Sánchez-Dehesa, J., Yeyati, A.L., Lambin, P., Khelif, A., Vasseur, J.O., Rouhani, B., Evidence of Fano-like interference phenomena in locally resonant materials. *Phys. Rev. Lett.*, 88, 225502, 2002.
66. Fang, N., Xi, D., Xu, J., Ambati, M., Srituravanich, W., Sun, C., Zhang, X., Ultrasonic metamaterials with negative modulus. *Nat. Mater.*, 5, 452–456, 2006.
67. Dühring, M.B., Laude, V., Khelif, A., Energy storage dispersion of surface acoustic waves trapped in a periodic array of mechanical resonators. *J. Appl. Phys.*, 105, 93504, 2009.
68. Berenger, J.-P., A perfectly matched layer for the absorption of electromagnetic waves. *J. Comput. Phys.*, 114, 185–200, 1994.

69. Achaoui, Y., Khelif, A., Benchabane, S., Robert, L., Laude, V., Experimental observation of locally-resonant Bragg b gaps for surface guided waves in a phononic crystal of pillars. *Phys. Rev. B*, 83, 104201, 2011.
70. Balaji, D., Ranga, J., Bhuvaneswari, V., Arulmurugan, B., Rajeshkumar, L., Manimohan, M.P., Devi, G.R., Ramya, G., Masi, C., Additive manufacturing for aerospace from inception to certification. *J. Nanomater.*, 2022, 1–18, 2022.
71. Wang, G., Wen, X., Wen, J., Shao, L., Liu, Y., Two-dimensional locally resonant phononic crystals with binary structures. *Phys. Rev. Lett.*, 93, 154302, 2004.
72. Vasseur, J.O., Deymier, P.A., Chenni, B., Rouhani, B., Dobrzynski, L., Prevost, D., Experimental theoretical evidence for the existence of absolute acoustic b gaps in two-dimensional solid phononic crystals. *Phys. Rev. Lett.*, 86, 3012, 2001.
73. Sigalas, M. and Economou, E.N.B., Structure of elastic waves in two dimensional systems. *Solid State Commun.*, 86, 141–143, 1993.
74. Medvedev, N. and Milov, I., Electron-phonon coupling in metals at high electronic temperatures. *Phys. Rev. B*, 102, 64302, 2020.
75. Vasseur, J., Pennec, Y., Rouhani, B., Vasseur, J., Larabi, H., Khelif, A., Choujaa, A., Benchabane, S., Laude, V., Acoustic channel drop tunneling in a phononic crystal. *Appl. Phys, Lett.*, 87, 261912, 2005.
76. Luk'yanchuk, B., Zheludev, N.I., Maier, S.A., Halas, N.J., Nordler, P., Giessen, H., Chong, C.T., The Fano resonance in plasmonic nanostructures and metamaterials. *Nat. Mater.*, 9, 707, 2010.
77. Ramesh, M., Balaji, D., Rajeshkumar, L., Bhuvaneswari, V., Manufacturing methods of elastomer blends composites, in: *Elastomer Blends and Composites, Principles, Characterization, Advances, and Applications*, S.M. Rangappa, J. Parameswaranpillai, S. Siengchin, T.B. Ozbakkaloglu, (Eds.), pp. 11–32, Elsevier, USA, 2022.
78. Liu, Z., Zhang, X., Mao, Y., Zhu, Y.Y., Yang, Z., Chan, C.T., Sheng, P., Locally resonant sonic materials. *Science*, 289, 1734–1736, 2000.
79. Kushwaha, M.S., Halevi, P., Martinez, G., Acoustic dispersion relation of periodic elastic composites. *Phys. Rev. Lett.*, 71, 2022–2025, 1993.
80. Khelif, A., Choujaa, A., Benchabane, S., Djafari-Rouhani, B., Laude, V., Guiding bending of acoustic waves in highly confined phononic crystal waveguides. *Appl. Phys. Lett.*, 84, 4400–4402, 2004.
81. Khelif, A., Aoubiza, B., Mohammadi, S., Adibi, A., Laude, V., Complete b gaps in two-dimensional phononic crystal slabs. *Phys. Rev. E*, 74, 46610, 2006.
82. Khan, A., Rangappa, S.M., Siengchin, S., Asiri, A.M., Ramesh, M., Maniraj, J., Kumar, L., Biocomposites for energy storage, in: *Biobased Composites*, A. Khan, S.M. Rangappa, S. Siengchin, A.M. Asiri (Eds.), pp. 123–142, Wiley, Germany, 2021.
83. Krödel, S., Thomé, N., Daraio, C., Wide b-gap seismic metastructures. *Extreme Mech. Lett.*, 4, 111–117, 2015.

84. Papageorgiou, C. and Smith, M.C., Laboratory experimental testing of inerters. *Proc. 44th IEEE Conf. Decis. Control*, pp. 3351–3356, 2005.
85. Ramesh, M., Kumar, L., Khan, A., Asiri, A.M., Self-healing polymer composites and its chemistry, in: *Self-Healing Composite Materials: From Design to Applications*, A. Khan, M. Jawaid, S.N. Raveendran, A.M. Asiri (Eds.), pp. 415–427, Elsevier, USA, 2019.
86. Zeng, Y., Cao, L., Wan, S., Guo, T., Wang, Y.-F., Du, Q.-J., Assouar, B., Wang, Y.-S., Seismic metamaterials: Generating low-frequency b gaps induced by inertial amplification. *Int. J. Mech. Sci.*, 221, 107224, 2022.
87. Ramesh, M., Kumar, L., Bhuvaneshwari, V., Bamboo fiber reinforced composites, in: *Bamboo Fiber Composites*, M. Jawaid, S.M. Rangappa, S. Siengchin (Eds.), pp. 1–13, Springer, Singapore, 2021.
88. Lin, Q., Zhou, J., Wang, K., Xu, D., Wen, G., Wang, Q., Cai, C., Low-frequency locally resonant b gap of the two-dimensional quasi-zero-stiffness metamaterials. *Int. J. Mech. Sci.*, 222, 107230, 2022.
89. Kumar, T.P., Kumar, S., Kumar, L., Jute fibers, their composites applications, in: *The Textile Institute Book Series*, M. Rangappa, S. Parameswaranpillai, J. Siengchin, S. Ozbakkaloglu, T. Wang (Eds.), pp. 253–282, Woodhead Publishing, USA, 2022.
90. Laghfiri, H. and Lamdouar, N., Periodic structures as a countermeasure of traffic vibration earthquake: A review. *Int. Conf. Adv. Technol. Humanit.*, pp. 359–373, 2021.
91. Ramesh, M., Rajeshkumar, L., D.B., Aerogels for insulation applications, in: *Aerogels II: Preparataion, Properties, Applications*, Inamuddin, R. Mobin, M.I. Ahamed, T. Altalhi (Eds.), pp. 57–76, Materials Research Forum, Millersville, USA, 2021.
92. Ramesh, M., Rajeshkumar, L., Balaji, D., Bhuvaneswari, V., Influence of moisture absorption on mechanical properties of biocomposites reinforced surface modified natural fibers, in: *Aging Effects on Natural Fiber-Reinforced Polymer Composites: Durability Life Prediction*, C. Muthukumar, S. Krishnasamy, S.M.K. Thiagamani, S. Siengchin, (Eds.), pp. 17–34, Springer, Singapore, 2022.
93. Huang, Y., Liu, S., Zhao, J., Optimal design of two-dimensional b-gap materials for uni-directional wave propagation. *Struct. Multidiscip. Optim.*, 48, 487–499, 2013.
94. Schurig, D., Mock, J.J., Justice, B.J., Cummer, S.A., Pendry, J.B., Starr, A.F., Smith, D.R., Metamaterial electromagnetic cloak at microwave frequencies. *Science*, 314, 977–980, 2006.
95. Cai, W., Chettiar, U.K., Kildishev, A.V., Shalaev, V.M., Optical cloaking with metamaterials. *Nat. Photonics*, 1, 224–227, 2007.
96. Chen, H. and Chan, C.T., Acoustic cloaking in three dimensions using acoustic metamaterials. *Appl. Phys. Lett.*, 91, 183518, 2007.
97. Reassen, E., Lazarov, B.S., Sigmund, O., Design of manufacturable 3D extremal elastic microstructure. *Mech. Mater.*, 69, 1–10, 2014.

98. Ramesh, M., Rajeshkumar, L., Balaji, D., Bhuvaneswari, V., 14 - Keratin-based biofibers their composites, in: *The Textile Institute Book Series*, S.M. Rangappa, M. Puttegowda, J. Parameswaranpillai, S. Siengchin, S. Gorbatyuk (Eds.), pp. 315–334, Woodhead Publishing, USA, 2022.
99. Wang, Y., Luo, Z., Zhang, N., Kang, Z., Topological shape optimization of microstructural metamaterials using a level set method. *Comput. Mater. Sci.*, 87, 178–186, 2014.
100. Zheng, X., Lee, H., Weisgraber, T.H., Shusteff, M., DeOtte, J., Duoss, E.B., Kuntz, J.D., Biener, M.M., Ge, Q., Jackson, J.A., Ultralight, ultrastiff mechanical metamaterials. *Science*, 344, 1373–1377, 2014.
101. Xia, L. and Breitkopf, P., Design of materials using topology optimization energy-based homogenization approach in Matlab. *Struct. Multidiscip. Optim.*, 52, 1229–1241, 2015.
102. Colquitt, D.J., Brun, M., Gei, M., Movchan, A.B., Movchan, N.V., Jones, I.S., Transformation elastodynamics cloaking for flexural waves. *J. Mech. Phys. Solids*, 72, 131–143, 2014.
103. Yu, X., Zhou, J., Liang, H., Jiang, Z., Wu, L., Mechanical metamaterials associated with stiffness, rigidity compressibility: A brief review. *Prog. Mater. Sci.*, 94, 114–173, 2018.
104. Ramesh, M. and Kumar, L.R., Bioadhesives, in: *Green Adhesives*, Inamuddin, R. Boddula, M.I. Ahamed, A.M. Asiri (Eds.), pp. 145–164, Wiley, Germany, 2020.
105. Chen, W. and Huang, X., Topological design of 3D chiral metamaterials based on couple-stress homogenization. *J. Mech. Phys. Solids*, 131, 372–386, 2019.
106. Li, H., Luo, Z., Gao, L., Walker, P., Topology optimization for functionally graded cellular composites with metamaterials by level sets. *Comput. Methods Appl. Mech. Eng.*, 328, 340–364, 2018.
107. Lakes, R., Foam structures with a negative Poisson's ratio. *Science*, 235, 1038–1040, 1987.
108. Guest, J.K. and Prévost, J.H., Design of maximum permeability material structures. *Comput. Methods Appl. Mech. Eng.*, 196, 1006–1017, 2007.
109. Diaz, A.R. and Sigmund, O., A topology optimization method for design of negative permeability metamaterials. *Struct. Multidiscip. Optim.*, 41, 163–177, 2010.
110. Bhuvaneswari, V., Rajeshkumar, L., Ross, K., Influence of bioceramic reinforcement on tribological behaviour of Aluminium alloy metal matrix composites: Experimental study analysis. *J. Mater. Res. Technol.*, 15, 2802–2819, 2021.
111. Evans, J.S.O., Negative thermal expansion materials. *J. Chem. Soc. Dalton Trans.*, 19, 3317–3326, 1999.
112. Ramesh, M., Rajeshkumar, L., Balaji, D., Priyadharshini, M., Properties characterization techniques for waterborne polyurethanes, in: *Sustainable Production Applications of Waterborne Polyurethanes*, Inamuddin, R.

Boddula, A. Khan (Eds.), pp. 109–123, Springer International Publishing, Cham, 2021.

113. Wu, S., Eichenberger, J., Dai, J., Chang, Y., Ghalichechian, N., Zhao, R.R., Magnetically actuated reconfigurable metamaterials as conformal electromagnetic filters. *Adv. Intell. Syst.*, 4, 2200106, 2022.
114. Kim, J., Lee, S., Brovman, Y.M., Kim, M., Kim, P., Lee, W., Weak antilocalization conductance fluctuation in a single crystalline Bi nanowire. *Appl. Phys. Lett.*, 104, 119901, 2014.
115. Ramesh, M., Rajeshkumar, L., Bhuvaneswari, V., Leaf fibres as reinforcements in green composites: A review on processing, properties applications. *Emergent Mater.*, 5, 833–857, 2022.
116. Ramesh, M., Balaji, D., Rajeshkumar, L., Bhuvaneswari, V., Saravanakumar, R., Khan, A., Asiri, A.M., Tribological behavior of glass/sisal fiber reinforced polyester composites, in: *Vegetable Fiber Composites and their Technological Applications*, M. Jawaid and A. Khan (Eds.), pp. 445–459, Springer Singapore, 2021.
117. Shah, S.I.H. and Lim, S., Review on recent origami inspired antennas from microwave to terahertz regime. *Mater. Des.*, 198, 109345, 2021.
118. Zhu, Z., Wang, H., Li, Y., Meng, Y., Wang, W., Zheng, L., Wang, J., Zhang, J., Qu, S., Origami-based metamaterials for dynamic control of wide-angle absorption in a reconfigurable manner Zhibi. *IEEE Trans. Antennas Propag.*, *46*, 1349–1352, 2022.
119. Valente, J., Ou, J.-Y., Plum, E., Youngs, I.J., Zheludev, N.I., Reconfiguring photonic metamaterials with currents magnetic fields. *Appl. Phys. Lett.*, 106, 111905, 2015.
120. Nagasaki, Y., Gholipour, B., Ou, J.-Y., Tsuruta, M., Plum, E., MacDonald, K.F., Takahara, J., Zheludev, N.I., Optical bistability in shape-memory nanowire metamaterial array. *Appl. Phys. Lett.*, 113, 21105, 2018.
121. Tang, J. and Sun, B., Reprogrammable shape transformation of magnetic soft robots enabled by magnetothermal effect. *Appl. Phys. Lett.*, 120, 244101, 2022.
122. Zhao, R., Kim, Y., Chester, S.A., Sharma, P., Zhao, X., Mechanics of hard-magnetic soft materials. *J. Mech. Phys. Solids*, 124, 244–263, 2019.
123. Gao, X., Yang, W.L., Ma, H.F., Cheng, Q., Yu, X.H., Cui, T.J., A reconfigurable broad polarization converter based on an active metasurface. *IEEE Trans. Antennas Propag.*, 66, 6086–6095, 2018.
124. Driscoll, T., Basov, D.N., Padilla, W.J., Mock, J.J., Smith, D.R., Electromagnetic characterization of planar metamaterials by oblique angle spectroscopic measurements. *Phys. Rev. B*, 75, 115114, 2007.
125. Wang, H.L., Ma, H.F., Chen, M., Sun, S., Cui, T.J., A reconfigurable multifunctional metasurface for full-space control of electromagnetic waves. *Adv. Funct. Mater.*, 31, 2100275, 2021.

8

Metamaterials for Cloaking Devices

M. Rizwan[1*], M. W. Yasin[1], Q. Ali[1] and A. Ayub[2]

[1]*School of Physical Sciences, University of the Punjab, Lahore, Pakistan*
[2]*Department of Physics, University of the Punjab, Lahore, Pakistan*

Abstract

Electromagnetic cloaking has gained the attention of scientists and researchers who are working on metamaterials. Exotic electromagnetic properties occur in artificial synthesis. The invisible cloak has attained huge attention due to its remarkable effects on the modern world. For flexible and malleability wave cloaks, a higher level of controlled features is provided by an active wave that is different from the standard structure. The cloaking structure contains polymethylioxane and piezoelectric patches and 16 different films, which are interlinked with the active control system. For the flexural wave cloak, the productive frequency range broadens from 900 Hz to 1200 Hz, attained by computational and experimental results of some elements. This type of cloaking is significantly used in shielding applications. In this chapter, we are going to discuss the basic working principles of metamaterials, especially those that explain the latest evolution in the electromagnetic cloaking domain and are used for cloaking applications. Due to its application in the network of transmission lines, it gained so much attention. With negative permittivity ($\varepsilon < 0$) and permeability ($\mu < 0$) long-distance transmission lines have special characteristics that do not arise in natural material.

Keywords: Transformation coordinates, metamaterials, electromagnetic cloak, cylindrical shape, ductility, camouflage, cloaking

8.1 Introduction

Invisible cloaks have attracted the huge attention of scientists and engineers. The construction process for the cloak mainly depends on the

**Corresponding author*: rizwan.sps@pu.edu.pk

Inamuddin and Tariq Altalhi (eds.) Electromagnetic Metamaterials: Properties and Applications, (165–184) © 2023 Scrivener Publishing LLC

transformation of coordinates. The invisibility gadgets must lead the light incoming from an object in such a way as nothing was there at that point from where the light is coming. Due to the dual nature of light, it is very difficult to make ideal invisibility devices. After a lot of research and experimentation, scientists can create a highly workable invisible cloak with the precision of geometrical optics. Artificially fabricated metamaterials have allowed us remarkable flexibility in handling electromagnetic radiation or wave and generating new functionalities, including the transformation of the coordination construction process [1]. All the cloaking processes have some limitations like these methods are not work perfectly on subwavelength objects. But scientists make some designs that provide macroscopic objects' invisibility by using the transformation method. Besides, this design is not sensitive to objects that are being cloaked. Like a cloak, design metamaterial is being used in many fields of electromagnetic after its inventions. Some exotic properties are shown by metamaterial that do not occur in natural materials. One of the most popular applications of metamaterial is cloaking. Bypassing an electromagnetic field around the object electromagnetic cloak hide the objects. Cloaking devices are very famous in armies because they are used in hiding weapons, tanks, and military aircraft. The transformation optics approach in cloaking is very useful even for hiding a large structure. On the other hand, this problem is solved by the scattering reduction technique for cloaking because it has better performance, flexibility, and simplicity reduction technique of cloaking creates opposite diploe moments. If the object does not have scattered waves, it can be hidden electromagnetically. In the scattering reduction technique for vanquishing the scattering, the metamaterial is used with negative permittivity. Almost all cloaks have a cylindrical shape [2].

By introducing a prescribed spatial variation in the constitutive parameter, a design was proposed in which EM wave is handled inside a material. At different frequency regions, this type of exciting application, metamaterial obtained huge attention from scientists and researchers [3].

8.2 What is Cloaking and Invisibility?

Scattering properties of vacuum are present in good invisibility cloaks made by metamaterials. For a certain frequency range, a device that makes the object invisible is called an electromagnetic cloak. Working on a cloak in the visible region of the spectrum is the most exciting thing about a cloaking device. If the wave does not scatter, or reflect by the object or if no shadow is formed, then an object is considered hidden or invisible. Hence, any type of power should not be absorbed by the object. In other words, we can say that the electric field that is present outside the object does not affect the object [4].

The term cloaking is referred to the vanquishing of total scattering of an object. The ratio of diverged power to the incoming power is known as scattered cross-section (SCS) of an object. But keep in mind that cloaking and stealth technology are not the same. In stealth technology, a small amount of power is reflected by the object and detected by radar. If observed from the left, right, and back side, the ideal stealth craft is also detectable [5]. Even a 50% scattering cross-section cannot be decreased by shaping the object and absorbing the covering [6].

8.3 Basic Concepts of Cloaking

The main purpose of cloaking is to hide the object from the detector. This can be obtained only when the electromagnetic wave does not reflect and is scattered by the object, and it means incident light should come out without interacting with an object. Transformation optics is used in cloaking to gain a spatially scattered set of constitutive factors a conformal coordinate transformation is operated to the Maxwell equation that defines the cloak. The material becomes spatially invariant, inhomogeneous, and anisotropic, we extract the permeability and permittivity in such a way that these properties are required to obtain cloaking [7].

Permittivity tensor:

$$\varepsilon' = \frac{\Lambda \varepsilon \Lambda^T}{det|\Lambda|} S \quad \text{(Eq. 8.1)}$$

Permeability tensor:

$$\mu' = (\Lambda \mu \Lambda^T / det|\Lambda|) \quad \text{(Eq. 8.2)}$$

Here, Λ is Jacobuan transformation ε is permittivity in free space and μ is permeability in free space.

$$\Lambda_{ij} = \frac{dx_i}{dx_j}$$

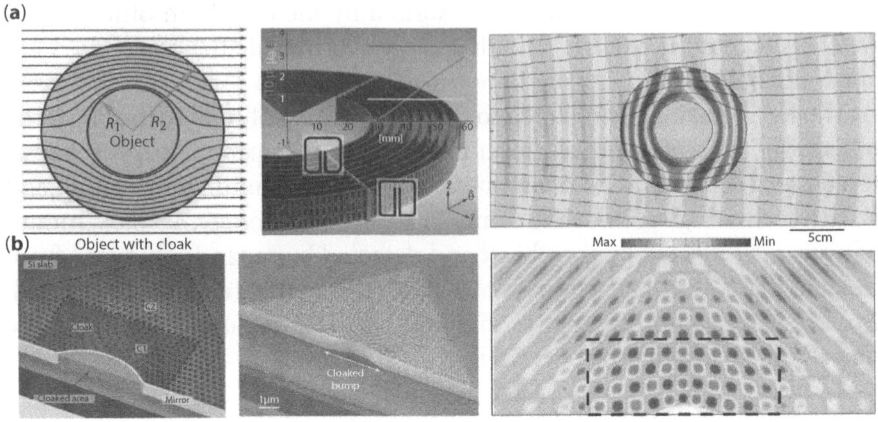

Figure 8.1 (a) Light ray trajectory within cloaking shell, (b) carpet cloaking [12].

Using a coordinate transformation, the volume of free space is converted into a shell-type region that can protect the inside of the shell from electromagnetic radiation interaction of incident waves as shown in Figure 8.1. The object to be cloaked is shown as a PEC cylinder, and the cloak is a multilayer structure of radius b [3, 8].

8.4 Design and Simulation of Metamaterial Invisibility Cloak

The analysis is very complex in the three-dimensional conformal multilayer structure and a systematical process is existing for the same. So for the construction and visualization of the invisible cloak, the numerical technique is compulsory. Inductor-capacitor representation of metamaterial, FDTD, FEM, LC-MTM, and FIT are four subsections of simulation categorized. FDTD and FIT are considered numerical techniques. On the other hand, LC MTM is a circuit analysis method [8].

An estimation solution is given to the integral equation and partial differential equation by numerical technique (finite-element method). With polynomial estimation operation, the finite element technique is an exceptional case of Galerkin's method. For the design and visualization of the invisible cloak, this method is mostly used. Schurig *et al.* purposed the first 2D practical metamaterial in 2006. The cloak consists of 10 parallel layers of cylinders made of copper and each has three unit cells set in horizontal lines 850 GHz is the frequency for the function of the cloak. Due to absorption and estimation, the invisibility was imperfect. Cylindrical region

$0 < r > b$ is compressed $a < r < b$ by transformation coordinate, and this transformation helps to obtain the objective [9].

$$r' = \frac{b-a}{b} r + a \qquad \text{(Eq. 8.3)}$$

$$\theta' = \theta \qquad \text{(Eq. 8.4)}$$

and

$$z' = z \qquad \text{(Eq. 8.5)}$$

For permittivity and permeability tensor

$$\epsilon_r = \mu_r = \frac{r-a}{r} \qquad \text{(Eq. 8.6)}$$

$$\varepsilon_\theta = \mu_\theta \left(\frac{b}{b-a} \right)^2 \frac{r-a}{r} \qquad \text{(Eq. 8.7)}$$

To decrease flaws in invisibility inside the precision of geometrical optics, Leonharaolt purposed a conformal mapping technique. The wave is not only refracted but also reflected, so perfect invisibility is very difficult to obtain proved by this technique, but reflection is decreased by anti-reflection coating. At optical frequency, for the design of a non-magnetic cylindrical cloak, Cai *et al.* used a method in mid of 2007 known as coordinate transformation. By squeezing the cylindrical region into concentric cylinders, this cloak hides microscopic objects. For operation, the wavelength is 632.8 nm and by using metal wires, sub-wavelength permittivity is obtained. COMSOL shows field distribution. In the absence of the cloak, the object emits shadow, and in the presence of the cloak, the wavefront moves around the cloaked region with little disorder as shown in Figure 8.1.

Two parameters are varied, and one parameter is kept constant, as a result, the constitutive parameter tensor of the cloak reduces. So permittivity and permeability are written as:

$$\varepsilon_r = \left(\frac{b}{b-a} \right)^2 \frac{r-a}{r} \qquad \text{(Eq. 8.8)}$$

$$\varepsilon_\theta = \left(\frac{b}{b-a}\right)^2 \qquad \text{(Eq. 8.9)}$$

$$\mu_z = 1 \qquad \text{(Eq. 8.10)}$$

The power of reflection is obtained as

$$\Gamma = \left(\frac{R_{ab}}{2-R_{ab}}\right)^2 \qquad \text{(Eq. 8.11)}$$

where $R_{ab} = \frac{b}{a}$. In infrared and microwave regions cloaking is obtained by this method [8].

8.5 Types of Cloaking

There are several types of cloaking optical cloaking, acoustic cloaking, and elastic cloaking.

8.5.1 Optical Cloaking

The aim of making the object invisible forced scientists in the optics field to explore and invent materials for cloaking technology [9, 10]. These artificial design metamaterials are made of a dielectric. Metamaterials play an important role in cloaking technology by moving incoming light on all side objects and making the object indistinguishable from the detector or observer as shown in Figure 8.2(a). To obtain permeability and permittivity and to create a space under coordinate transformation in-variance of the Maxwell equation is used in the first proposal to obtain cloaking for the metamaterial, that will guide incident light to move around the object or voids as shown in Figure 8.2(a). A spherical shell is the main part of the recommended invisible cloak that protects the object in such a way that no incident electromagnetic wave impinges on the object and makes the object hidden from the detector by maintaining the external field unchanged. Obtaining cloaking by resonance satisfied the requirement of permeability and permittivity by splitting the resonator the first cylindrical metamaterial cloaking shell made after the theoretical work of seminal after the center of the metamaterial shell copper cylinder of internal

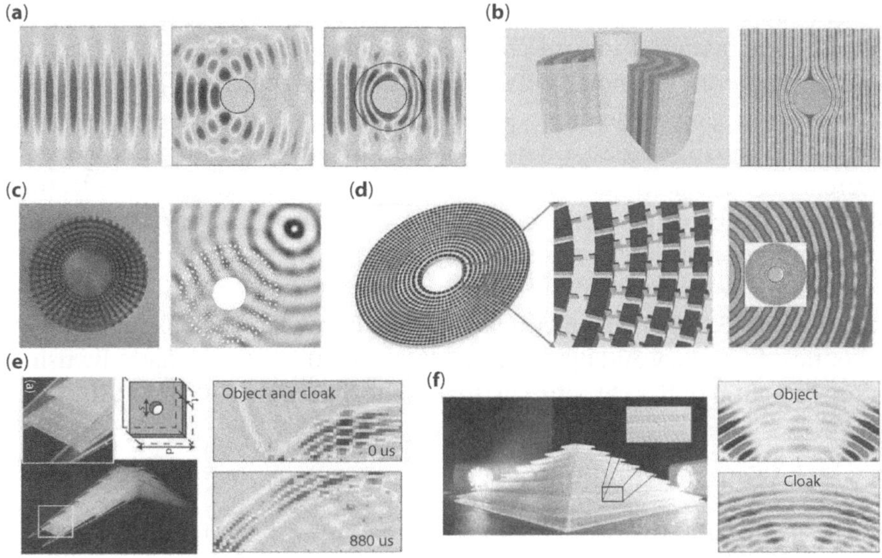

Figure 8.2 (a) Without an object the propagation of waves under stress and pressure. (b) By using the two-layered material, cylindrical shell is made. (c) Cylindrical shell for cloaking an object in a fluid by surface wave. (d) Cylindrical shell for cloaking an object in water by using ultrasound. (e) Acoustic wave for ground cloak. (f) For sound there is a three-dimensional cloak which is of acoustic ground [12].

radians copper cylinder of internal radius R_1 and external radius radius R_2 and function in the microwave region $0 < r < R_0$ into angular region $R_1 < r < R_2$ the coordinate transformation to object is achieved here (r', θ', z') are transformed coordinates and (r, θ, z) are original cylindrical coordinates such transformation is written as [9].

$$r' = \frac{R_2 - R_1'}{R_2} r + R_1, \ \theta' = \theta, \ z' = z \qquad \text{(Eq. 8.12)}$$

For the cloaking shell permittivity and permeability, the tensor component is achieved by applied transformation. The tensor component of permeability and permittivity is given by polarization experimentally [11].

$$\mu_r = \left(\frac{r - R_1}{r}\right)^2, \quad \mu_\theta = 1, \quad \epsilon_z = \left(\frac{R_2}{R_2 - R_1}\right)^2 \qquad \text{(Eq. 8.13)}$$

After an illustration of a cloaking device at microwave frequency, scientist moves their intention toward creating a cloaking device that functions in optical frequencies region. Metamaterial made of split resonator does not effective at higher frequencies because SRR decreased the resonance efficiency on a small scale. So, a new theoretical design or cloaking purposed to overcome this limitation A random boy that is based on the planar reflecting surface is ids by using carpet cloaking. A carpet cloak is very easy to construct and it works on optical frequencies. The copper-clad printed circuit is the main component of the carpet cloak and experimentally this cloak functions at 13 to 16 GHz microwaves. As we know carpet cloak functions at an optical frequency and operated at 1400 to 1800 nm wavelength. By using transformation optics carpet cloak makes the object invisible [12, 13].

8.5.2 Acoustic Cloaking

In the beginning, the ordinary electrodynamic wave equation is studied, and found that it is not constant with the transformation coordinate. By converting the acoustic equation to the Maxwell equation Cummer and Schurig proposed an acoustic cloaking in 2 dimensional in 2007. An acoustic cloak wraps an object in such a way that sound coming from all directions does interact with the object and is presented as an object that was not there [14]. For mapping a cloaked region using a change of variable method and transformation, an acoustic theory is developed with scattering strength reduction. But, the parameter of the acoustical present in the cloak should be anisotropic. For single bulk modulus fluid if the rigidity is isotropic, at the inner surface the inertial density must be finite. But need a massive cloak. By using anisotropic rigidity with finite mass the perfect cloaking is achieved [15]. Pentamode material is the general class of anisotropic material. The parameters of penta mode are categorically certain or its characteristics depend on tensor (stress) if the transformation slope or gradient is harmonic which satisfied the equilibrium condition. In pseudo-acoustic waves the speed and phase of the cloak are unique but the composition is not distinctive for a given mapping transformation. A fluid having properties of scalar bulk modulus and anisotropic density is needed for an acoustic cloaking solution reported by Cummer and Shurig and these variables depend upon the axial distance to the hidden object. The required functional form is

$$\frac{\rho_r}{\rho_b} = \frac{r}{r - R_1} \qquad \text{(Eq. 8.14)}$$

$$\frac{\rho_\theta}{\rho_b} = \frac{r - R_1}{r} \quad \text{(Eq. 8.15)}$$

$$\frac{B}{B_b} = \left(\frac{R_2 - R_1}{R_2}\right)^2 \frac{r}{r - R_1} \quad \text{(Eq. 8.16)}$$

Outer radii for cloaking shell is R_2 and inner radii for cloaking the shell is R_1 the bulk modulus for the shell is B.Mass density tensor diagonals components are ρ_θ and ρ_b. Like fluid and gas those quantities that surround the background are denoted by subscripted b. The acoustic parameter of acoustic metamaterial should change as a function of distance from layer to center of the shell and this metamaterial is used to make multilayered structures in acoustic cloaking Homogenized 2D sonic crystal consists of two fully elastic cylinders per unit cell by using this we can obtain radial dependence practically [12].

8.5.3 Elastic Cloaking

In solid obtaining the cloaking for elastic waves was the next logical step. Furthermore, elastic waves are more challenging to conceal than acoustic ones, as wave equations derived by strokes do not often preserve their shape through coordinate changes. It was necessary to find transformation-based solutions for simpler scenarios to enable the construction of cloaking devices for elastic waves, allowing them to catch up to the state-of-the-art in acoustic and electromagnetic cloaking. With a cylindrical cloak, the very first hypothetical study to accomplish elastic wave cloaking in a two - dimensional solid at 40 Hz frequencies for in-plane fixed stress and tangential elastic waves was proposed [16]. The transformation coordinate t required that the cloaking shell have both a mass density(a scalar quantity) and a completely non-symmetric stiffness tensor. Following this breakthrough, researchers started looking into cloaking in thin elastic plates to account for transverse waves, also known as bending or flexural waves. Bi-harmonic Coordinate transformations of the mathematical statement were used to derive the theoretical characteristics of a cylinder-shaped cloak with outward variable scalar mass isotropic density and radial variable and ductility stiffness [12].

Subsequently, a simple elastic ultrawide band cloaking was hypothetically presented utilizing slender plates that govern flexible elastic waves, which helped to overcome the restrictions brought on by the complicated material requirements. The cloak, composed of isotropic elastic concentric multilayers, preserved the phase of a 250 Hz incoming plane wave, demonstrating quantitatively almost flawless elastic wave cloaking. Theoretical studies were followed by the first experimental illustration of an elastic wave cloak, which made use of twenty cylindrical concentric multilayers, sixteen of which had a different effective elastic modulus than the others [17].

Excellent elastic wave cloaking was attained from the left side by sending a plane wave of the cloak at frequencies between 200 and 450*Hz*. These findings provide the first experimental evidence of a broadband elastic cloak, which might be used to shield infrastructure from destructive seismic waves [18].

Pentamode elastic metamaterials are a new type of material structure that has been studied for their potential to allow the construction of structures with tunable shear (G) and bulk modulus (B), allowing for the creation of 3D elastic wave cloaks. This breakthrough motivated the development of the first practical elastic cloak made possible by pentamode metamaterials. For the inner-shell configuration to be contract- and shear-insensitive, a Penta-mode metamaterial was placed around an empty cylinder with high bulk moduli and shear moduli. After subjecting the system to compression and shear, side images were taken and the film was recorded. Electrostatic cloaking is achieved by exerting force on the structure from above using a stamp, demonstrating the potential of pentamode metamaterials for elastic cloaking [18]. Figure 8.3 provides the configuration details of a pentamode based elastic cloaking device.

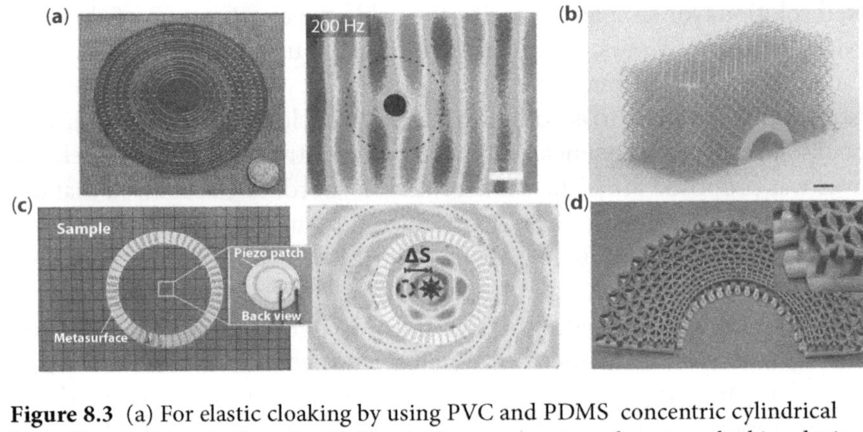

Figure 8.3 (a) For elastic cloaking by using PVC and PDMS concentric cylindrical layers like cloaking shell made. (b) For elastic wave pentamode carpet cloaking devices made by using hollow solid cylinder. (c) The point source is moved by $\delta S = 25$ mm at frequency 12 KHz experimentally measured using elastic wave. (d) Top view of polar mechanical cloak [12].

Recently, metamaterial elastic cloaks have also been used to guide seismic waves. At first, we presented a plate prototype for concealing flexible and elastic waves at the air-to-soil boundary, and we established experimentally that these waves could be shielded at frequencies of about 50 Hz. To redirect surface elastic waves in soil with frequencies between 3 and 10 Hz, Luneburg lenses were used, a technique borrowed from the field of transformation optics [18, 19].

8.5.4 Thermal Cloaking

An ultimate thermal hallucination process known as thermal cloaking is the result advance heat exploitation or handling with metamaterials. Without affecting the ambiance thermal environment heat is guided to move around the object smoothly. All preceding thermal metamaterials have defects like they were massive, lacking the flexibility of changing geometries and lacking the functionality of switching on/off. To pump heat from one side to another side of the invisible object thermoelectric components are used in thermal cloaking and all procedure is controlled by electric voltage. A real heat radiant is transformed into a pre-controlled-on in thermal camouflage cloaking. Due to distinctive scattering radiant in different fields like optics, thermotic, and acoustic, the object can be identified. A real opinion converts into pre-controlled perception in thermal camouflage like we dd analogous to ave dynamic hallucinations legitimize unparallel application in thermal cloaking. To obtain 2D and 3D thermal cloaking using two layers of bulk isotropic material is used in thermal cloak fabrication depending on scattering cancellation [20]. A polystyrene cylindrical layer neighbored by nickel-based alloy is used to create a bilayer thermal cloak that is used 2D cloaking an object at the center of the bilayer cloak [21]. Maintaining the heat distribution outside and maintaining stable heat protection to objects inside the thermal cloak an excellent performance is observed by the bilayer thermal cloak. By cloaking the sphere from the outer thermal flux field and using 100μm thickness spherical copper layer the experimental exhibition is conducted, in 3D. For making ultrathin spherical copper thermal cloaking 3D is used [19, 22]. Figure 8.4 explains the configuration details of thermal cylindrical cloak.

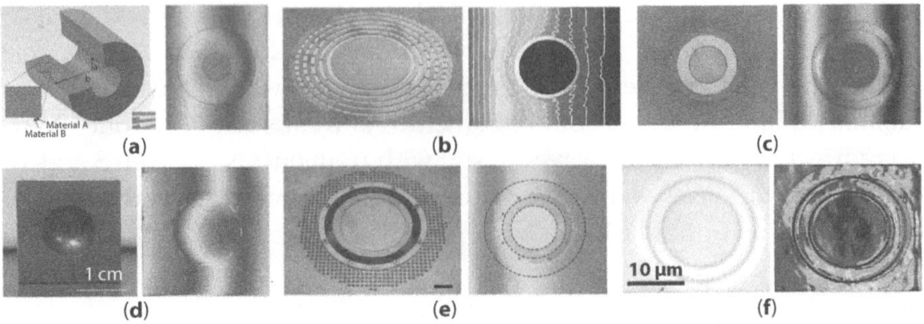

Figure 8.4 (a) Thermal multilayer cylindrical cloak is manufactured by latex rubber and silicone elastomer. (b) For the transient regime, by using PDMS and copper thermal cloak is manufactured. (c) Nickel-based alloy layer neighboring by a polystyrene 2D bilayer thermal cloak is manufactured. (d) By shaping a thin copper sphere into the block of stainless steel thermal 3D cloak is manufactured. (e) Thermal cloak (with zero-index) work in the background of copper. (f) Four Layered thermal cloaks at the microscale [12].

8.5.5 Mass Diffusion Cloaking

Mass diffusion is an adjacent field discussion that contributes to resembling heat conduction. Mass diffusion cloaking is known as metamaterial that leads a mass to move around the object while controlling the external mass concentration incline. Ions are prevented from approaching the block of metal inside and concrete and cloaking chloride ions, for an increased lifetime of tangible foundation mass diffusion cloaking is introduced experimentally [23]. Satisfying the required parameters for cloaking, by varying cement and glue amounts of six different types of concrete were made to design or fabricate a multilayered mass diffusion cloak. At the center, steel bars are located and ions are redirected to move towards them and the concentration of chloride ions was measured. Where it is arduous to separate different species mass diffusion control plays a vital role here [12].

8.5.6 Light Diffusion Cloaking

We pay attention to cloaking light propagation in a diffuse regime in this comprehensive section where light shows great resemblance with mass diffusion and heat. By collecting motivation from thermal cloak researchers experimentally reported cloaking of diffusive light. To serve as a reflector, a thin acrylic white-coated hollow stainless steel cylinder is used in the cloak as the core [24]. Tank having white wall pint and deionized water, the core-shell system is submerged into it and the shell was made by doping 10µm dielectric particles. A remarkable cloaking is noted when the cylindrical diffuse cloak reveals white light. Inside the diffusive shell, the phase is terminated in diffuse light cloaking in contrast to the white light case. The flexibility of coordinate transformation clearly shows in diffuse light experimentally [12].

8.5.7 Multifunctional Cloaking

The quick success in cloaking extended the research to make cloaking devices that cloak more than physical phenomena at a time. To control heat and electricity theoretically first multifunctional metamaterial is purposed and similarly to control heat flux and electricity current first multifunctional cloaking is purposed. The impinging electric and heat flux is redirected to move around an air cavity force by a bilayer cloak made o silicon. While regaining the external flux bilayer cloak strongly scatters both electric and thermal fields [25]. When sited under thermal and electric fields bifunctional device shows excellent cloaking as shown in Figure 8.5(a). In the center of devices to concentrate on both thermal and electric fields, bifunctional devices show excellent cloaking. ABS thermoplastic eight different wedges and aluminum is used to design a fan-shaped structure [26].

Both conductions filed are presided over by the Laplace equation according to this situation electrical-thermal cloaking is made. However, the Helmholtz wave equation is also used to describe electromagnetic and acoustic water waves. Coupled thermal conduction and radiation

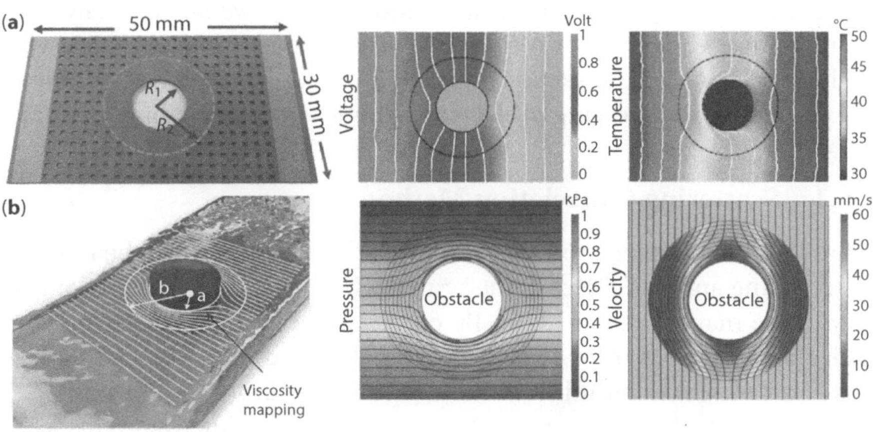

Figure 8.5 (a) Thermal and electric bifunctional silicon-based cloak promoting inner and outer radius of the shell. (b) Hydrodynamic cloaking demonstration showing fluid flow on all sides of cloak (around) of cloaking shell and dark object [12].

bio-function form of cloaking is reported. For the Roseland diffusion equation for thermal radiation Stefan-Boltzmann's approximation law is used. It is also used to control these thermal fields. By using these approximations, diffusion of the photon is considered thermal radiation which guides to conductivity-type multilayer structure for cloaking made of isotropic material. Recently theoretically by the transformation of the coordinate of the thermal conductivity a permeability cloaking device is become able to cloak heat conduction and fluid [12, 27].

8.6 Cloaking Techniques

Scientific researchers are becoming increasingly interested in electromagnetic cloaking, particularly those who are working to create "metamaterials"—artificial composites with unusual electromagnetic properties. The concept of a tool that renders the body hidden from the naked eye has old the past, beginning with different nationalities of the folklore: we have all heard stories of several "invisibility hats" and "invisibility cloaks," like in Harry Potter the cloak worn by J. K. Rowling's [1].

A device that hides the object from electromagnetic radiation is known as an electromagnetic cloak. Of course, the most intriguing uses for cloaks that operate in the visible spectrum may be imagined. The object is considered invisible if it does not produce any shadow, does not scatter or diverged waves, and also does not reflect the light toward the source. These circumstances imply that any type of power present in the system is not absorbed by the system. To put it in another way, we say that any type of interference of an object with an electromagnetic field is not observed. Reducing the total cross-sectional to zero is cloaking o object, according to the physics of the diverging of electromagnetic waves, including light [2].

8.6.1 Scattering Cancelation Method

It has long been understood that scattering from an object can be reduced by including another object in the system whose scattering is complementary to the main scatterer [3–5]. By covering the primary scattering item with one or more layers of dielectric material, for instance, one can achieve this form of scattering minimization [6–8]. The idea of employing plasmonic materials for transparency has recently sparked interest in this technique [9].

The scattering cancellation principle is seen in Figure 8.6. Here, a shell of dielectric material with a permittivity that is less than that of the

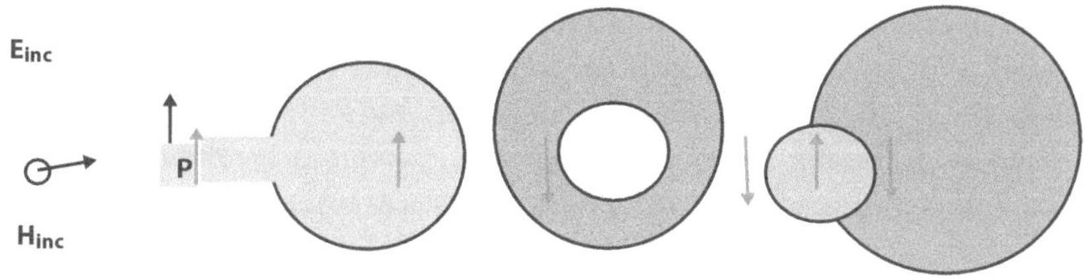

Figure 8.6 Scattering cancellation technique.

surrounding medium is placed over a sphere of dielectric material. Since the dipole moment can be induced in the shell having an opposite sign, shell width will be selected so that the core scattering and the shell scattering are equal and can annul each other. Although higher modes may potentially exist adding in the modes dipolar, it is demonstrated that effective intangible will be attained even by scattering the dipolar being suppressed alone. Higher modes can be suppressed, although doing so always complicates the design [10].

The advantages of this technique include easy construction also design (as long as the materials with the necessary qualities are available), as well as the ability to achieve covert with homogenous and isotropic substances. Taking into account the disadvantages, relying on whether penetrable or impenetrable things must be made impalpable, are the realization of metamaterials with the necessary properties in the absence of plasmonic materials, bandwidth restrictions present in different doable (resonant) metamaterials, and fundamental restriction on the velocity energy when cloaking impenetrable things having passive cloaks in the free space [11].

8.6.2 Coordinate Transformation Technique

Leonhardt [10] and Pendry *et al.* [28] described the use of metamaterials for cloaking, which enables the construction of electromagnetic field-free volumes in the inner side of a device made of these materials. These methods rely on coordinate transformations, such as in physical space when the point in EM space is converted into a sphere, creating a sphere-shaped volume in which EM fields do not exist but are rather directed around the volume.

This technique necessitates the use of metamaterials which are lossless anisotropic with some part of the effective relative permittivity (ε_r) and/or permeability (μ_r), which are less in values than these values in free space to cape objects (or like medium having similarity to free space). The simple theoretical design and the verity that it is not dependent on the shape and/or constituent stuff of the shrouded item are advantages of the coordinate transformation technique. Disadvantages include challenges in realizing materials with the necessary characteristics, particularly when large bandwidths and signal masking from energy pulses are needed [3].

8.6.3 Transmission

Not long ago, we put forth a wrapping method that significantly varies from those covered in the earlier parts. This method is based on the utilization of variometer structures made up of transmission-line networks in two or three dimensions. The electromagnetic fields in these structures spread inside the transmission lines, thereby obscuring the space between them [29].

A link layer is required to join the meadow (fields) between the local medium and the network because fields coming into the cloak from the surrounding medium must be "squeezed" into the coax lines. Although the primary idea of transmission-line networks for cloaking is quite straightforward, it cannot get around the following fundamental limitation. To completely conceal an item in outer space, the velocity of the wave in the transmission lines needs to be faster than light. Because a single coax line perceives all other coax lines as regular loads, the network itself "slows" the wave. By adding periodic reactive loads to such a network, optimal wave numbers can be attained even when hiding objects in space. The inherent problems of the complexity of design and substantial dispersion of frequency are present in this type of system, though. Therefore, it has been determined that even when the velocity propagates in the cloak is not perfect, using straightforward unloaded transmission-line networks is superior for real applications that call for wide bandwidths and/or concealing from signals. The straightforward design, straightforward production, and assembly processes, as well as wide-band operating, are advantages of the coax-line technology. The restriction on the dimensions of the shrouded things in this approach's biggest flaw, especially when collated with the methods earlier mentioned [30].

8.6.4 Other Cloaking Techniques

There are further cloaking methods as well. Kildal and his co-authors first proposed one such method more than 10 years ago. The principle behind this method was to cover a metal object with a purported "hard surface" that had an extended design in the direction of the propagation of the wave. This has been demonstrated that this method can be used to successfully lower the next scattering from these objects, which lowers the total scattering cross-section. The masking (cloaking) effect, which relies heavily on the angle at which the EM wave is impinging, is this technique's obvious flaw. This follows naturally from the "cloaking" device's elongated shape rather than its asymmetrical design. However, in common applications where the incidence angle is known, such as lowering dispersion from antenna support struts, this technique can be employed for cloaking. Milton, Nicorovici, and their co-authors proposed another extremely unusual cloaking method that relies on the deployment of a purported "super lens" to conceal scattered objects that are positioned adjacent to this device. The basis for this cloaking phenomenon is the anomalous localized resonance which can be excited on the super lens surface This cloaking method's ability to achieve cloaking for things placed outside the "cloaking device" is one of its unique features [31].

8.7 Conclusion

In this chapter, the complete process of cloaking is discussed from design to methods and analysis. Previous studies only covered only design or simulation but here other aspects are elaborated. Now, we can design a cloak that can hide other external objects as well, which mean it hides the material and surrounding by using the anti-object. Researchers gave theoretical designs for electromagnetic cloaks, by using metamaterial scientists made electromagnetic and carpet cloaks in the lab. Many fields of metamaterial are still undiscovered containing quantum systems so the future of cloaking is very bright. For various physical fields, mostly all work is concentrating on higher-level metamaterial cloaking devices. New fundamental physics evaluation is based on atomic length scale but in multifunctional integrated cloaking metamaterial work at nanometer length and millimeter length. Optical cloak design is based on transformation coordinates. In three-dimensional space, gradient magnetic metamaterial construction becomes so easy due to our design's non-magnetic nature. In the proposed cloak due to impedance mismatch, the obtained invisibility is not perfect. We also note that the wavelength is six times less than the size of the cloak while the wavelength is 20 times less than the simulated area.

References

1. Ning, L., Wang, Y.-Z., Wang, Y.-S., Active control cloak of the elastic wave metamaterial. *Int. J. Solids Struct.*, 202, 126–135, 2020.
2. Cui, T.J., Smith, D.R., Liu, R., *Metamaterials*, Springer, Boston, MA, USA, 2010.
3. Alitalo, P. and Tretyakov, S., Electromagnetic cloaking with metamaterials. *Mater. Today*, 12, 22–29, 2009.
4. Fleury, R. and Alu, A., Cloaking and invisibility: A review. *Forum Electromagn. Res. Methods Appl. Technol.*, 1, 1–25, 2014.
5. Ufimtsev, P.Y.P., New insight into the classical Macdonald physical optics approximation. *IEEE Antennas Propag Mag.*, 50, 11–20, 2008.
6. Chen, P.Y., Soric, J., Alù, A., Invisibility and cloaking based on scattering cancellation. *Adv. Mater.*, 24, OP281–OP304, 2012.
7. Schurig, D., Mock, J.J., Justice, B., Cummer, S.A., Pendry, J.B., Starr, A.F., Smith, D.R.J.S., Metamaterial electromagnetic cloak at microwave frequencies. *Sci.*, 314, 977–980, 2006.
8. Choudhury, B. and Jha, R., Materials and continua: A review of metamaterial invisibility cloaks. *Comput. Mater. Contin.*, 33, 275–303, 2013.
9. Pendry, J.B., Schurig, D., Smith, D.R., Controlling electromagnetic fields. *Science*, 312, 1780–1782, 2006.
10. Leonhardt, U., Optical conformal mapping. *Sci.*, 312, 1777–1780, 2006.
11. Schurig, D., Pendry, J., Smith, D.R., Calculation of material properties and ray tracing in transformation media. *Opt. Express*, 14, 9794–9804, 2006.
12. Martinez, F. and Maldovan, M., Metamaterials: Optical, acoustic, elastic, heat, mass, electric, magnetic, and hydrodynamic cloaking. *Mater. Today Phys.*, 27, 100819, 2022.
13. Cai, W., Chettiar, U.K., Kildishev, A.V., Shalaev, V.M., Optical cloaking with metamaterials. *Nat. Photonics*, 1, 224–227, 2007.
14. Milton, G.W., Briane, M., Willis, J.R., On cloaking for elasticity and physical equations with a transformation invariant form. *J. New Phys.*, 8, 248, 2006.
15. Norris, A.N., Acoustic cloaking. *Acoust. Today*, 11, 38–46, 2015.
16. Brun, M., Guenneau, S., Movchan, A.B., Achieving control of in-plane elastic waves. *Appl. Phys. Lett.*, 94, 061903, 2009.
17. Farhat, M., Guenneau, S., Enoch, S., Ultrabroadband elastic cloaking in thin plates. *Phys. Rev. Lett.*, 103, 024301, 2009.
18. Norris, A.N. and Shuvalov, A.L., Elastic cloaking theory. Elsevier, *Wave Motion* 48, 525–538, 2011.
19. Nguyen, D.M., Xu, H., Zhang, Y., Zhang, B., Active thermal cloak. *Appl. Phys. Lett.*, 107, 121901, 2015.
20. Fan, C., Gao, Y., Huang, J., Shaped graded materials with an apparent negative thermal conductivity. *Appl. Phys. Lett.*, 92, 251907, 2008.
21. Han, T., Bai, X., Gao, D., Thong, J.T., Li, B., Qiu, C.-W., Experimental demonstration of a bilayer thermal cloak. *Phys. Rev. Lett.*, 112, 054302, 2014.

22. Han, T., Bai, X., Thong, J.T., Li, B., Qiu, C.W., Full control and manipulation of heat signatures: Cloaking, camouflage and thermal metamaterials. *Adv. Mater.*, 26, 1731–1734, 2014.
23. Guenneau, S. and Puvirajesinghe, T., Fick's second law transformed: One path to cloaking in mass diffusion. *J. R. Soc. Interface*, 10, 20130106, 2013.
24. Guenneau, S., Amra, C., Veynante, D.J., Transformation thermodynamics: Cloaking and concentrating heat flux. *Opt. Express*, 20, 8207–8218, 2012.
25. Li, J., Gao, Y., Huang, J., A bifunctional cloak using transformation media. *J. Appl. Phys.*, 108, 074504, 2010.
26. Fujii, G. and Akimoto, Y., Optimizing the structural topology of bifunctional invisible cloak manipulating heat flux and direct current. *Appl. Phys. Lett.*, 115, 174101, 2019.
27. Mei, J.-S., Wu, Q., Zhang, K.J., Multifunctional complementary cloak with homogeneous anisotropic material parameters. *Opt. Soc. Am.*, 29, 2067–2073, 2012.
28. Rahm, M., Schurig, D., Roberts, D.A., Cummer, S.A., Smith, D.R., Pendry, J., Nanostructures-fundamentals, and applications, design of electromagnetic cloaks and concentrators using form-invariant coordinate transformations of Maxwell's equations. *Photonics Nanostruct.*, 6, 87–95, 2008.
29. Tretyakov, S., Analytical modeling in applied electromagnetics, Artech, Boston, 2003.
30. Milton, G.W., Nicorovici, N.-A.P., McPhedran, R.C., Opaque perfect lenses. *Phys. B*, 394, 171–175, 2007.
31. Edwards, B., Alù, A., Silveirinha, M.G., Engheta, N., Experimental verification of plasmonic cloaking at microwave frequencies with metamaterials. *Phys. Rev. Lett.*, 103, 153901, 2009.

9

Single Negative Metamaterials

M. Rizwan[1]*, U. Sabahat, F. Tehreem[1] and A. Ayub[2]

[1]School of Physical Sciences, University of the Punjab, Lahore, Pakistan
[2]Department of Physics, University of the Punjab, Lahore, Pakistan

Abstract
Single negative metamaterials have been deliberated on in this chapter along with their properties and potential applications. Left-handed metamaterials, which have negative refractive indices and are used in microwave and optical range structures are rereviewed. This chapter demonstrates the presence of an acoustical metamaterial with a negative effective density and bulk modulus, demonstrating that it is an effective medium in the most precise sense. For those unfamiliar, negative-index materials (NIMs) are a subset of metamaterials distinguished by an effective negative index, which results in peculiar wave phenomena, such as reverse negative refraction. We examined how metamaterials respond to electromagnetic behavior, these are manufactured media are manufactured materials that differ from those of their components. We also have discussed the history of negative index materials, the primary design approaches, and some potential applications, such as sub-wavelength resonant cavities. Several real-world applications of microwave technology are also analyzed. The most promising possibilities for future metamaterials study are highlighted.

Keywords: Left-handed metamaterials, negative refraction, EM response, antenna miniaturization, negative permittivity, permeability, electromagnetic metamaterials

9.1 Introduction

An artificial substance called a metamaterial is created to have properties that "may not be readily found in nature." The properties in these materials are typically derived from structure instead of composition, with the

**Corresponding author*: rizwan.sps@pu.edu.pk

Inamuddin and Tariq Altalhi (eds.) Electromagnetic Metamaterials: Properties and Applications, (185–204) © 2023 Scrivener Publishing LLC

incorporation of minute inhomogeneities enforcing significant macroscopic behavior. In recent years, metamaterials have emerged as a mainstream tool in electromagnetics. This category of materials covers a wide range of artificial polymers with adjustable permittivity and permeability that are either unattainable or difficult to get from natural sources [1]. The metamaterials appear similar to natural crystals because they are constructed from the unit cell along a side length of an array in a periodic fashion [2]. Metamaterials are defined by their unique and desirable characteristics that arise because of their certain design and manufacturing. Electromagnetic waves specifically alter a composite media's macroscopic effective permittivity and permeability because they affect the parts that make up the medium, producing electric and magnetic moments. Because created inclusions can be incorporated into a preset host media to create metamaterials. Because of this, the designer has access to a wide variety of controllable variables including host material characteristics, inclusion size, and morphology. All these design variables may significantly affect the final product. These offer a novel opportunity for metamaterial synthesis due to the shape of their inclusions [1].

The electromagnetic characteristics of metamaterials are defined by the way light interacts with metals. It is mentioned that metamaterials are substances in which both the electric permittivity and the magnetic permeability are concurrently negative. The designing of these metamaterials and, consequently, their electromagnetic responses may now be done with a considerable amount of freedom because of advancements in modeling and fabrication technology. Since metamaterials could be used to modify the electromagnetic responses of materials, researchers have taken a keen interest in them for a wide range of p ractical purposes [3]. The effective ε_{eff} and μ_{eff} of these materials could be adjusted, allowing for a wide range of possible applications that would be impossible or difficult to achieve with naturally occurring substances. The effective permittivity and permeability are given by:

$$\varepsilon_{eff} = \varepsilon_o \varepsilon_r \quad \text{(Eq. 9.1)}$$

$$\mu_{eff} = \mu_o \mu_r \quad \text{(Eq. 9.2)}$$

For such a formulation to be introduced, artificial resonators with a size significantly less than the wavelength are needed. If this condition holds, then the medium reaction is assumed to be local, meaning that the effective permittivity and permeability averaged over a given unit cell are independent of the values at surrounding unit cells. Thus, in this type of medium, it is acceptable to disregard the consequences of spatial dispersion [4].

If we consider that the magneto-electric coupling is small, then the behavior of the material will be isotropic for a certain polarization of field and over a specified frequency range. In the lack of impressed sources, the time-harmonic form of Maxwell's equations appears as:

$$\nabla \times E = -j\omega\mu H \quad \text{(Eq. 9.3)}$$

$$\nabla \times H = -j\omega\varepsilon E \quad \text{(Eq. 9.4)}$$

where

ε = Electric permittivity
μ = Magnetic permeability

These quantities become extremely complicated if the losses are considered here. Mostly natural materials may be described in terms of these two parameters because of the way electromagnetic waves interact with them, resulting in numbers that fit within the constraints imposed by Re $[\varepsilon] \geq \varepsilon_o$, Re $[\mu] \geq \mu_o$, $\text{Im}[\varepsilon] < 0$ and $\text{Im}[\mu] < 0$. This denotes that the substance is inert and has a refractive index larger than or equivalent to the value in a vacuum [5].

To be more precise, the real components of the ε and μ of an inactive material can, in essence, produce any real value at a given frequency, as long as the appropriate temporal dispersion respects constraints specified by causality [6].

9.2 Classification of Metamaterials

The schematic in Figure 9.1 [5] can be used to classify materials whose fundamental relations are represented by the simple equation in equations 1.3 and 1.4. The vast majority of naturally occurring substances referred to as "double-positive (DPS) media" are those in which ε and μ each have positive real parts [8]. In contrast, materials that have both of these values in the negative range (shown by the third quadrant in Figure 9.2) are referred to as "double-negative" (DNG). Engineers and physicists have taken a keen interest in these materials because of their anomalous wave refraction. The second quadrant contains "ε-negative (ENG) media", so called because the

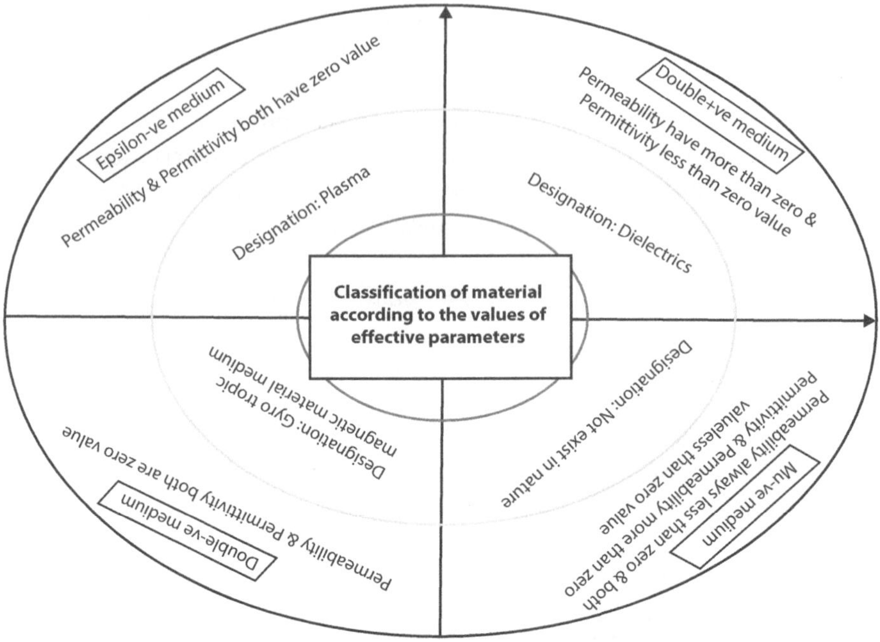

Figure 9.1 Classification of materials as per their effective parameter values (μ and ε): [7].

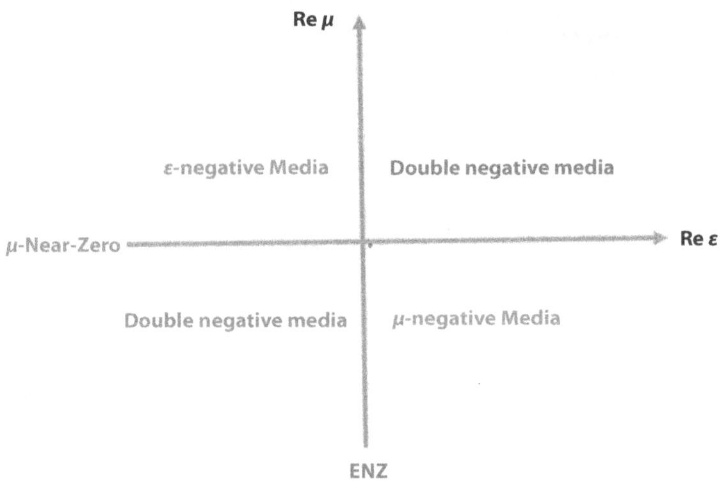

Figure 9.2 The names of metamaterials based on the values of actual components of their μ and ε [5].

real component of their permittivity is negative but their permeability is positive. Under their plasma frequencies, they are classified as plasmonic materials, which include noble metals as well as certain polar dielectrics. There has been extensive research into the potential uses of plasmas and other ε-negative materials for centuries. The μ-negative (MNG) materials occupy the fourth quadrant. These can be achieved with ferromagnetic materials or created by incorporating appropriate inclusions into a host material. It is impossible to create double negative materials without the use of artificially realized, μ-negative materials. Single-negative (SNG) media can be thought of as an analog to double-negative (DNG) media, and comprise both ε-negative and, μ-negative media [8].

9.3 Types of Metamaterials

The term "metamaterial" encompasses a broad category of materials. Thus, there are different types of metamaterials discussed in this chapter. Furthermore, there are several distinctive types of metamaterials, and these classes are described as follows:

9.3.1 Electromagnetic Metamaterials

Metamaterials, as their name implies, are created to outperform conventional hybrids in ways that were previously considered impossible [9]. In recent years, research into electromagnetic (EM) materials that have been artificially constructed has exploded in popularity due to the promise of developing materials with unique EM responses that cannot be found in nature. These substances often referred to as metamaterials (MMs), exhibit electromagnetic activity because of the oscillation of electrons in highly conducting metals like Au or Cu, which produces a precisely regulated resonant output of the electrical permittivity or magnetic permeability. There is a long history of understanding in EM theory of continuous media with negative constants, namely negative ε_r or μ_r [10].

Metamaterials with distinct electromagnetic characteristics already contain a wide range of structural components. Some of the structural elements of metamaterials have a negative dielectric permittivity, while others have a negative magnetic permeability or very high magnetic permeability within a given frequency range. Even though they are made of paramagnetic materials, these constituents have these characteristics. There are underlying structures in metamaterials. For a metamaterial to affect electromagnetic waves, its structural characteristics must be smaller than the wavelength of the electromagnetic radiation it meets. An electromagnetic metamaterial provides an electromagnetic response that is more precisely controlled than what is found in nature. It can have a periodic, quasi-periodic, non-periodic, or fractal structure, or it can have an irregular structure. Fabricated dielectrics were the earliest examples of electromagnetic metamaterials. Periodic electromagnetic metamaterials include photonic crystals, electromagnetic crystals, and left-handed metamaterials [11].

Optical metamaterials are electromagnetic metamaterials that are designed to transmit or reflect electromagnetic waves in the ultraviolet, visible, or infrared spectrum.

9.3.2 Negative Refractive Index

Left-handed metamaterials (LHMs) are substances with a negative effective refractive index value. It is said that they are composite sub-wavelength artificial structures whose ε_{eff} and μ_{eff} have been artificially manipulated to create negative values of their real component. To distinguish them from chiral metamaterials, which are also frequently referred to as "left-handed media," these are also designated as negative refractive index materials (NRM) [12].

Generally, an object's complex refractive index is defined as the velocity of an electromagnetic wave in the medium divided by the velocity of the same wave in a vacuum, hence it can be denoted as: $n^2 = \mu\varepsilon$, here μ and ε are magnetic permeability and electric permittivity correspondingly. If permittivity and permeability together are negative over some range of wavelengths then we can write: $\mu = |\mu| exp\ (i\Pi)$, also in a similar manner: $\varepsilon = |\varepsilon| exp\ (i\Pi)$

$$\begin{aligned} n &= \sqrt{|\varepsilon||\mu|\, exp(2i\Pi)} \\ &= \sqrt{|\varepsilon||\mu|}\sqrt{exp(2i\Pi} \\ n &= -\sqrt{|\varepsilon||\mu|} \end{aligned} \qquad \text{(Eq. 9.5)}$$

In other words, if both ε and μ are negative in the same material, then that medium must also have a negative refractive index.

Since there is currently no such thing as a material with both negative permeability and permittivity, an NRM metamaterial is a combination of two materials, each of which exhibits $\varepsilon < 0$ and $\mu < 0$ separately [13].

The alteration of Snell's law, known as negative refraction, is arguably the most consequential and well-studied effect of NRM. A plane electromagnetic wave traveling through a vacuum (or air) is shown being refracted by a material with a greater electromagnetic thickness illustrated in Figure 9.3 [13], when the real component of effective refractive index is positive (on top) or negative (below). When the refractive index is positive, the wave vector and the Poynting vector are tangential, but when the index is negative, they are antiparallel. So, light traveling in a beam is refracted back to the same side it arrived from when it encounters the boundary between the positive and negative index media. This might not make sense at first. Therefore, when $d_1 + d_2 = d$, a planar parallel slab of a negative refractive

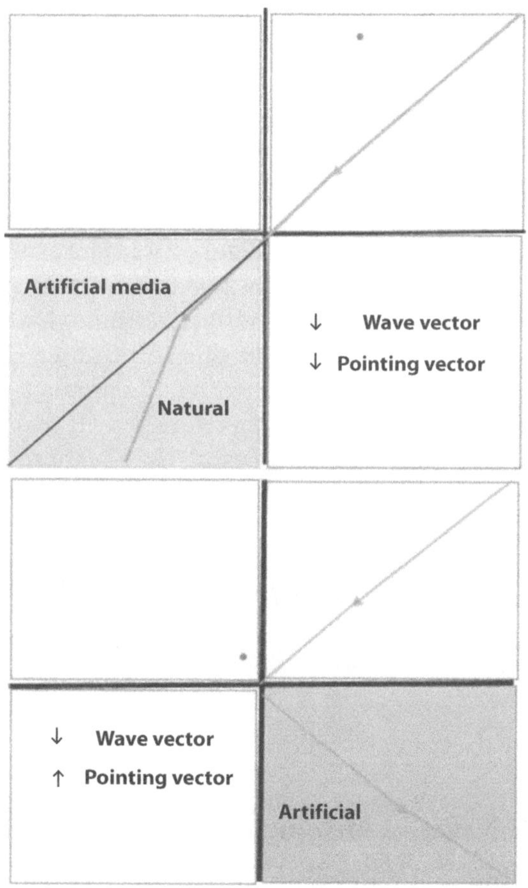

Figure 9.3 Schematic compares the refraction that occurs in normal materials with that which occurs in negative refractive index media [13].

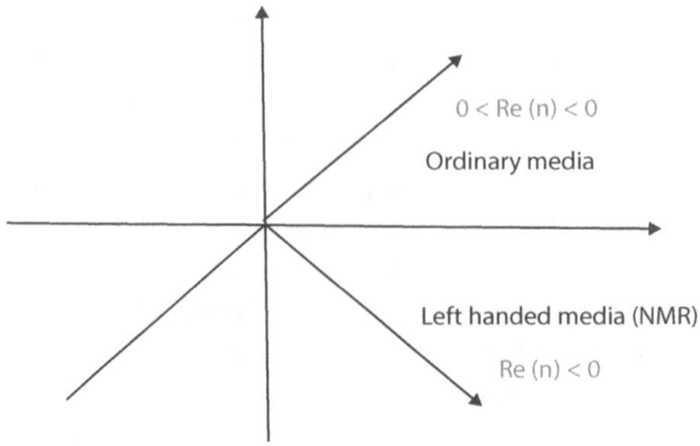

Figure 9.4 The schematic diagram of refraction ordinary media versus NMR material [15].

material will concentrate rays coming from a distance d_2 rather than scattering them.

As an alternative explanation, a convex lens will cause a planar wave to diverge, whereas a concave lens will cause the wave to converge. It is worth noting that the NRM has no effect on the beam's angle after reflection [14].

The simplest schematic diagram of the refraction in a left-handed media and an ordinary media is depicted in Figure 9.4 [15]. Permittivity and permeability values are expected to be positive for all transparent non-metamaterial media. Conventionally, the refractive index is expressed as the square root of a positive number. Meanwhile, $\varepsilon < 0$ and $\mu < 0$ can be found in certain artificial metamaterials. Given that the product of $\varepsilon\mu$ is positive, we know that n is non-negative and therefore real. If this is the case, one must calculate the refractive index by using the square root of minus one as explained in equation 9.5 [15].

9.4 Different Classes of Electromagnetic Metamaterials

Electromagnetic metamaterials are classified into various types of metamaterials that are described as follows:

9.4.1 Double Negative Metamaterials

Due to the negative permittivity and permeability, double negative metamaterials (DNG) have a negative index of refraction (NIM) [16]. Veselago developed the theory of negative refraction in an EM wave, and this idea has lately been demonstrated using the concept of metamaterials. Composites having intrinsic resonance structures are used to generate EM metamaterials, and these materials have an effective negative value of μ and ε under specified frequency ranges. Due to their interesting properties, such as negative refraction and sub-wavelength imaging, these "double negative" media have attracted a lot of attention. It is important to keep in mind that negative refraction and "double negativity" are similar but distinct phenomena [17].

The negative refractive index of Veselago medium in electromagnetic metamaterials is the result of the material's negative permittivity as well as its negative permeability. It is possible to express both continuity and Newton's second law in terms of an acoustical wave, which can be written as:

$$\begin{aligned} \nabla.v &= \frac{-i\omega}{k}\rho = 0 \\ \nabla.\rho &- i\omega\rho v = 0 \end{aligned} \quad \text{(Eq. 9.6)}$$

where

$\rho(v)$ = Pressure of the field

The refractive index n should be defined by taking into account a plane-wave solution with wave vector 'k' within a homogeneous media of uniform density and bulk modulus and could be expressed as:

$$k = |n|\frac{\omega}{c} \quad \text{(Eq. 9.7)}$$

Consequently, we need either a positive density and bulk modulus or a negative ρ and bulk modulus for a plane wave to propagate within the media. Furthermore, the pointing vector is defined as:

$$S = \frac{i}{2\omega\rho}\rho\nabla\rho^* = \frac{\rho^2}{2\omega\rho}k \quad \text{(Eq. 9.8)}$$

If an acoustic wave were to pass through the Veselago medium, we would expect S and k to be in opposing directions. To put it in physical terms, at certain frequencies the medium responds anomalously by expanding when compressed (negative k) and shifting left when pushed right (negative density).

9.4.2 Single Negative Metamaterials

In single negative (SNG) metamaterials, only one of the two properties— ε or — μ is a negative value. Here we will talk about 2 types of metamaterials, epsilon negative metamaterials, and μ negative metamaterials. Integrating two SNG sheets into a single metamaterial has led to some intriguing experiments. As so, they constitute yet another variety of DNG metamaterial. Experiments on the reflection of waves have been performed using a connected slab of ENG and MNG materials. This allowed for the manifestation of characteristics including resonances, anomalous tunneling, transparency, and zero reflection. Since SNGs are inherently dispersive, their — ε and μ, index of refraction all change with frequency, just as they do in Double negative metamaterials [15].

9.4.3 Chiral Metamaterials

Chiral metamaterials are those that do not exhibit any planes of mirror symmetrical. When the electric and magnetic fields are connected, the medium is said to be chiral in electromagnetic theory. When trying to characterize the optical response of a generic chiral media, the constitutive relations are then employed:

$$D = \varepsilon_O \epsilon E + \frac{i}{c_o} \chi H \quad \text{(Eq. 9.9)}$$

$$B = -\frac{i}{c_o} \chi E + \mu_O \mu H \quad \text{(Eq. 9.10)}$$

Here,

ϵ = permittivity
μ = permeability
χ = chirality tensor

In addition, the ε_o, μ_o are the permittivity and permeability in free space accordingly. Right (+) and left-handed (-) circular polarization have distinct values of refractive index and are expressed as:

$$n_{\pm} = \sqrt{\varepsilon\mu} \pm \chi \quad \text{(Eq. 9.11)}$$

Equation 9.11, shows that waves of opposite-handedness will accumulate phase differences in different ways as they pass through a chiral material. Efficient material parameters, which may be obtained from scattering characteristics using the retriever method, are most frequently used to characterize the properties of metamaterials. Even so, this strategy may only be used if the complex constitutive parameters are already known and stored in a tensor form. The former quality greatly increases the difficulty of studying complicated optical effects including anisotropy, no locality, and spatial dispersion. The optical response provides a more accurate description of metamaterial characteristics [18].

9.4.4 Hyperbolic Metamaterials

Based on the shape of the iso-frequency plane, we refer to these metamaterials as hyperbolic metamaterials (HMMs). The spherically surface of iso-frequency can be expressed by the following expression:

$$k_x^2 + k_y^2 + k_z^2 = \frac{\omega^2}{c^2} \qquad \text{(Eq. 9.12)}$$

where k is the propagation vector

ω = frequency of the light wave

c = speed of light in free space

In case of a special wave (TM polarized) propagating over a uniaxial media, this iso-frequency relationship changes to:

$$\frac{k_x^2 + k_y^2}{\varepsilon_{zz}} + \frac{k_z^2}{\varepsilon_{xx}} = \frac{\omega^2}{c^2} \qquad \text{(Eq. 9.13)}$$

It is important to note that uniaxial media is specified by a dielectric sensitivity represented via a tensor:

$$\varepsilon = \{\varepsilon_{xx}, \varepsilon_{yy}, \varepsilon_{zz}\}s \qquad \text{(Eq. 9.14)}$$

Due to this occurrence, the material must act like metal in one dimension and as an insulator in other. The latter is difficult to do in nature at optical frequencies, but it is possible to do so by using metamaterials, which are fabricated nanostructured electromagnetic media. The behavior of waves with a large fraction of wave vectors is taken into consideration as the most important aspect of such media. However, at the ideal limit, waves with infinitely high wave vectors can propagate because of the iso frequency surface's open shape in hyperbolic materials. This special characteristic of high-k wave propagation offers a wealth of device possibilities for hyperbolic media [18].

a. Semiconductor Metamaterials

Negative-index materials (NIMs) as a concept have been around for a while, but it has only recently that genuine metamaterials with NIM features have been created. A workable NIM was first proposed, and then it was implemented in the microwave area, using nanowires to create electric and magnetic resonance, yielding the values, $\varepsilon < 0$ and $\mu < 0$. Recently, novel designs have been manufactured and characterized that show negative refraction. These materials still no longer employ the aforementioned architecture; rather rely on the overlapping resonance of ε and μ. The difficult design and production of this double-resonance technique make it difficult to produce a 3D metamaterial from pre-existing structures. Making a material with only one resonance can substantially simplify the structure, diminish the losses, and provide a large range of metamaterials. To distinguish these materials from naturally occurring resonant materials, negative refraction must be produced for these materials by using anisotropic, birefringent, or chiral materials. A semiconductor metamaterial that displays negative refraction for all incident angles in the long-wave infrared portion of the spectrum can be used to overcome the drawbacks of double-resonance metamaterials [19].

b. Quantum and Atomistic Metamaterials

The study of metamaterials is expanded to the quantum level by quantum metamaterials (Atomistic metamaterials). They use the principles of quantum physics to manipulate electromagnetic radiation. In general, a quantum metamaterial is one whose behavior is determined by quantum mechanical equations since its certain quantum characteristics must be taken into consideration. Its behavior confirms the presence of electromagnetic waves as well as matter waves. Depending on the frequency range, the components must either be at a tiny or nanoscopic scale (e.g., optical or microwave).

An atomistic metamaterial must exhibit quantum dynamics, put it more precisely. A device like this can regulate electromagnetic wave propagation in additional ways by acting as a geographically extended controlled quantum object [20].

A more specific definition of quantum metamaterials is an optical medium that:

1. Is made up of artificial quantum coherent unit components.
2. Display tunable quantum states.
3. Maintain coherency for some time greater than the navigational time of a relevant EM signal.

To regulate and alter electromagnetic radiation, quantum metamaterials, therefore, integrate quantum coherent states. These materials integrate the study of metamaterials with quantum information processing (periodic artificial electromagnetic materials). The unit cells can be regarded as quantum bits that maintain quantum resonance "sufficiently for the EM pulse to go through. The material's separate cells enable the quantum state. The entire system maintains quantum coherence when each cell interacts with the electromagnetic pulse that is spreading [20].

There are numerous kinds of metamaterials being researched. Quantum dots, placed in periodic nanostructures, can serve as the structure's unit cells or fabricated atoms in nanowires. It is easy to construct and exhibits a negative refractive index and magnetic properties. The wavelength of interest that is radiated has a substantially bigger diameter than its constituent. Another form makes use of ultra-cold gases to create cold atom cells that are spaced out frequently. As a quantum system, this structure demonstrates controllability and stability, as well as a photonic bandgap. Research is now being done on MM prototypes based on superconducting devices with and without Josephson junctions.

c. Properties of Metamaterials

i. Central Equations Describing Electromagnetic Behavior

Let us review Maxwell's equations to describe the fundamental characteristics of metamaterials.

$$\nabla \times E = -\mu_0 \mu_r \frac{\partial H}{\partial t}$$

And

$$\nabla \times H = \varepsilon_0 \varepsilon_r \frac{\partial H}{\partial t} \quad \text{(Eq. 9.15)}$$

Where, μ_r and ε_r are relative permeability and permittivity, respectively, and $\nabla = \left[\frac{\partial}{\partial x}, \frac{\partial}{\partial y}, \frac{\partial}{\partial z}\right]$. One can get the wave equation from the aforementioned formulae;

$$\nabla^2 E = -\varepsilon_r \mu_0 \varepsilon_r \mu_r \frac{\partial^2 E}{\partial t^2} \quad \text{(Eq. 9.16)}$$

When the signs of ε_r and μ_r are simultaneously altered, the wave equation remains unchanged if losses are ignored and ε_r and μ_r are thought of as real values [21].

d. Development of Metamaterials

The study of the properties of novel materials has recently been replaced by the engineering of materials in the discipline of material science. Metamaterials (MMs) have made important advancements in the study of electromagnetism by connecting artificial media's tiny structures to their ultimate properties. MMs have physical properties not seen in natural materials, such as a negative refractive index that permits the cloaking of materials, reflection cancellation from a particular surface, and the concentration of electromagnetic energy in the tightest possible space. However, at this time, once an MM is made, its parameters cannot be changed programmatically. Therefore, with little to no reusability, each unique application necessitates a particular MM design and manufacturing method.

e. Software Developed Metamaterials (SDM)

An SDM is a 2D surface that allows for the offset and programmed fabrication of unique metamaterial (MM) building blocks. In particular, it is made up of active, ultra-low-power electronic components that serve as control mechanisms and passive components called substrates. Since nano networks offer the highest spatiotemporal control granularity, they are the optimal choice for the control agent. The control agents act on the SDM substrate after receiving programmed instructions and transmitting them over their network, producing unique patterns. The origin of the directives can be from within the SDM or from outside. In the former situation, a regular system can communicate with the SDM, however, in the latter scenario, the control agents are allowed to work. For instance, they could automatically assess their surroundings and respond cooperatively. The network control agent plus an external hardware interface makes up the control module [22].

The SDM interface, for the instance, plays two unique roles:

1. Transmit the information from the outside world to the SDM control agents. These could be sensor readings or programming directives.
2. Handles all deferrable, computationally demanding tasks centrally, unloading the network of control agents whenever it is practical.

A task that requires precise floating-point calculations and is computationally taxing is manipulating digital signal processing. Massively parallel computer architectures are excellent choices for implementing SDM interfaces because

(a) There are likely to be many control agents.
(b) They all carry out a limited set of identical tasks

A Software Developed Metamaterial (SDM) comprises the following components:

1. Communicating module
2. A Central Processing Unit (CPU)
3. Actuator

A bidirectional connection is unquestionably preferable since it increases the number of SDM applications available and enables node-collaborative sensing and actuation over substrates as well as network and SDM monitoring.

f. Development of Graphite-based Metamaterial for X-band Applications
Modern electronic systems and wireless communication both require electromagnetic shielding that is easy to make, reliable, thin, and flexible, and exhibits adequate absorption. Metamaterials with designed sub-wavelength structures have recently been the subject of in-depth research as electromagnetic absorbers. Metamaterial absorbers (MAs) are thinner, lighter, and, most critically, independent of the foundation material than conventional absorbers. Moreover, by simply altering the unit cell design, MAs' absorption frequency can be changed. The majority of the documented research on MAs uses metallic conducting layers on rigid substrates like FR4, silicon wafers, or thin films made of vanadium oxide, etc. (mainly copper). The MA cannot be used to cloak aircraft or electronic devices, isolate electromagnetic (EM) waves for body-worn applications,

or use non-planar surfaces like radomes, due to their rigid substrates. The copper-based ground planes and periodic unit cell architectures of MAs prevent them from being used in any practical context. Additionally, a ground plane with a metal backing is less flexible. The conformability problems might be resolved by using flexible polymer materials in place of stiff substrates. An excellent candidate for MA substrate materials is linear low-density polyethylene (LLDPE). Furthermore, carbon nanotubes and graphene, which have conductivities of almost the same order as metallic unit cells but are non-corrosive, can be used in place of metallic unit cells. However, the production of carbon nanotubes and graphene is a time- and money-consuming process. Instead, expanded graphite (EG), which can be produced easily and affordably, can be used. Its conductivity is 106 S/m, making it lighter and less corrosive than copper. An EG layer can even take the place of an MA's metal ground plane [23].

g. Single Negative Metamaterials

Negative permittivity or negative permeability characterizes single-negative metamaterials (SNM). Wires and metallic split ring resonator (SRR) arrays are incorporated into a host material, to create single negative metamaterials [15].

i. Negative Permittivity

It is generally known that metals exhibit an electric permittivity at optical frequencies that vary with the frequency following Drude relation

$$\varepsilon(k) = \varepsilon_0 \left[1 - \frac{\omega_p^2}{\omega(\omega + i\gamma)}\right] \quad \text{(Eq. 9.17)}$$

Where $\omega_p^2 = \dfrac{Ne^2}{m\varepsilon_0}$ = plasmonic frequency

The electronic density, charge, and mass are denoted by N, e, and m, respectively, and γ is the rate at which the amplitude of the plasma oscillation is decreasing [24].

ii. Negative Permeability

Magnetic permeability governs how materials react to an applied EM field. In ordinary materials, magnetic permeability is positive. Negative magnetic permeability values are not present, hence exploring negative-index materials has not been particularly motivated. Split-ring resonator (SRR) structures were proposed by Pendry *et al.* to achieve negative permeability values. The capacitive components (gaps and splits) are the cause of the resonant behavior of SRRs, which in turn causes quite large positive and negative values of permeability close to the magnetic resonance frequency (ω_m) [25].

iii. Design of Single Negative Metamaterials

One unit cell of the metamaterial is put within a waveguide with sidewalls that have PEC and PMC boundary conditions set to support a transverse electromagnetic wave with a constant profile while the design is being made. At the waveguide's entrance and exit are waveguide ports that act as the source and detector, respectively. The direction of the entering electromagnetic wave is x, while the magnetic and electric fields are maintained in y and z, respectively. When the magnetic and electric energy stored in the inductor and capacitor are equal, the resonance is seen in the SRR. The resonant frequency is given by:

$$f_0 = \frac{1}{\sqrt{LC}} \qquad \text{(Eq. 9.18)}$$

The gap capacitance (C_g), coupling capacitance (C_c), created between the ring and its neighbors, and surface capacitance (C_s), formed between the ring and the substrate, make up the majority of the total capacitance. It is intended to link these three capacitors in parallel.

A modified metamaterial made of aluminum is used to replicate the transmission spectrum for a variety of thin film thicknesses to study the characteristics of dielectric sensing. With a permittivity of $\varepsilon_r = 1\ 1.9$, the silicon thin film thickness affects the spectrum. The resonance shifts down by increasing the thicknesses. The SRR resonates at 151 GHz for thicknesses of 2 m. It has been discovered that when thin film thickness increases, the peak location of the resonant mode switches from the frequency range of 139 GHz to 165 GHz to a lower frequency range. Since the peak location shifts with the thickness of the thin film, it is important to notice that the cutoff mode depends on the resonance [26].

9.5 Applications

There is no question that the applications made possible by materials with unexpected characteristics will likewise be unusual. One particularly intriguing application is for stoplights. We would want to draw attention to several important recent developments, such as the development of "perfect lenses" that surpass the diffraction law's fundamental limit and the potential for developing an optical equivalent of a black hole. Some of the applications of metamaterials in various fields are stated below:

i. Super Lens
Metamaterials are employed by super lenses to achieve a resolution that exceeds the diffraction limit. The fineness of the resolution of traditional lenses and microscopes is constrained by the diffraction limit, which depends on the light wavelength and the objective lens's numerical aperture (NA) [26].

ii. Cloaking Devices
The foundation for trying to create a workable cloaking device is metamaterials. The cloaks refract the microwave beams so that they pass around a "hidden" object inside with minimum distortion, giving the impression that nothing is there at all. Such a gadget normally entails enclosing the target object in a shell that blocks light from getting close to it [26].

iii. Acoustic Metamaterials
Sonic, infrasonic, or ultrasonic waves may arise in gases, liquids, or solids. Acoustic metamaterials are materials that are intended to regulate, direct, or alter the sound in these different waveforms.

iv. Seismic Metamaterials
Seismic metamaterials are types of metamaterial that are intended to mitigate the negative impacts of seismic waves on built-up areas on or near the earth's surface. This is a type of metamaterial created to mitigate the damaging effects of seismic waves on man-made structures found on or close to the earth's surface

v. Sub-Wavelength Apertures for Transmission Enhancement
To enhance the transmission rate within the diffractive region, ε-near-zero and μ-near-zero coverings are used. The power transmission efficiency of an isolated sub-wavelength aperture, which is proportional to the electrical size's fourth power, is frequently subpar. The grounded values of ε-near-zero and μ-near-zero layers allow an increase in transmitted power. The impinging power can be gathered at the screen's entrance side, tunneled through the tiny aperture, helped to tunnel once more at the exit side, and then molded to increase the beam's directivity in the desired direction. These combined impacts can greatly boost the transmitted power [15].

9.6 Conclusion

We have discussed the single negative metamaterials in this chapter. Classification of metamaterials has been discussed which shows that these devices are studied and are used widely consequently. Different types of metamaterials such as electromagnetic, chiral, semiconductor, and atomistic metamaterials have been studied and distinguished in this chapter so that their uses and purposes must be found familiar. In this chapter, we offer a robust approach for systematically extracting the effective constants for practical single-negative media. Here, we report on the development of a metamaterial with a single layer of connected plasmonic coaxial light waves that effectively modifies the refractive index. Metamaterials can be used as cloaking devices, for transmission purposes and to enhance acoustical processes. Various types and applications of metamaterials listed in this chapter demonstrate that these materials have vast applications and can be considered promising materials for Nano-structural materials in the future.

References

1. Turpin, J.P., Bossard, J.A., Morgan, K.L., Werner, D.H., Werner, P.L., Reconfigurable, and tunable metamaterials: A review of the theory and applications. *Int. J. Antennas Propag.*, 2014, 429837, 2014.
2. Noginov, M.A. and Podolskiy, V.A., *Tutorials in metamaterials*, CRC Press, Boca Raton, 2011.
3. Engheta, N. and Ziolkowski, R.W., *Metamaterials: Physics and engineering explorations*, Wiley-IEEE Press, New Jersey, 2006.
4. Sihvola, A.J.M., Metamaterials in electromagnetics,. *Metamaterials*, 1, 1, 2–11, 2007.
5. Alù, A., Engheta, N., Erentok, A., Ziolkowski, R.W., Single-negative, double-negative, and low-index metamaterials and their electromagnetic applications. *IEEE*, **49**, 23–36, 2007.
6. Zhang, S., Fan, W., Malloy, K., Brueck, S., Panoiu, N., Osgood, R.M., Near-infrared double negative metamaterials. *Opt. Express*, **13**, 4922–4930, 2005.
7. Kumar, R., Kumar, M., Chohan, J.S., Kumar, S., Overview on metamaterial: History, types and applications. *Mater. Today Proc.*, **56**, 3016–3024, 2022.
8. Srinivasan, K., Ali, N.B., Trabelsi, Y., Rajan, M.M., Kanzari, M., Design of a modified single-negative metamaterial structure for sensing application. *Optiks*, **180**, 924–931, 2019.
9. Fan, K. and Padilla, W.J., Dynamic electromagnetic metamaterials. *Mater. Today*, **18**, 39–50, 2015.

10. Walser, R.M., Electromagnetic metamaterials, in: *Complex Mediums II: Beyond linear isotropic dielectrics. Proc: SPIE*, vol. 4467, 2001.
11. Tao, H., Padilla, W.J., Zhang, X., Averitt, R.D., Recent progress in electromagnetic metamaterial devices for terahertz applications. *IEEE*, **17**, 92–101, 2010.
12. Zharov, A.A., Shadrivov, I.V., Kivshar, Y.S., Nonlinear properties of left-handed metamaterials. *Phys. Rev. Lett.*, **91**, 037401, 2003.
13. Jakšić, Z., Dalarsson, N., Maksimović, M.J., Negative refractive index metamaterials: Principles and applications. *Mikrotalas. Rev.*, 12, 36–49, 2006.
14. Padilla, W.J., Basov, D.N., Smith, D.R., Negative refractive index metamaterials. *Mater. Today*, **9**, 28–35, 2006.
15. Mendhe, S.E. and Kosta, Y.P., Metamaterial properties and applications. *IJITKM*, **4**, 85–89, 2011.
16. Engheta, N. and Ziolkowski, R.W., A positive future for double-negative metamaterials. *IEEE*, **53**, 1535–1556, 2005.
17. Li, J. and Chan, C.T., Double-negative acoustic metamaterial. *Phys. Rev. E*, **70**, 055602, 2004.
18. Wang, B., Zhou, J., Koschny, T., Kafesaki, M., Soukoulis, C.M., Chiral metamaterials: Simulations and experiments. *J. Opt.*, **11**, 114003, 2009.
19. Naik, G.V. and Boltasseva, A., Semiconductors for plasmonics and metamaterials. *RRL*, **4**, 295–297, 2010.
20. Hoffman, A.J., Alekseyev, L., Howard, S.S., Franz, K.J., Wasserman, D., Podolskiy, V.A., Narimanov, E.E., Sivco, D.L., Gmachl, C., Negative refraction in semiconductor metamaterials. *Nat. Mater.*, **6**, 946–950, 2007.
21. Plumridge, J., Clarke, E., Murray, R., Phillips, C., Ultra-strong coupling effects with quantum metamaterials. *Solid State Commun.*, **146**, 406–408, 2008.
22. Liaskos, C., Tsioliaridou, A., Pitsillides, A., Akyildiz, I.F., Kantartzis, N.V., Lalas, A.X., Dimitropoulos, X., Ioannidis, S., Kafesaki, M., Soukoulis, C.J., Design and development of software defined metamaterials for nanonetworks. *IEEE*, **15**, 12–25, 2015.
23. Borah, D. and Bhattacharyya, N.S., Design and development of expanded graphite-based non-metallic and flexible metamaterial absorber for X-band applications. *J. Electron. Mater.*, **46**, 226–232, 2017.
24. Wartak, M.S., Tsakmakidis, K.L., Hess, O., Introduction to metamaterials. *Phys. Can.*, **67**, 30–34, 2011.
25. Ozbay, E., Guven, K., Aydin, K., Metamaterials with negative permeability and negative refractive index: Experiments and simulations. *J. Opt.*, **9**, S301, 2007.
26. Cui, T.J., Smith, D.R., Liu, R., *Metamaterials*, Springer, London, 2010.

10

Negative-Index Metamaterials

Rajesh Giri[1] and Ritu Payal[2*]

[1]Department of Physics and Electronics, Rajdhani College, University of Delhi, Delhi, India
[2]Department of Chemistry, Rajdhani College, University of Delhi, Delhi, India

Abstract
A negative-index metamaterial, also known as a negative-index material (NIM), is defined as an artificially engineered structure (metamaterial) with a negative refractive index for electromagnetic radiations across a certain frequency range. Early on, NIMs were a contentious topic, with academics disputing their existence. However, there has been strong evidence over the last two decades that some specific sporadic assemblies can have an effective negative refractive index across a restricted range of frequencies that is absent in naturally available materials. NIMs have recently attracted scientific importance, showcasing metamaterials' extraordinary perspective to enable new electromagnetic breakthroughs. In this chapter, we have reviewed different aspects of NIMs and explored their potential applications in different domains allowing access to new dimensions of material response.

Keywords: Metamaterial, permittivity, electromagnetism, refraction, permeability

10.1 Introduction

The word "metamaterial" was coined for Negative-index metamaterials (NIMs) which are made of designed inclusions displaying outlandish and exceptional electromagnetic characteristics which are neither intrinsic in the discrete integral components nor encountered in nature [1, 2]. These materials are engineered to have negative permittivity as well as permeability resulting in mysterious behavior ranging from negative refractive index

**Corresponding author*: ritupayal.10@gmail.com

Inamuddin and Tariq Altalhi (eds.) Electromagnetic Metamaterials: Properties and Applications, (205–218) © 2023 Scrivener Publishing LLC

values to subwavelength focusing. Victor Veselago, an eminent researcher from the Institute of Physics and Technology, Moscow hypothesized in 1968 that material with simultaneously negative permittivity (ε) and permeability (μ) might exist without violating any physics laws [3].

Veselago referred NIMs as "left-handed" or "double negative" because the cross product obtained by the two vectors namely electric and magnetic fields is antiparallel contrary to the conventional right-handed. He also speculated the optical properties of negative-index materials or these left-handed materials to be the polar opposite of the other media, namely air, glass, and other translucent media. These materials were supposed to have capricious and surprising characteristics, such as deflection or bending light in the negative direction, backward Cherenkov radiation, and Doppler shift reversal. However, it took 33 years for the first real metamaterial to be built, and it also yield the concepts of Russian researcher Veselago [1, 4, 5]. A decade later in 1978, Professor Efimov of Bauman Moscow State Technical University discovered an unanticipated phenomenon in wave refraction theory. His study includes the elementary properties of Maxwell's equations. He discovered the parameter of an anisotropic media, which is a completely non-reflecting crystal, is crucial in the development of metamaterial ideas [6, 7].

NIMs are made up of unit cells, which are periodic basic parts that are typically much smaller compared to the externally supplied wavelength of electromagnetic radiation. The initial NIMs' unit cells were made of circuit board material, dielectrics, and wires, and were experimentally tested. These synthetic cells either piled or planar are designed in a defined repetitive configuration to constitute a discrete NIM. The early NIMs' unit cells, for example, were layered vertically and horizontally, resulting in a repetitive and designed architecture.

In this article, we survey the basic concept and theory of metamaterials, some specific applications of metamaterials, and predict a feasible scientific expansion of this field shortly.

10.2 The Journey from Microwave Frequency to Electromagnetic Radiation

NIMs were initially validated for microwave frequencies but designing NIMs for the optical range has proven mammoth, limiting them to optically attenuated samples due to strong energy dissipation in metals as well as considerable manufacturing hurdles [6, 7]. These attenuated structures are similar to the self-assembly of atoms, thus, making it a herculean task to allocate them the bulk parameters like the index of refraction. After demonstrating promising results at microwave frequencies, intensive efforts have been made to extend metamaterials to the infrared, terahertz, and visible bands. Many research initiatives have been dedicated to understanding and developing higher-frequency NIMs [8]. Although the field of NIMs as a whole is still new, artificial structure scaling is validated for a broad spectrum including millimeter-wave [9], radio frequencies [10], near-infrared [11], mid-infrared [12, 13], and far infrared [14] wavelengths, covering approximately seven orders of frequency scales.

These days, NIMs are technologically advanced to control electromagnetic energy in a novel approach. For instance, electromagnetic and optical features of natural materials are usually modified using chemistry by altering the configuration of the unit cells. These constituents are arranged in such configurations so that their dimensions are fractions of the radiated electromagnetic wave's wavelength. As a result of the combined effects, the electromagnetic wave yields, a broad range impact on the material. Consequently, transmission can also be manipulated by altering the size, shape, and alignments of the unit cells. As a result, material properties like magnetic permeability and permittivity can be controlled, which in turn regulates the proliferation of electromagnetic radiation in a matter. The manipulation of both permittivity and permeability values governs the value of the refractive index as negative, zero, or positive. Everything is dependent on the envisioned applications or anticipated outcomes. Consequently, optical qualities can be enhanced afar what mirrors, lenses, and other traditional materials can do. The negative index of refraction is also one of the most studied effects.

As discussed above the two essential parameters in electromagnetism that determine a medium's electromagnetic quality are electric permittivity (ε) and magnetic permeability. Permittivity and permeability are significant physical properties that define how an electric and magnetic field, respectively are stimulated and exaggerated by a medium and are estimated by a material's tendency to disperse electric and magnetic fields in reaction. In terms of electromagnetic characteristics, an illustration (Figure 10.1) has been used to represent different types of materials. Most dielectric materials fall within Region I in the upper-right quadrant, which includes constituents with positive values of both ε and μ. Ferroelectric materials, metals, and doped semiconductors having negative permittivity at specific frequencies fall within Quadrant II (lower than the plasma frequency). Region IV contains a few ferrite materials manifesting negative permeability; however, their magnetic responses diminish promptly beyond microwave frequencies. Quadrant III, when values of both permittivity and permeability are negative, is the most fascinating region in the material parameter space. There is no such material in nature [15, 16].

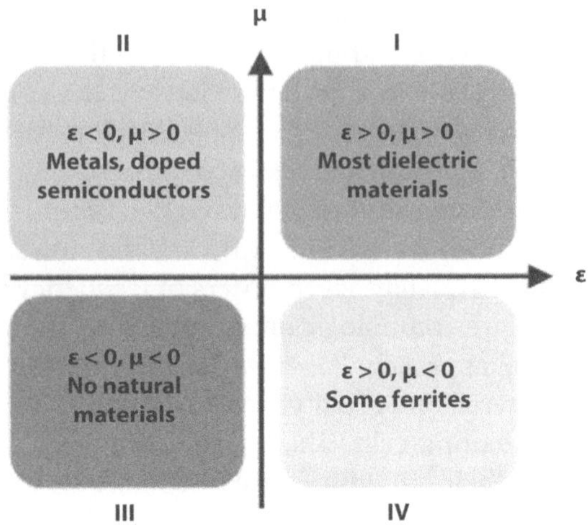

Figure 10.1 Quadrants representing different types of materials.

10.3 Experimentation to Justify Negative Refraction

The debate over negative refraction open the doorways for global study in this field, which came to an end with the findings of the experiments carried out by Houck *et al.* and Parazzoli *et al.* These experiments back up the original findings and advocate Veselago's idea. Houck *et al.* [17, 18] at the MIT Media Laboratory, Massachusetts used wedge-shaped samples of both positive and negative-index materials using a planar waveguide arrangement and recorded the configuration of transmitted microwaves. They utilized a wedge of positive-index material Teflon (Figure 10.2a), as a control sample, which deflects the direction of a microwave beam and exits the sample by making a positive angle with an antiparallel surface, as per Snell's law. On the other hand, a metamaterial wedge constructed of a network of rings and cables caused the emerged beam to exit at a negative angle for specific frequencies (Figure 10.2b). The researchers employed two distinct wedge samples with the incidence of wave on the metamaterial at two unlike angles to establish that the emerged beam was triggered by the intrinsic negative index of the wedge (Figure 10.2b) rather than a lossy or diffractive sample artifact. In both examples (Figure 10.2a and b), negative refraction was observed at angles that were compatible with samples with apparent negative index, thereby confirming Snell's law. The researchers demonstrated that a rectangular-shaped metamaterial may divert photons from a proximate antenna, fulfilling one more of Veselago's numerous speculations.

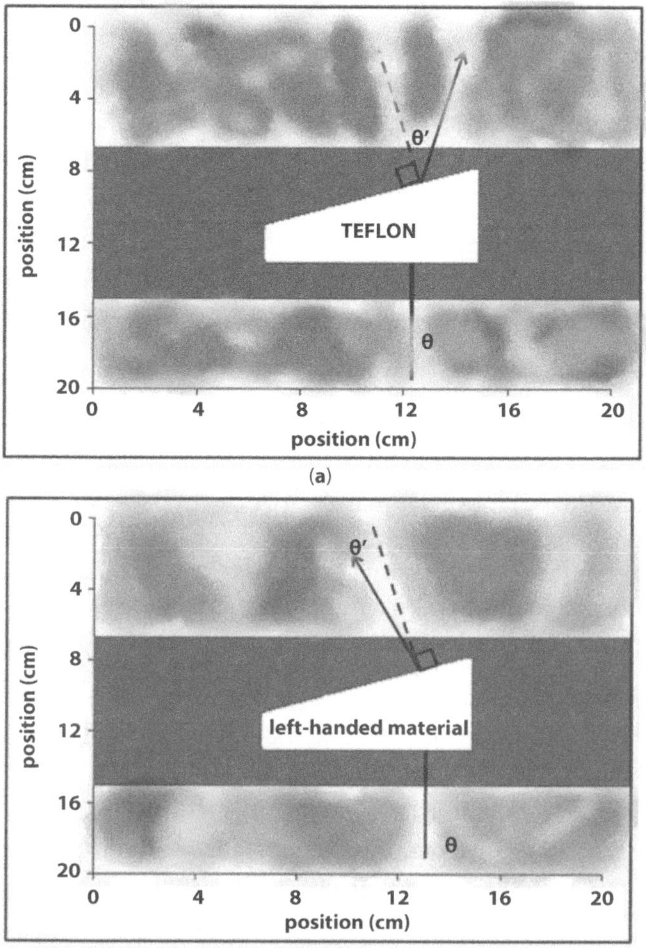

(a)

(b)

Figure 10.2 Refraction in a conventional material with a positive refractive index (a) versus a negative-index metamaterial (b). The incident beam (θ) comes in through the air and deflects in a usual (θ') or metamaterial (-θ') refraction pattern.

Parazzoli *et al.* at Boeing's Phantom Works division fabricated a wedge sample suitable for free-space measurements by taking a different methodology (Figure 10.3).

The Boeing sample, which is based on a ring and cable arrangement, visibly validates negative refraction [19]. The negatively refracted beam does not decline in any unusual way as a function of distance, even when measured to the extent of 28 wavelengths and nearly 66 cm away from the

Figure 10.3 Ring and cable assembly depicting negative refraction.

sample, as in the MIT experiment performed by Houck *et al.* [4]. Using electromagnetic simulations, the Boeing team calculated the metamaterial's refractive index. The resulting value is in perfect corroboration with the refraction data and exhibits similar frequency-dependent fluctuation as the experimentally measured index.

10.3.1 Reverse Propagation

The transmission of an electromagnetic wave is reversed when the index of refraction is negative. It is now possible to achieve resolutions below the diffraction limit. Subwavelength imaging is the term for this type of imaging. Another capability is the capacity to transmit a light beam through an electromagnetically flat surface. Conventional materials, on the other hand, are frequently curved and can't resolve the diffraction boundary. Furthermore, changing the direction of electromagnetic waves in a material and combining it with some other common material (such as air), reduces the losses that would otherwise occur. The negative index of refraction is also shown by the reversal of the electromagnetic radiation, which has a perpendicular phase velocity [1, 3, 4].

10.3.2 Properties of NIMs

Designer materials with a variety of unique and switchable electromagnetic properties at practically configurable wavelength bands are becoming a reality as NIMs design and fabrication methods become more common. Metamaterials investigations have exploded in today's modern era because of the incredible capabilities they provide for manipulating light. Artificial chirality [20], negative refraction [21], optical magnetism [22], remarkable optical transmission [23], and super absorption [24] are examples of exceptional properties. Out of the various published designs for achieving NIMs at optical frequencies the multilayer fishnet design (3D structures) emerges as the most propitious architecture in terms of the figure of merit, tuneability, and strong negative values [25].

Moreover, NIMs are specialized composites that are fabricated to achieve a specific goal and can be combined to provide optical qualities that are rarely found in nature. The composite material's qualities are derived from its framework assembled from constituents that are smaller than the imposing electromagnetic wavelength and also manage the distance smaller than the imposing electromagnetic wavelength. Strange and counterintuitive features are currently being used to manipulate electromagnetic radiation in wireless communication systems. The study still going on in different areas of the electromagnetic scale, such as visible light [1, 4].

10.4 Electromagnetic Response of Materials

To comprehend NIMs, one must first grasp the concept of electromagnetic wave sensitivity in general. Electromagnetic response in homogeneous materials is primarily ruled by two factors. The first characteristic, $\varepsilon(\omega)$, specifies a material's response to the electric component of the beam of polarized light (or another electromagnetic wave), while the second factor, $\mu(\omega)$, describes its response to the magnetic field component at a frequency. $\varepsilon(\omega)$ and $\mu(\omega)$ are frequency-reliant complicated values, therefore providing a total of four parameters that fully explain the response of an isotropic material to electromagnetic radiation at a particular frequency value as:

$$\varepsilon(\omega) = \varepsilon_1(\omega) + i\varepsilon_2(\omega) \tag{10.1a}$$

$$\mu(\omega) = \mu_1(\omega) + i\mu_2(\omega) \tag{10.1b}$$

The parameters ε and μ are the most significant and convoluted parameters for most of the materials, dictating the response between light and matter. Nevertheless, there are appreciable electromagnetic characteristics that explain wave propagation in other fields of study that are associated with the material parameters presented in Eqs 10.1a and 10.1b by simple arithmetic relationships; for instance, absorption or refraction processes of a material characterized by ε and μ. The index of refraction is an often used EM parameter that measures the EM wave's speed as it propagates through a material and is demarcated as:

$$n(\omega)^2 = \varepsilon\,(\omega)\mu(\omega) \tag{10.2}$$

Furthermore, the refractive index is used to calculate the deflection of a light beam as it passes through an interface amid two materials with unlike refractive indices. Willebrord Snell [26] offered a quantitative measure of this bending in 1621, demonstrated that:

$$n_1 \sin\theta_1 = n_2 \sin\theta_2 \tag{10.3}$$

where n_1 and n_2 specify the first and second media's index of refraction, respectively, θ_1 and θ_2 mentions the angles that the light waves make, with the normal surface of each medium. A basic ray tracing illustration (Figure 10.4) depicts the emergence of waves from a point source in vacuity, incidence on a slab of positive refraction index material.

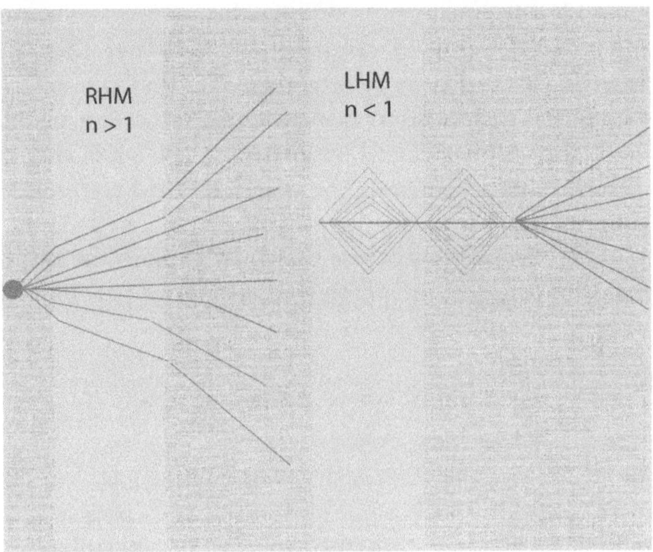

Figure 10.4 A point source illuminates a conventional slab of glass (pink) on the left (brick red lines). Eq. 10.2 depicts Snell's Law showing the divergence and refraction of rays at the interface. A flat slab of NIM is depicted on the right, with rays from a point source (brown lines) incident upon it. The rays refract at the interface in this example, according to Snell's law, but with an index of n = -1.

When the rays from the source reach the slab's surface, they deflect at the contact between the glass and free space at an angle specified by eq 10.2. The refractive index is always considered to be positive in almost all educational institution's transcripts based on electricity, optics, and magnetism. Nature, on the other hand, has kept a colossal secret from us, which Russian physicist Victor Veselago uncovered first. He understood that if he identified a material with negative values for both the electric ($\varepsilon(\omega) < 0$) and magnetic response functions ($\mu(\omega) < 0$), the value of $n(\omega)$ would also be negative. Even though Veselago speculated that naturally occurring negative refractive index materials would be discovered or produced, however, such materials have never been observed. Artificially structured NIMs, on the other hand, can have systematized electric and magnetic responses over a broad spectrum of the frequency range, therefore, it is realistic to attain the conditions $\varepsilon < 0$ and $\mu < 0$ in these aggregates thus validating the fabrication of Veselago's proposed material.

10.5 Application of NIMs

NIMs are emerging materials and will linger to have control over the domains of engineering, materials science, nanotechnology, optics, physics, and many other divisions of study by allowing access to new worlds of material response. NIMs are being combined with traditional devices that transmit, broadcast, shape, or accept electromagnetic signals traveling across wires, cables, or air. The qualities of the materials, equipment and systems used in this project could be changed or enhanced. As a result, commercially available metamaterial antennas and accompanying equipment are already doing this [27]. Furthermore, these metamaterial apparatuses are still being explored in the wireless sector. Other applications are being investigated as well. These include electrically small resonators, radar-microwave absorbers, waveguides that can work beyond the diffraction limit, enhanced electrically small antennas, phase compensators, and advances in microwave lenses [28–31]. The evolvement of superlens makes it possible to obtain imageries below the diffraction limit in the optical frequency regime. Optical nanolithography, nanotechnology circuits, subwavelength photolithography, and near-field superlens for biomedical imaging are all possible applications for negative-index metamaterials [32].

NIMs also find use in various optoelectronic devices like Electro-optical meta devices [33–35], Liquid-crystal meta devices [36, 37], Phase-change meta devices [38–40], Superconducting meta devices [41, 42], Ultrafast photonic meta devices [43, 44], Nonlinear meta devices with varactors [45,

46], meta devices driven by electromagnetic forces [47, 48]. Researchers are continuously working on NIMs to explore their potential applications in different domains allowing access to new dimensions of material response.

10.6 Conclusions

Over the last two decades, there has been a surge in interest in metamaterials research. NIMs are rationally built composites made up of customized building blocks using one or more constituent bulk materials that have qualities that are superior to their constituents. NIMs are gaining traction in a variety of engineering fields, including mechanical, acoustic, transport characteristics, materials science, nanotechnology, optics, physics, and a variety of other fields. NIMS showing a negative refractive index have been designed in the broad wavelength range of 1500 nm - 200 THz - GHz to visible light, Infrared, and others. Because of their ease of production and reduced design complexity, 1D and 2D model architectures paved the way for 3D microstructures and nanostructures.

Acknowledgments

The authors are thankful to the Library and central computer laboratory of Rajdhani College, Raja Garden, West Delhi for assessing their library and internet facilities.

References

1. Helby, R.A., Smith, D.R., Shultz, S., Experimental verification of a negative index of refraction. *Science*, 292, 5514, 77, 2001.
2. Padilla, W.J., Basov, D.N., Smith, D.R., Negative refractive index metamaterials. *Mater. Today*, 9, 7-8, 28, 2006.
3. Veselago, V.G., The electrodynamics of substances with simultaneously negative values of \epsilon and. *Sov. Phys. Usp.*, 10, 509, 1968.
4. Engheta, N. and Ziolkowski, R.W., Introduction, history, and selected topics in fundamental theories of metamaterials, in: *Metamaterials: Physics and Engineering Explorations*, N. Engheta and R.W. Ziolkowski (Eds.), pp. 5–42, Wiley & Sons, Inc., USA, 2006.
5. Lezec, H., NIST, NIST, US, 2009, https://www.nist.gov/programs-projects/nanoplasmonics-and-three-dimensional-plasmonic-metamaterials.

6. Efimov, S.P., Compression of electromagnetic waves by anisotropic media. *Radiophys. Quantum Electron.*, 21, 2, 916, 1978.
7. Efimov, S.P., Compression of waves by artificial anisotropic medium. *Acoust. Phys.*, 25, 2, 234, 1979.
8. Smith, D.R., Padilla, W.J., Vier, D.C., Nemat-Nasser, S.C., Schultz, S., Composite medium with simultaneously negative permeability and permittivity. *Phys. Rev. Lett.*, 84, 18, 4184, 2000.
9. www.nanotechnology.bilkent.edu.tr/research%20areas/documents/mm-waveleft-handed.htm.
10. Wiltshire, M.C.K., Pendry, J.B., Young, I.R., Larkman, D.J., Gilderdale, D.J., Hajnal, J.V., Microstructured magnetic materials for RF flux guides in magnetic resonance imaging. *Science*, 291, 5505, 849, 2001.
11. Enkrich, C. *et al.*, Magnetic metamaterials at telecommunication and visible frequencies. *Phys. Rev. Lett.*, 95, 20, 203901, 2005.
12. Linden, S., Enkrich, C., Wegener, M., Zhou, J., Koschny, T., Soukoulis, C.M., Magnetic response of metamaterials at 100 terahertz. *Science*, 306, 5700, 1351, 2004.
13. Zhang, S., Fan, W., Panoiu, N.C., Malloy, K.J., Osgood, R.M., Brueck, S.R.J., Experimental demonstration of near-infrared negative-index metamaterials. *Phys. Rev. Lett.*, 95, 13, 137404, 2005.
14. Yen, T.J., Padilla, W.J., Fang, N., Vier, D.C., Smith, D.R., Pendry, J.B., Basov, D.N., Zhang, X., Terahertz magnetic response from artificial materials. *Science*, 303, 5663, 1494, 2004.
15. Liu, Y. and Zhang, X., Metamaterials: A new frontier of science and technology. *Chem. Soc. Rev.*, 40, 2494, 2011.
16. Liu, Z., Du, H., Li, Z.Y., Fang, N.X., Li, J., Nano-kirigami metasurfaces by focused-ion-beam induced close-loop transformation. *APL Photonics*, 3, 10, 100803, 2018.
17. Houck, A.A., Brock, J.B., Chuang, I.L., Experimental observations of a left-handed material that obeys snell's law. *Phys. Rev. Lett.*, 90, 13, 137401, 2003.
18. Smith, D.R., Department of Physics, University of California, San Diego, US, 2003, https://physicsworld.com/a/the-reality-of-negative-refraction/#:~:-text=Negative%20reaction&text=This%20means%20that%20refraction%20at,no%20single%20negatively%20refracted %20wave.
19. Parazzoli, C.G., Greegor, R.B., Li, K., Koltenbah, B.E.C., Tanielian, M., Experimental verification and simulation of negative index of refraction using snell's law. *Phys. Rev. Lett.*, 90, 10, 107401, 2003.
20. Liu, Z., Du, H., Li, J., Lu, L., Li, Z.Y., Fang, N.X., Nano-kirigami with giant optical chirality. *Sci. Adv.*, 4, 7, eaat4436, 2018.
21. Valentine, J., Zhang, S., Zentgraf, T., Ulin-Avila, E., Genov, D.A., Bartal, G., Zhang, X., Three-dimensional optical metamaterial with a negative refractive index. *Nature*, 455, 7211, 376, 2008.

22. Many, V., Dézert, R., Duguet, E. *et al.*, High optical magnetism of dodecahedral plasmonic meta-atoms. *Nanophotonics*, 8, 4, 549, 2019.
23. Rodrigo, S.G., de Leon-Perez, F., Martin-Moreno, L., Extraordinary optical transmission: Fundamentals and applications. *Proc. IEEE*, 104, 12, 2288, 2016.
24. Molet, P., Garcia-Pomar, J.L., Matricardi, C., Garriga, M., Alonso, M.I., Mihi, A., Ultrathin semiconductor superabsorbers from the visible to the near-infrared. *Adv. Mater.*, 30, 9, 1705876, 2018.
25. Soukoulis, C.M., Linden, S., Wegener, M., Negative refractive index at optical wavelengths. *Science*, 315, 5808, 47, 2007.
26. Grzegorczyk, T.M., Nikku, M., Chen, X., Wu, B.I., Kong, J.A., Refraction laws for anisotropic media and their application to left-handed metamaterials. *IEEE Trans. Microw. Theory Tech.*, 53, 4, 1443, 2005.
27. Slyusar, V.I., Metamaterials on antenna solutions, in: *7th International Conference on Antenna Theory and Techniques ICATT'09*, Lviv, Ukraine, pp. 19–24, 2009.
28. Engheta, N. and Ziolkowski, R.W., A positive future for double-negative metamaterials. *IEEE Trans. Microw. Theory Tech.*, 53, 4, 1535, 2005.
29. Beruete, M., Navarro-Cia, M., Sorolla, M., Campillo, I., Planoconcave lens by negative refraction of stacked subwavelength hole arrays. *Opt. Express*, 16, 13, 9677, 2008.
30. Alu, A. and Engheta, N., Guided modes in a waveguide filled with a pair of single-negative (SNG), double-negative (DNG), and/or double-positive (DPS) layers. *IEEE Trans. Microw. Theory Tech.*, 52, 1, 199, 2004.
31. Shalaev, V.M., Optical negative-index metamaterials. *Nat. Photonics*, 1, 1, 41, 2007.
32. Pendry, J.B., Negative refraction makes a perfect lens. *Phys. Rev. Lett.*, 85, 18, 3966, 2000.
33. Chen, H.T. *et al.*, Active terahertz metamaterial devices. *Nature*, 444, 597, 2006.
34. Papasimakis, N. *et al.*, Graphene in a photonic metamaterial. *Opt. Express*, 18, 8353, 2010.
35. Ju, L. *et al.*, Graphene plasmonics for tunable terahertz metamaterials. *Nat. Nanotechnol.*, 6, 630, 2011.
36. Wang, X. *et al.*, Tunable optical negative-index metamaterials employing anisotropic liquid crystals. *Appl. Phys. Lett.*, 91, 143122, 2007.
37. Zhang, F. *et al.*, Magnetic control of negative permeability metamaterials based on liquid crystals. *Appl. Phys. Lett.*, 92, 193104, 2008.
38. Zheludev, N., All change, please. *Nat. Photonics*, 1, 10, 551, 2007.
39. Driscoll, T. *et al.*, Dynamic tuning of an infrared hybrid-metamaterial resonance using vanadium dioxide. *Appl. Phys. Lett.*, 93, 024101, 2008.
40. Appavoo, K. and Haglund Jr., R.F., Detecting nanoscale size dependence in VO_2 phase transition using a split-ring resonator metamaterial. *Nano Lett.*, 11, 1025, 2011.

41. Tsiatmas, A., Fedotov, A.V., Zheludev, N.I., Low-loss terahertz superconducting plasmonics. *New J. Phys.*, 14, 11, 115006, 2012.
42. Tsiatmas, A. *et al.*, Superconducting plasmonics and extraordinary transmission. *Appl. Phys. Lett.*, 97, 111106, 2010.
43. Nikolaenko, A.E. *et al.*, Carbon nanotubes in a photonic metamaterial. *Phys. Rev. Lett.*, 104, 153902, 2010.
44. Ren, M.X. *et al.*, Nanostructured plasmonic medium for terahertz bandwidth all-optical switching. *Adv. Mater.*, 23, 5540, 2011.
45. Shadrivov, I.V., Kozyrev, A.B., van der Weide, D., Kivshar, Y.S., Nonlinear magnetic metamaterials. *Opt. Express*, 16, 20266, 2008.
46. Shadrivov, I.V., Kozyrev, A.B., van der Weide, D.W., Kivshar, Y.S., Tunable transmission and harmonic generation in nonlinear metamaterials. *Appl. Phys. Lett.*, 93, 161903, 2008.
47. Lapine, M., Shadrivov, I.V., Powell, D.A., Kivshar, Y.S., Magnetoelastic metamaterials. *Nat. Mater.*, 11, 30, 2012.
48. Lapine, M., Shadrivov, I.V., Powell, D.A., Kivshar, Y.S., Metamaterials with conformational nonlinearity. *Sci. Rep.*, 1, 138, 2011.

11

Properties and Applications of Electromagnetic Metamaterials

Km. Rachna* **and Flomo L. Gbawoquiya**

School of Basic Sciences and Research, Department of Environmental Sciences, Sharda University, Greater Noida, India

Abstract

It is not unexpected given the recent quick advancements in the realm of metamaterials that the concept of metamaterials was uncovered as quickly as the foundations for macroscopic characterization were established and a strong similarity to optical crystals was made clear. There has been a significant amount of interest in and growth in electromagnetic materials. Using SPR sensors, metamaterials, a new family of artificial electromagnetic (EM) substances with subwavelength regularity that is strongly influenced by indices of refraction, are helpful. During the year 2000, production of them within the microwave range of frequencies started. Notable cases of unusual phenomena like immaculate focusing, negative index of refraction, and transparency cloaking have sparked a lot of interest in them. The emerging area of electromagnetic metamaterials has given rise to novel phenomena like the negative index of refraction and technological innovations like the electromagnetic cloak. Engineered electromagnetic materials have advanced significantly within a decade of studying metamaterials, from microwave to optical frequencies. The various kinds of metamaterials were covered in this chapter, along with their characteristics and uses. The theory, experimentation, and uses of metamaterials are briefly highlighted in this chapter. The primary emphasis is placed on the meta-atoms that make up matter, how light is transformed, how negative refraction works in practice, and other topics. The many different procedures used to create metamaterials with dynamic behavior are logically divided into different groups and are then graphically presented. Metamaterial technology has advanced for several applications in the last ten years. Countless innovative metamaterial types with unusual electromagnetic properties, such as negative index, optical magnetism, etc., have been discovered.

**Corresponding author*: km.rachna@sharda.ac.in

Inamuddin and Tariq Altalhi (eds.) Electromagnetic Metamaterials: Properties and Applications, (219–260)

The field of electromagnetic metamaterials has experienced remarkable expansion and attention given the recent demonstration of unique phenomena.

Keywords: Metamaterials, single negative metamaterials, double positive media, permittivity and permeability

11.1 Introduction

The development of artificial dielectrics in microwave engineering immediately following World War II marks the beginning of the history of metamaterials. However, just before the climax of the 19th century, pioneering research into artificial materials for regulating electromagnetic waves was conducted. Based on the qualities of concern, it can have a myriad of starting locations. Beginning in 1904, similar initial wave research continued for most of the first part of the 20th century. The phase velocity to group velocity link as well as the wave vector to pointing vector connection were both the subject of early investigation. Metamaterials have been described as artificial recurrent formations whose fretwork constants are smaller as compared to the wavelength of the radiation that enters. Consequently, a negative index of refraction features is delivered. Metamaterials are a mixture of meta and material. According to Sihovola, the Greek term "meta" signifies "beyond," "altered," "changed," or "advanced." The electromagnetic properties of metamaterials can be accurately altered to a non-natural status. They are typically composed of synthetic materials. Thus, a theoretical investigation is offered. To show the effectiveness of synthetic materials, a microwave test has been performed on a perverted framework. Subsequently, theoretical research by physicists was reported and artificial electric plasma was created by Pendry *et al.* in [1] using a synthetic wiring medium with negative permeability. Following that, negative permeability magnetic plasma was created in 1999 using split-ring resonators (SRR). Smith and colleagues found a substance with a gradient refractive index in 2004 that possessed the capability of curving waves with electromagnetic status. Metamaterial opens up an entirely new, interesting realm favoring the scholarly community. Since most people are now familiar with the generalized idea of a negative refractive index, research is now focusing on applications in real-world situations. To produce an optimal combination of two or more reactions to a particular stimulation that is not possible naturally, metamaterials are categorized as tiny composites with anthropogenic, periodic cellular structures, and three dimensions. As a result, the development of metamaterials can be boiled down to the development of a few artificial materials that interact at radio, microwave, and eventually optical frequencies [2]. Conventional waves propagate energy but not matter through the medium (material). As an example, in a pond, waves don't move the molecules of water around; instead, the wave's energy is carried via water which helps to keep the molecules in position.

Additionally, as electric charges like electrons and protons move, they generate electromagnetic fields that convey electromagnetic radiation, often known as light. An altered magnetic field will result in an altered electric field, and vice-versa. These varying fields produce electromagnetic waves. Electromagnetic waves, as opposed to mechanical waves, can move without the need for a medium. This suggests that all these waves can travel into void space in addition to solid materials like air and rock.

Figures 11.1 and 11.2 represent how metamaterials and functional surfaces work; negative-index metamaterials configuration, respectively.

After the successful creation of the very first negative index metamaterial as seen in Figure 11.2, the subject of metamaterials has become an increasingly prevalent topic across several professions. Owing to a certain remarkable new use in industries, including solar cells, cloaking gadgets, lightweight antennae, and field super lenses. Recently published multiple monographs offer details on the metamaterials' complete outline and many noteworthy references [2]. Considering recent developments and the fascinating prospective uses of electromagnetic metamaterials, precise mathematical evaluation and simulation of metamaterials are necessary. In actuality, the creation of new metamaterials and the finding of novel metamaterial phenomena have both depended heavily on computer simulations of metamaterials. However, a lot of

Figure 11.1 Functional surfaces of metamaterials.

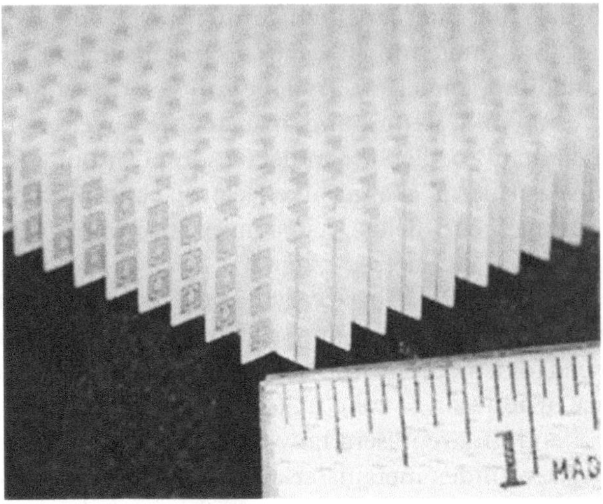

Figure 11.2 Configuration of negative index metamaterials.

simulations have been done up to this point utilizing at some point the traditional limited-difference time-domain (FDTD) technique or tools at the commercial level (*i.e., COMSOL and HFSS*). Although the FDTD method is popular among engineers, it produces the so-called staircase effect [3] when applied to a domain with complicated geometry. Even though the FDTD technique is well-liked amongst engineers, it results in the infamous staircase effect when employed on a domain with intricate geometry. To adequately address Maxwell's equations, the finite element technique (FEM) is commonly utilized, especially in domains with complex geometry. It is worth noting that several kinds of literature on very smaller elemental techniques for Maxwell's equations concentrate on medium with free space, with the possible exception of a handful of studies on dispersive media. The finite element approach and the applicability of the equations formulated by Maxwell concerning metamaterials in the domains of time and frequency have been the subject of several investigations [4]. It would be better to find an appropriate technique to transform the time-to-domain metamaterial concerning Maxwell's equations into a type of vector wave problem, with only one unknown, since it would enable a considerably faster approach. What follows is motivated by the orbit of two artificial satellites A and A+ around planet P. We assume that they move in a synchronized manner and distribute a circular orbit on the horizontal plane for the arc length l of the orbit between them to remain constant at all times. Presume that l is smaller than the orbit's | | half-length. After that, the triangle 1APA+ can be created. The three interior altitude lengths are important. Our calculations of these lengths will be important for constructing communications networks as well as other defensive tactics, among other things, because they rely on the orbit and can be calculated in a variety of ways. In the ideal scenario described earlier, it is not difficult to achieve the required approximations or bounding constants. However, in the grand scheme of things, we might be accounting for different orbital types, such as elliptical ones, or the fact that point P is not the middle of the earth, etc. The corresponding predictions are then hard to find.

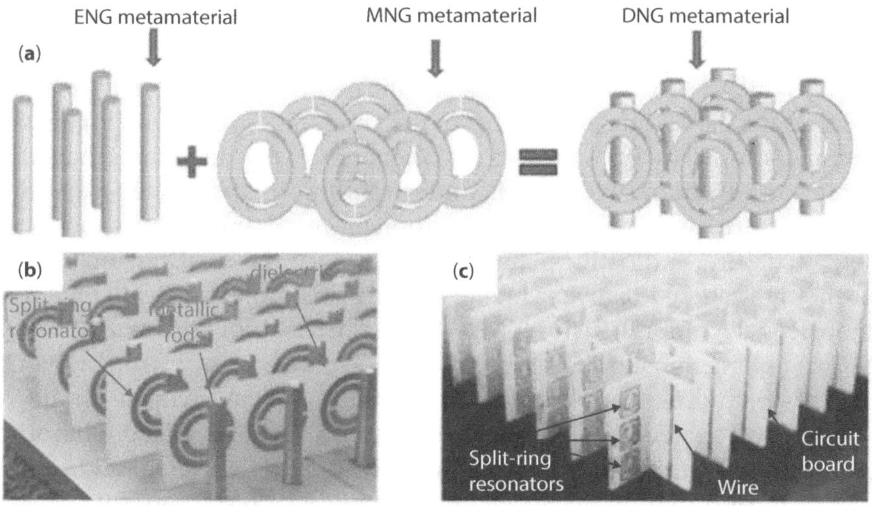

Figure 11.3 (a) ENG metamaterial. (b) MNG metamaterial. (c) DNG metamaterial.

Metamaterials are unnatural materials that might not be easily encountered in "nature" in their original form. Usually, the features of these materials come from their architecture instead of their makeup, and they leverage tiny inhomogeneities to create effective macroscopic behaviors. Metamaterials are currently a component of the general field of electromagnetics. The unique and desirable characteristics that metamaterials display as a reflection of their distinctive structure and design are what define them. The macroscopic functional permittivity and permeability of a specific composite medium is affected by the interaction of electromagnetic waves with inclusion to produce electric and magnetic moment. Considering metamaterials can be produced using purposefully made inclusions that are inserted in specific host media. Inclusion mixtures, shape, size, and host material properties are just a few of the independent factors the designer can access as a result (Figure 11.3). These design features could have a big influence on the result. Each of these designs has inclusions with a morphology that gives metamaterials more processing choices. Because of their unique acoustical, electromagnetic, optical, and mechanical properties, they have a great deal of ability for a variety of applications. The challenge of supplying the innovative capabilities offered by metamaterials employing conventional design is also another factor contributing to the rising interest in the advancement of metamaterials. Recent studies have shown that the metamaterial phenomenon can be used to build energy recovery technology, particularly in the area of lower-intensity energy scavenging. Using algorithms to order fundamental aspects at the Techniques are utilized to establish the correct sequence of response to incident energy at the sublevel of the micron scale. Additionally, there are many opportunities for harvesting energy due to the simplicity of customizing metamaterials in harmony with sources of energy like acoustic, mechanical, optical, and microwave. A full understanding of their categorization, manufacture, and potential for customization characteristics, such as size, structure, and lattice planes, is required for the creation and choice of the right energy recovery metamaterial. This project seeks to give you that deeper comprehension. Together with numerous ideas and tests, the use of metamaterials to exhibit and evaluate energy from various sources including sound waves, solar waves, and mechanical motions is also investigated. As the science of materials has advanced, photonic substances, that employ the photon of light as the principal additional source of information, have been developed. As a result, photonic crystals were developed, and in the early years of the millennium, it was demonstrated that the refraction of metamaterials which is negative as regards index would likely function in the microwave (10.5 Gigahertz) and optical ranges. Six years later, yet again in the microwave realm, the first metamaterial cloaking proof of concept—which obscures an object from view—was shown. As the field of materials science developed, photonic materials, which use the photon of light as their main information carrier, were developed. As a result, photonic crystals were developed, and concrete evidence experiments for functional metamaterials with negative indices of refraction in the optical and microwave (10.5 Gigahertz) bands were conducted at the beginning of the millennium. The first metamaterial cloaking proof of concept, which hides an object from view, was shown about six years later. There are various types of electromagnetic metamaterials as indicated below.

a) Negative refractive index

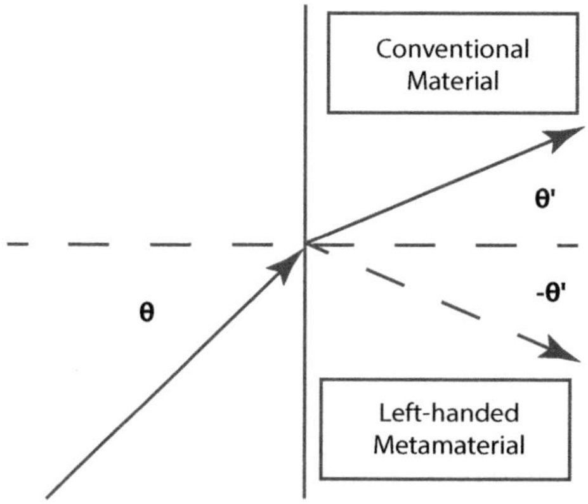

Figure 11.4 A left-handed comparison of metamaterial refraction in normal material.

b) Those metamaterials that possess an index but are negative can be distinguished by said (negative index) refraction (NIM). NIMs are often referenced as media with something like a negative index considered refractive, media with a reverse wave, and that which is regarded as left-handed. Double negative materials (DNG) are the other nomenclature for NIMs in which Negative permeability and unfavorable permittivity combine to produce negative indices of refraction [5].

c) Negative Single (NS)
Only negative relative permeability can be found in single negative metamaterials (r). It is claimed that they serve as metamaterials when they are amalgamated with second, alternative single negative metamaterials and act as a double negative metamaterial as a whole. Rather than displaying a negative r, the negative to epsilon's media exhibited a positive r. Many plasmas feature this character. Gold or silver are well-known metals. Metals are ENG for both the visible and ultraviolet spectral regions, including gold and silver. Positive εr and negative μr are seen in other media which have been explained [6]. Materials regarded as gyromagnetic have got this feature and the term gyrotropic describes a material that has undergone modifications as more than just a product of output in a stable equilibrium. In this type of substance, the rates of left and right-hand circulating elliptical polarisation propagation can vary. The Faraday impact occurs when light pierces a surface substance deemed magneto-optic. A Faraday rotator can then be made by rotating the polarisation plane (Figure 11.4). The rotation

orientations of these two primary polarizations are reversed in optical isomers, which are two gyrotropic materials.

11.2 Hyperbolic Metamaterials

Due to the severe anisotropy caused by both the negative and positive permittivity tensor components of hyperbolic metamaterials generate, they simultaneously serve as a defined path for light transfer and an insulator for the opposite. The metamaterial is considered to be hyperbolic if the transfer function in wavevector space has a hyperbolic form. Because of the high anisotropy of hyperbolic metamaterials light propagates both inside and outside in particular directions. In addition to improved reflection modulation, plasmon resonance effects, optical signal steering, and imaging, Hyperbolic metamaterials have shown a variety of other potential applications [7].

d) Bandgap

Metamaterials with electromagnetic bandgaps control how light travels (EBG or EBM). To do this, left-handed materials known as photonic crystals (PC) are used (LHM). Computers can fully halt the spread of light. Both classes may make bandgaps at selected frequencies and enable light to travel in planned, desired paths. Electromagnetic bandgaps produce interruption that is simultaneously productive and detrimental since a sizeable chunk of their wavelength is taken up by their period. PC can be differentiated from subwavelength structures like in the case of tunable metamaterials because they get their attributes from the characteristics to obtain from bandgap. In contrast to other metamaterials that reveal wavelength at the substructural level, photonic crystals are proportioned to the light's wavelength. PCs also use light scattering to power their operations. Nevertheless, metamaterial does not employ diffraction. Occasional inclusions in PCs generate harmful scattering interference that prevents waves from propagating. Thanks to their photonic bandgap property, computers are the electromagnetic counterpart of electronic semi-conductor crystals.

e) Medium that is Double-Positive

Nature has double positive media, just like natural sources of dielectrics. The dissemination of wave is pushed further ahead while permittivity and magnetic permeability are also both positive. Materials with intentionally blended DPS, ENG, and MNG features have already been created.

f) Bi-isotropic and bianisotropic
Usually, it is believed that metamaterials have two distinct responses (i.e., magnetic and electric) magnetic responses and these responses can be classified as dual, single, or dual positive. When the electric field regularly leads to magnetic polarisation while the magnetic field frequently produces it, this is a phenomenon termed the coupling of magnetoelectric. Bi-isotropic media are those in question. Several metamaterial structures are anisotropic and exhibit magnetoelectric interaction in bi-anisotropic media [8, 9]. Four material properties control the intrinsic magnetoelectric coupling of bi-isotropic media. The flux densities for electricity (E), magnetism (H), electricity (D), and magnetism (B) are exactly that. The letters, and, represents permittivity, permeability, chirality strength, and the Tellegen parameter, respectively, are used to represent these variables. Modifications in a revolving coordinate system of measurements do not affect the material properties in this type of medium. In this sense, they are scalar or invariant.

g) Chiral
Since the phrase left- and right-hand are used, there is a contestation for this phrase in the literature on metamaterials, metamaterial handedness may be misunderstood. The first refers to either of the two circular polarization oscillations that serve as the mechanisms of transmission in chiral media. The second discusses the trio of various phenomena (i.e., magnetic field, electric field, etc). that appears in media with negative refractive indices, which are often nonchiral. An electromagnetic response that is chiral and/or bianisotropic is typically caused by 3D geometrical chirality: 3D-metamaterials, which display a phenomenon of mirror images related to polarization impacts such as optical activity and dichroism that is rotational, are synthesized in a host medium. According to the notion of 2D mirror images, a flat object is considered chiral if it is unable to be superposed onto its mirror counterpart without being elevated off the plane. Circular conversion dichroism causes anisotropic and this has been It has been shown that lossy 2D-chiral metamaterials transmit (reflect, absorb) circular polarization waves in a directionally asymmetric manner.

11.3 Properties of Metamaterials

The idea of metamaterials is without a doubt one of the most significant recent developments in the study of complicated media and unusual materials. Researchers that investigate difficult media have already coined the word to describe a class of synthetic materials having peculiar electromagnetic properties that are absent in natural materials [11]. This idea has led to the use of the Greek word "meta" to describe the "uprising" of unique features from a specific annexing of materials that exhibit traditional electromagnetic characteristics [12]. The definition of the prefix "meta" in the context of metamaterials that is currently most frequently used is "beyond" as in the phrase "metaphysics."

Veselago has described the most intriguing features of DNG materials, which we shall briefly examine. Let's pretend that an electromagnetic field is moving above an unconstrained DNG medium as plane waves. Because all of the domains (including the wave's vector, magnetic fields, and electric fields) follow the left-hand principle, the wave's vector displays the reverse of the Poynting vector. This allows any portion in a recurring series of changes to expand along a similar direction like in the case of the energy but with the inverse direction. As a result, the plane-wave transfers energy as it would in a conventional material, but phase-front propagation is in the opposite direction. This is a novel finding when comparing this result to those derived from the electrodynamics of common materials. However, we must point out that the electrodynamics of common materials also displays the property of having distinctive routes for the energy circulation and the velocity stage. Phase velocity and Poynting vector in anisotropic biaxial materials form an angle and are not comparable. This results in a 180° angle amongst the Poynting and waves vectors for a plane vibrating system in DNG metamaterials, which is unusual for a macroscopically homogeneous and isotropic material. Veselago saw a fresh and unexpected result that has other fascinating unintended consequences. For example, the Doppler shift in a DNG medium that has no limits differs from that in ordinary materials in the reverse directions [13]. Whenever the person or place or thing moves a distance apart from the viewer while the viewer remains static, the frequency perceived increases, and it decreases when the source advances to the viewer. This attitude is rather strange and contradicts logic. Taking into consideration the case that is a little more difficult of a consistent plane wave infringing at a certain angle on an unending planar interaction between two materials—a regular and DNG material—to see how it might affect the results. We also should note that alternative techniques can also produce negative refraction, which does not necessarily result in refraction with a negative index. Without being able to determine an index of refraction, some EBG materials, for instance, can allow negative refraction (or, better yet, a negative direction of the beam), to arise from these [14, 15]. The exposition of all the aforementioned traits first gave off the impression to be intellectually fascinating but without a direct correlation to applicability. Also not known at the time were the practical applications for such unusual materials. Before talking about these two components, we'd want to examine some other aspects of metamaterials (real implementation and potential applications). First off, due to energy considerations, a material with constant and real values of negative of the property of dielectric medium and the permeability at any frequency cannot produce a macroscopic reaction [16]. In essence, if it were possible, the energy content of the electromagnetic field would've been negative, which is clearly against the principles. This simple presumption forces extra prohibitions on the frequency behavior of the permittivity and permeability functions and prohibits them from acting randomly. As a result, materials with negative constitutive characteristics must be dispersive. The interested party can learn more about this topic by doing further research [17]. Furthermore, causality prevents a material with frequency-dependent constitutive properties from being completely lossless. Kramers-Kronig correlations, which are originally derived by enforcing causation, relate the actual and fictitious components of a dispersion medium's fundamental features [17]. These first considerations lead to the realization in the process of modeling a metamaterial with negative values for its constitutive parameters, propagation and loses must be taken into consideration. From the standpoint of applicability, material losses and propagation also cause components based on metamaterials to operate narrowband and perform less efficiently. As a result, these two aspects in particular need continual attention. Materials that concurrently display negative real portions of permittivity and permeability are considered to be the first and most well-known category of metamaterials. These substances are also known as "left-handed substances," "dual"-negative (DNG) substances, "negative index substances," and "reverse (BW) media." These materials (i.e., ENG, DPS, DNG, and MNG) each exhibit unique characteristics, as seen in (Figure 11.5). These names have their genesis here.

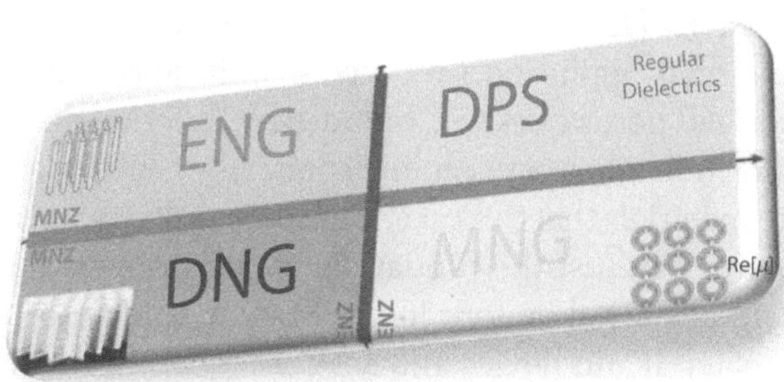

Figure 11.5 Different terms in the study of metamaterials.

11.4 Application of Metamaterials

Metamaterials provided alternate and untapped perspectives that provided exciting answers, quickly addressed challenging problems, and overcame the inherent constraints of both conventional and cutting-edge technologies because of their abnormal contact with the electromagnetic field. Since the first experimental confirmation of metamaterials [17], it is now common practice in electromagnetics to consider materials with extreme (i.e., very low and very high) permittivity and permeability, negative refractive index, negative permeability, and permittivity. Therefore, this became possible to revisit every notion from traditional electrodynamics and practical electromagnetics from a completely fresh perspective, leading to the discovery of novel electromagnetic material characteristics, the expansion and generalization of electromagnetic concepts, the circumvention of the inherent limits of traditional materials, and the enhancement of the effectiveness of existing devices through novel designs encompassing the new metamaterial properties. Following Martin's [13] findings and Veselago's presentation of them, Pendry [18] put out the idea for the ideal lens as the first and presumably easiest use of a DNG metamaterial. A flawless lens is composed of an incredibly long flat surface of a DNG substance bathed in a vacuum. Essential features of the DNG, that have a constant thickness d, are in opposition to that of the vacuum (relative permittivity and permeability which are both equal to –1). The anomalous effects at the two interfaces between the slab and the vacuum allow the DNG planar slab to create a perfect image of a source placed at the interval of time d/2 before the lens as well as d/2 from the lens. To create a perfect image, the dissimilating and disappearing portions of the source spectrum must point to the image's plane. Due to the interfacial events that occur at the contact points of the ideal lens, the subwavelength components of a picture can be precisely focused on the focal plane. Conventional lenses can only reconstruct the dissimilating portions of an image. Anyone with curiosity can obtain information regarding the potential uses of perfect lenses that operate at different frequencies [19–21]. Now, utilizing the ideal lens outcome, we will present a conventional, application-focused definition of metamaterials. The term "metamaterials" here refers to a brand-new functionalizing layer in the technology platform that lies between the concepts of "material" and "device." Because they are materials that can carry out some tasks independently, metamaterials are more than just the fundamental components for producing a device. Metamaterials are therefore located in the middle of the two basic conceptions of material and device and are consequently closer to the action of a device than either of the two. The DNG sheet of the perfect lens serves as a buffer, an activator, and an enabler of lens action in this manner. We are going to briefly discuss a few ideas that serve as the foundation for several uses of metamaterials across microwave frequencies to learn more about the physics underlying the operation of the ideal lens. Both an insulating medium's feature and a quantitative measure of a material's degree of magnetism exhibit a sign inversion while moving through two contact points between the DNG surface and vacuum. When the permittivity, as well as permeability substances sign flips, the tangential magnetic field's ratio of the interval and horizontal distances between the points changes across the interface. The slope of the tangential electric field also varies with the sign flip of the permeability, and it also changes with the sign flip of the permittivity.

Whereas, if multiple substances' variables are specially selected, a reverberating method that is restricted to the interaction can be thrilled, as is the situation at specific wavelengths at the interaction among a plasmonic material (such as a noble metal) and a frequent dielectric or vacuum. The higher intensity of the subwavelength constituents and consequently the ideal lens efficiency is caused by these interfacial resonances. Only permeabilities that are transverse-electric (TE), are significant. However, only permeabilities matter when the source field is transversely magnetic. Therefore, depending on the polarity, the best lens can be made from either ENG substance or MNG substance. This is a crucial subject since a real DNG perfect lens is challenging to develop and manufacture. In actuality, there exist similar difficulties as we moved beyond high-performing to real ones. The smallest characteristic that the lens can detect should ideally be contained by inclusion size and separations. Reductions are severe and could prevent perfect-lens functioning from occurring, logically speaking.

The fact that the interface resonance produced by the combination of a DPS and a DNG material is unaffected by the total thickness of the sets of two specimens is another noteworthy finding [37]. Because of this, it is feasible to drastically reduce the size of both samples of the two substances while still maintaining a pattern resonance in the interconnection. The 1D subwavelength resonator that was suggested as the first instrument to harness these physical phenomena blends a substance with a standard dielectric DNG that contains the same real values of the variables thought to be constitutive.

The longitudinal dimension of the resonator is hypothetically unlimited in the second instance. Naturally, if the appropriate modifications are taken into account for its practical implementation. The actual dimension of the additions and relative interval, considering Li *et al.* [38] as well as the level of the innovation into account, determine the smallest diameter that can be achieved. The intriguing finding is that by employing metamaterials, it is possible to make the longitudinal diameter of the resonators independent of wavelength and hence even lower than the conventional half-wavelength size. The breakthrough of phase compensation has the potential to minimize the size of any constituent and build innovative, more efficient technologies. It is impossible to generate this impact using standard substances. Unlike traditional equipment, whose least parameters are ascertained by the operating wavelength, the use of metamaterials has made it possible to create surface and metallic waveguides, dipoles, monopoles, microstrip antennas, absorbers, etc. with circumferential facets, lengths, and thicknesses that are considerably smaller than the correct wavelength [22–25].

Numerous guided-wave metamaterial techniques have been developed, such as strict coupled-line phase, coupling-level impedance defined by impulse, remarkable multiple surface compact formations, resonators with zeroth sequence having a uniform field dispersion, and dual-band and augmented components. Figure 11.6's illustration of the CRLH impedance coupler demonstrates how a new concept works.

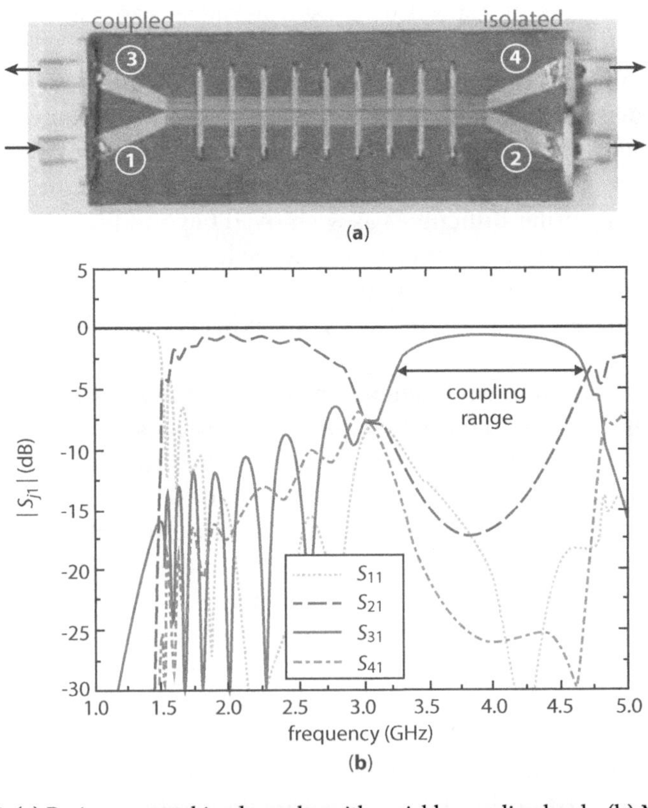

Figure 11.6 (a) Resistance combined coupler with variable coupling levels. (b) Microstrip-enabled shunt stub inductors including an interdigital series capacitor.

11.5 Single Negative Metamaterials

Noble metals and some dielectrics have been discovered to display a fascinating and anomalous electromagnetic reaction in the visual spectrum. These materials contain electric resonances in the microscopic molecule domain that generally results in a negative electric property of a dielectric medium for huge quantities medium at optical frequencies. Similar to something like this, further research logically suggested and constructed metamaterials by appropriately putting matched additions in a particular host material. These metamaterials had non-standard ratios of their inherent features. These metamaterials operate at lower frequencies to mimic the chemical processes that give rise to these aberrant resonances. The construction of these metamaterials and, consequently, their electromagnetic reactions, can be carried out with a rather broad degree of freedom in today's world thanks to advancements in modeling and fabrication technology. Metamaterials have drawn a great deal of interest due to their ability to manipulate a material's electromagnetic processes for a wide range of uses. It is interesting to note that recent advances in nanotechnology and molecular bioengineering have sparked debate among academics about the viability of attempting to bring back these metamaterial principles to the noticeable frequencies and about the ideal method for creating a synthetic arrangement of the molecule to make unnatural optical metamaterials to regulate their electromagnetic attributes at both the level of visible and infrared frequencies. Maxwell's equations, that is, the time-harmonic, with an $e^{j\omega t}$ time dependence with non-existence, inspired sources are represented simply as:

$$\nabla \text{ X } \mathbf{E} = j\omega\mu \text{ } \mathbf{H}; \nabla \text{ X H} = j\omega\mu \text{ } \mathbf{E}$$

If we presume that the magnetoelectric coupling is minimal and that the substance reaction is at minimum isotropic for a specific band of frequencies and polarities of the fluxes, where ε and μ denote the local magnetic permeability and permittivity, respectively, and are complex numbers when losses are being considered. The majority of materials in nature are described by these two quantities based on their electromagnetic wave interactions. These two quantities produce values that are following the constraints $\text{Re}[\varepsilon] \geq \varepsilon_0$, $\text{Re}\ [\mu]\ ] \geq \mu_0$, $\text{Im}\ [\varepsilon] < 0$, $\text{Im}\ [\mu] < 0$, which imply that the material is passive and has a greater refraction index than or equivalent to equal value of space that is free. However, assuming a preferred dispersion meets with the restrictions imposed by accident, the genuine segments of a passive permeability and permittivity can potentially yield any real value at a given frequency [26]. The majority of naturally occurring substances, sometimes known as "double-positive" (DPS) media, possess positive discrete components for both permittivity and permeability. When both of these properties are negative, however, the associated substances are referred to as "double-negative" media. These materials were experimentally realized by the UCSD group in 2000, but because of their anomalous wave refraction, the engineering and physics sectors have been very interested in them. There, media having a positive permeability but a real permittivity component with a negative value is distinguished and given the abbreviation "-negative (ENG)." They incorporate plasmonic and plasma materials, such as some polar dielectrics and noble metals, below their plasma frequencies.

The use of plasmas and other -negative materials has been investigated for a long time in a variety of fields. As a result of the advancement of nanotechnologies, there has been a rush in research into the plasmonic resonances related to subwavelength particles and surfaces. We have -negative (MNG) materials that can be synthesized from ferromagnetic materials or artificially synthesized by adding the proper inclusions to a host material. The synthetically produced, negative substances are the basic elements of double negative substances. The terms "negative" and "negative substances" can also be employed to describe single-negative media, much like double-negative substances.

11.6 Hyperbolic Metamaterials

Metamaterial advancements have occurred in the last 10 years for several applications, including faultless absorption, cloaking, and high-resolution imaging [27, 28]. Many new metamaterial classes with unusual electrical characteristics have been discovered, including those with negative index [29], optical magnetism [30], giant chirality [31], epsilon-near zero [32], and bianisotropy. The creation of a medium with a cell unit even significantly less than the wavelength is the core concept behind all metamaterials. The unit cell's unique resonances, which are governed by its structure, composition, and cell-to-cell interaction, produce a specified macroscopic electromagnetic reaction. Hyperbolic metamaterials are one type of synthetic media that has drawn a lot of interest [33–45]. They are known by this term because of the isofrequency curve's unusual shape, which is hyperbolic instead of circular like in regular dielectrics. They have received a lot of interest because of their merit figure, nonresonant reactivity that id broad-band, tunability of wavelength, a huge response with three dimensions, and excellent figure of merit despite their comparatively simple nanofabrication (Figure 11.7).

Numerous uses, such as negative index waveguides, are possible for hyperbolic metamaterials, subdiffraction photonic funnels [36], nanoscale resonators, and more [37]. The most feasible synthetic media for relatively close and transparent wavelength applications are hyperbolic metamaterials. hyperbolic metamaterials can be likened to uniaxial metacrystals since they have an extreme anisotropic dielectric field. Specifically, ε = diag [εxx, εyy, εzz] such that εxx = εyy and εzz. εxx <0. The frequency surface of exceptional waves in this medium is best studied to better understand the characteristics of hyperbolic metamaterials:

$$k_x^2 + k_y^2 / \varepsilon_{zz} + k_z^2 / \varepsilon_{xx} = (\omega / c)^2.$$

When εzz. εxx < 0, the aforementioned equation denotes a hyperbola, with a surface that is easily approachable as opposed to the confined spherical distribution in an isotropic material. The occurrence of propagating waves with this dispersion relation's immediate physical consequence is that they have.

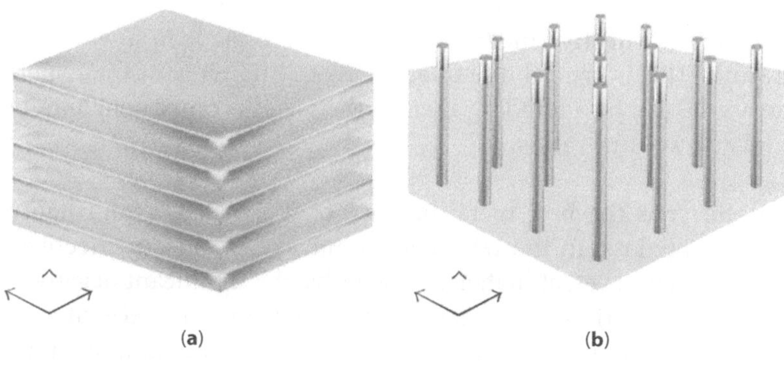

Figure 11.7 (a) Form of multi-layer hyperbolic metamaterials with subwavelength strata of metal and dielectric. (b) Nanorod of metal r in a dielectric medium as host.

High-k waves are large wavevectors but are impermanent in traditional media. Features of these high-k waves are related to many technological applications and processes that are physical in metamaterials that are termed hyperbolic. Moreover, it has been suggested by Jacob *et al.* [38] and Smolyaninov & Narimanov [39] that these states result in a wide disparity in the photonic density of states, the controlling physical principle of many phenomena like spontaneity and thermal. This supposition led to multiple computational [40, 41] and pragmatic [38] attempts to validate the notion and look into the recommended layout of this phenomenon. The emphasis of this discussion is on terminology that categorizes the two kinds of hyperbolic metamaterials within categories based on how many of its constituents contain negative charges in the dielectric tensor [42]. Remember that the outcome is an efficient metal that stops waves from passing through it because all three of its constituents are negative. Type I: If the tensor contains only one negative constituent, or zz 0, we referred to that metamaterials as type I hyperbolic metamaterials. Despite being challenging to accomplish in actuality, they have a low loss because they are primarily dielectric. Type II: Type II HMMs are described as consisting of two negative components, or $xx = yy < 0$, in the dielectric tensor. Because they are primarily metallic, they exhibit significant loss and impedance mismatches with the vacuum.

Two widely utilized techniques for new standard hyperbolic media are identified. The first has alternating sections of metal and dielectric, and its layer thicknesses are noticeably thinner than wavelengths. The second method employs metal nanorods and is based on a permeable anodic alumina dielectric host. These two methods provide the strongly anisotropic reaction that is predicted by Maxwell-effective Garnett's medium model [43, 44]. It's crucial to remember that the necessary reaction's non-resonant makeup allows effective medium theory to forecast it in a wide spectral bandwidth. This is crucial due to absorption. The reply of the hyperbolic metamaterial can be altered by altering the metal's component and dielectric's relative volume ratios.

1. Silver is the best metal because of its small UV and light supply loss. In the UV area, alumina (Al_2O_3) is an effective dielectric; yet, at higher wavelengths, the significant negative actual portion of the metallic dielectric constant necessitates the employment of high-index dielectrics. Titanium dioxide is a great choice because of its high pace since it just allows for the possibility of impedance that corresponds to a vacuum [45]. Near-IR: Conventional plasmonic metals like gold and silver do not make good choices for hyperbolic metamaterials at near-IR wavelengths. This is because metals exhibit strong reflection and increased mismatch with the media around them at frequencies reduced as compared to the plasma frequency. Considering their plasma velocity can be changed to be situated close to the -IR, the most advanced preferred

plasmonic metals depending on oxides and nitrides are perfect for usage in hyperbolic metamaterials [46].

2. Doped semiconductors are a potential source of metallic basic components for hyperbolic metamaterials at mid-infrared wavelengths. Employing phonon-polaritonic materials, like silicon carbide, whose Reststrahlen group resides in the mid-IR range, provides an alternative [47].

11.7 Classes of Metamaterials

The characteristics of the components that comprise a system determine how that system reacts to the existence of an electromagnetic field. By characterizing the macroscopic characteristics of these substances, permeability and permeability are employed to describe these qualities. The division of metamaterials into several categories is as followed; Figure 11.8 shows the division of the medium graphically.

Media that already have permittivity and permeability indices that are both above zero (e > 0, > 0) are referred to as double positive (DPS) media. The bulk of naturally produced media, including dielectrics, fits this definition. A medium is considered to be epsilon negative (ENG) if its permittivity is less than zero and its permeability is more than zero (e 0, > 0). Many plasmas have these characteristics in particular frequency ranges. Media that is Mu negative (MNG) are those that have permittivity values greater than 0 and permeability values less than 0 (e > 0, 0). In specific frequency domains, this property is found in some gyrotropic materials. Double Negative media are defined as having an attribute of a dielectric medium and a lead to a

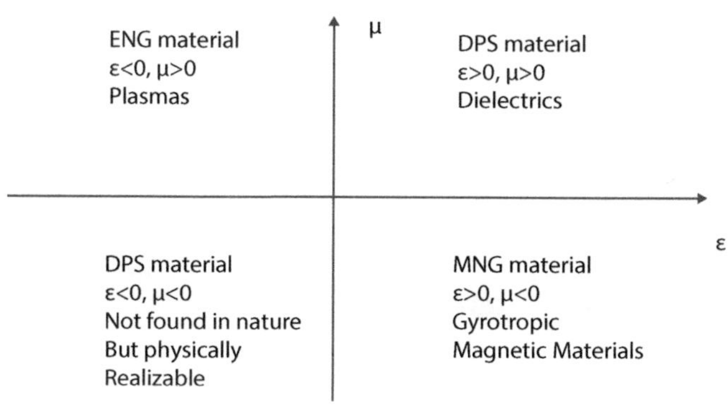

Figure 11.8 Categorization of metamaterials.

number of the intensity of magnetization both being less than zero (e 0, 0). This category of substances has been only illustrated by artificial constructs.

Various types of metamaterials exist, they are:

A. Electromagnetic metamaterials
   i. Double negative metamaterials
   ii. Single negative metamaterials
   iii. Electromagnetic bandgap metamaterials
   iv. Bi-isotropic and bianisotropic metamaterials
   v. Double positive medium
   vi. Terahertz metamaterials
   vii. Photonic metamaterials.
   viii. Tunable metamaterials

B. Nonlinear metamaterials.

C. Metamaterial absorber

11.8 Electromagnetic Metamaterials

The desire to build or produce something that is not present in the natural environment is one of the noteworthy elements of human civilizational development [48]. At first, people tried to reorganize their surroundings. As they continued to alter the form and composition of the things, separate and combine them, humanity eventually developed its first tools (Figure 11.9).

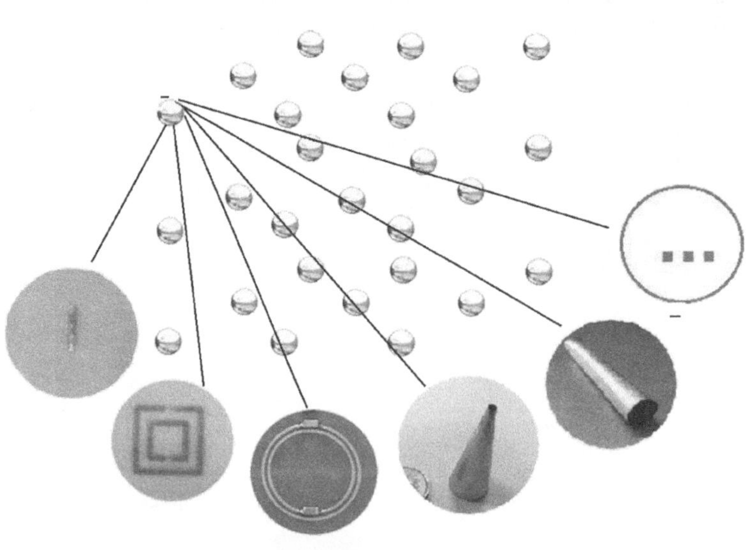

Figure 11.9 Metamaterial concept.

As beneficial compounds began to be recovered from natural resources, methods became more sophisticated to produce materials with specific qualities. As a result of the advancement of molecular and atomic-level manipulation techniques, the materials have since been synthesized. Moreover, altering specific structures can yield materials with amazing properties, giving rise to a variety of composites.

Each material is a composite in some way, even if its constituent atoms and molecules are. Thus, moving from materials to metamaterials requires little more than replacing the original concept's atoms with larger-scale structures. Metamaterials are a combination of synthetic structural elements intended to produce beneficial and unique features. Shivola *et al.* [49] provide a thorough analysis of the terms. As illustrated in Figure 11.8, most metamaterials are periodic lattices of elements that are similar, which are analogous to crystals. The components that make up a metamaterial are termed metal molecules or meta-atoms. Recognition of wire meshes, splitter ring resonators, conical Swiss rolls, and Swiss rolls through the meta-atoms are presented in Figure 11.8, in-depth examined in the study and further research of this intriguing domain permitting new forms of constitutive elements. Amorphous substances are made up of random or crooked arrangements of synthetic structural components.

$$a * \text{wavelength } (\lambda)$$

Here, λ represents the wavelength of the incident electromagnetic waves. Efficient permittivity and permeability are concepts that apply in this range.

The field of metamaterials has recently emerged in physics and electromagnetic (especially optics and photonics). They are used in microwave and optical processes, such as (brand-new beam steerers, modulators, lenses, couplers of microwave, and antenna randoms). Forms are a component of metamaterials. A metamaterial can affect electromagnetic waves by having elemental shapes that are smaller compared to the wavelength of electromagnetic waves it interfaces with. The greatest potential use of metamaterials is to build such a construct because no naturally produced material has a negative refractive index. Glass and water are just two examples of substances for use in optics that have good attributes for both permeability (μ) and permittivity (e). Nevertheless, a few metals exhibit negative (e) attributes at optical wavelengths, namely silver and gold. A substance with one (not both) negative e or ions cannot be penetrated by electromagnetic radiation. Refractive index (n) is commonly used, even though parameters e and μ adequately describe the optical properties of a transparent material.

Figure 11.8 depicts the refraction in left-handed metamaterial and regular material. Positive e and are present in all known transparent media that are not metamaterials. To determine whether n is affirmative, often, the square root is employed. Certainly, manufactured metamaterials do, nonetheless, exhibit e < 0 and 0 μ < values. N is authentic because the outcome of eμ is favorable. It is critical to determine n's negative square root in these circumstances. A physicist named Victor Veselago showed that some substances may transmit light. A reduced index of refraction results from the negative correlation between double negative metamaterial permeability and permittivity. Double negative metamaterials can also be called negative index metamaterials. The terms "left-handed media," "reversed-wave media," and "media having an index of refraction that is negative" are also used to describe double-negative metamaterials.

In single negative metamaterials, both permeability and permittivity are negatively associated. The ENG and MNG metamaterials indicated below are those specific ones. Several intriguing studies have integrated two SNG layers into a single metamaterial. This is effective in creating a different type of double negative metamaterial. To investigate wave reflection, a piece of ENG substance and a piece of MNG substance have also been connected. As a result, features like resonances, anomalous tunneling, opacity, and zero reflection were demonstrated. Akin to double negative metamaterials, single negative metamaterials are innately nonuniform, hence variations in frequency will affect their quantification and level of magnetization å, features of the dielectric medium μ, and refraction index n. Metamaterials with electromagnetic bandgaps control light transmission. This is accomplished by using either left-handed materials, a separate class, or photonic crystals (PC), a type of metamaterial (LHM) Both control and have an impact on the propagation of electromagnetic waves and are members of a new class of purposely constructed structures (light).

Double positive media (DPS), like naturally occurring dielectrics, are present in nature. Permittivity and permeability that is magnetic are both positive with the wave dispersion being put forward. Metamaterials are frequently categorized as double, single, or double positive substances based on the concept that they have distinct electrical and magnetic reactions, as represented by the variables. Nevertheless, magnetoelectric coupling happens frequently in electromagnetic metamaterials, where a magnetic field creates an electrical polarity and the electric field produces a magnetic polarity. These media are known as bi-isotropic media. Bi-anisotropic media are those that are anisotropic and also show magnetoelectric coupling.

11.9 Terahertz Metamaterials

Electromagnetic metamaterials have experienced unprecedented growth and attention as an outcome of the recent demonstration of unique phenomena such as perfect concentrating, negative refractive index, and invisible cloaking [28, 50]. In actuality, metamaterials provide an electromagnetic which has been demonstrated and proven impractical in substances that are found in nature. Metamaterials are manufactured items designed to respond to electromagnetic energy in a specific way. These substances usually possess subwavelength metallic additions incorporated inside or atop a substrate material.

Microwave radiation has been employed extensively in metamaterials research, in part because it is simple to create frameworks of subwavelength at these frequencies. Examples here include early microwave frequency depictions of media possessing a refractive index that is negatively made of metamaterial constituents that are negative quantifiable measures of the extent of magnetization (μ) and negative properties of a dielectric medium (ε). As a result, significant attempts have been made to expand the responsiveness of metamaterials concerning terahertz (1 THz=1012 Hz) [55], near-infrared, and visible frequencies [51–53]. Therefore, these designer materials have been proven to span a sizable percentage of the electromagnetic spectrum. They also expand and augment natural material response. Despite optical wavelengths being the focus of most metamaterials research, some work has focused on THz frequencies [54, 55]. Because of the relatively unusual reaction from naturally existing materials and the conspicuous lack of high-power areas, good detectors, as well as traditional device elements in this domain, there are numerous potential applications for metamaterials.

The frequency response range of metamaterials, which exhibit resonant electromagnetic activity, is rigidly constrained by their geometry and property of a dielectric medium component. A wider bandwidth is needed to be used in other ways while a narrower bandwidth is preferred for others, which poses different difficulties. For several possible uses, it could also be advantageous to build metamaterials that exhibit external monitoring active, dynamic, and/or changeable responses. Thus, due to the scarcity of materials that occur naturally and the prospect of huge rewards, the frequency regime of terahertz is a rich field for research into future metamaterial-based devices. Additionally, subwavelength elements can be created using normal photolithography at terahertz wavelengths (1 THz 300 m).

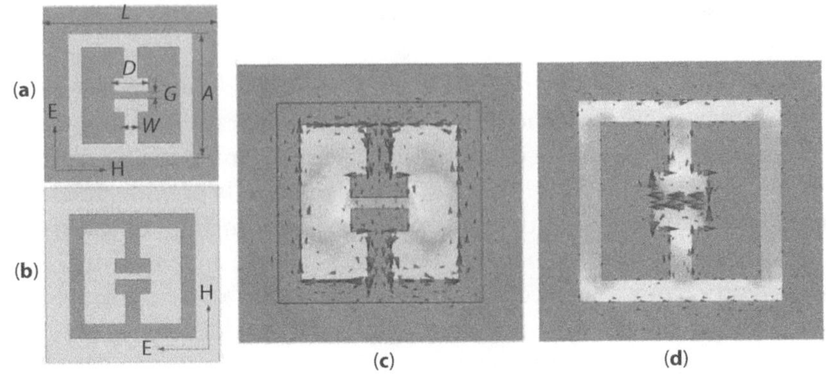

Figure 11.10 Architectures of cell's unit of the conventional (a) and (b); typical electric field of the conventional metamaterials, respectively (c) and (d) exhibits 11(a) and (b) show the cell unit designs of the basic and supplemental metamaterials, accordingly. Figures (c) and (d) show, correspondingly, the passage of electricity in the interface and electric field indicative of traditional and analogous metamaterials (d). A = 36 μm, G = 2 μm, D = 10 μm, W = 4 μm, and the lattice parameter L = 50 μm are the parameters of the structure.

By using divided ring resonators, metamaterials can be made to provide an entirely electric resonant output, analogous to magnetic resonance [50, 56, 57]. Figure 11.10 illustrates a flat metamaterial unit cell comprised of electric divided ring resonators (a). Two separated ring resonators are placed face-to-face in a harmonious arrangement across the building.

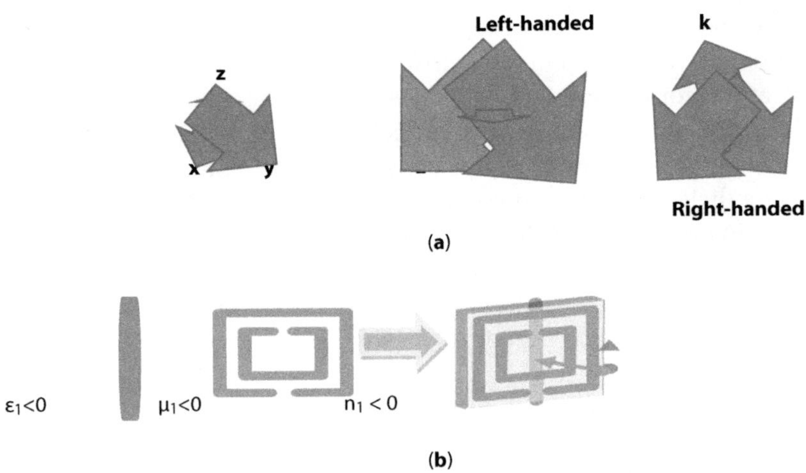

Figure 11.11 Illustration to establish a key unit of magnetic MM having a reflective index that is negative.

The counter-circulating currents, which are fuelled by the electric field of the conveyed electromagnetic waves, cancel out the opposing magnetic dipoles. The exclusively electric resonant reaction is the result of a numerical model employing finite elements, and it is shown in Figure 11.10(c). The observations of the architecture depicted in Figure 11.10 are chosen to result in a resonant response at THz frequencies. They are organized in an asymmetrical square pattern on a semi-insulating gallium arsenide.

11.10 Photonic Metamaterials

A purposefully manufactured subwavelength, periodic construction intended to mix with optical frequencies is known as a photonic metamaterial. It can be distinguished from the difference in the energy between two allowed ranges in a solid configuration thanks to a metamaterial's photonic subwavelength duration. A new class of unnatural electromagnetic materials with subwavelength periodicity is known as metamaterials. Around 2000, research on metamaterials in the microwave frequency spectrum began. One characteristic of MMs is that they exhibit peculiar EM approaches, which are infrequently, if ever, observed in currently known hard substances. The main goal was to achieve a strong magnetic response at frequencies greater than microwave from the beginning.

The condensed explanation that explained an increase in the amplitude of an oscillation in man-made metallic roll frameworks as a magnetic reaction was put forth in [58] which encouraged the creation of magnetic frameworks for microwaves. The manufactured roll architecture functions essentially as an inductance-capacitance circuit for electricity; the resonance is understood to be a magnetic reaction and is characterized by relative permeability. The early metamaterials scheme was straightforward. In the event wherein an individual could manufacture isotropic unnatural constituents displaying both electric and magnetic reactions such as $\varepsilon = -1$ (ε: relative permittivity) and $\mu = -1$, then the refractive index n would equal -1 because of the relation $n = -\sqrt{\varepsilon\mu}$. It is worth noting that, for $\varepsilon < 0$ and $\mu < 0$, the EM waves are left-handed as compared to the common right-handed EM waves. To simplify this, the equations formulated by Maxwell must be transformed.

A pointed illustration of a metamaterial with a negative refractive index has been explained. Figure 11.11 shows the basic building block of a metamaterial with an actual negative refractive index (b). The negative characteristic of a dielectric medium is represented by a copper (Cu) bar, and the negative quantitative assessment of the extent of magnetization in the microwave medium is represented by a divided (split) ring resonator. The components were set up like an electric circuit. In the experiment, the two parts were divided by a glass board that was 250μm thick and formed a 5 x 5 mm^2 unit cell. In the strictest sense, for the incident polarisation equally distant to the metal bar in Figure 11.11(b), the real components of ε and μ, denoted as ε1 and μ1, accordingly, are negative. a regular arrangement of meta-atoms operating at 10.5 GHz and made up of an MM with a negative refractive index (n1 < 0) (3 cm wavelength).

The finding of negative refraction sparked a thorough investigation into metamaterials. Increase frequency ranges, such as THz and optical frequencies, have now been added to the functional framework (PHz). Despite the reality that the transition in working frequencies forward into optical frequencies was preceded by a significant fall in|Re(μ)|, split ring resonators were examined in-depth as a constituent with a magnetic reaction that permitted an efficient negative at about 0.1 PHz. Scaled-down split ring resonator's ability to maintain the magnetic response at optical frequencies has been studied theoretically and numerically. The research all came to the same conclusion: at optical frequencies, the magnetic response associated with negativity decreases and vanishes within 0.3 PHz (1μm wavelength).

11.11 Tunable Metamaterial

Since the initial work by Smith *et al.* [59], proving the existence of such media, metamaterials, particularly negative index metamaterials [60], has received a lot of interest. Most recently, studies have concentrated on a variety of design methodologies and cutting-edge electromagnetic characteristics [61, 62] and applications [63]. Nevertheless, owing to the inherent and constrained resonance frequency range of the preponderance of metamaterials, it is unsuitable for use in certain significant operational frequency areas. Metamaterials, particularly negative index metamaterials [59], have subsequently received a great deal of attention following the first Smith *et al.* paper [59], that depicted the presence of these media. In the most recent studies, the diverse design methodologies [59, 64]; the unique electromagnetic properties, and the applications have been the main topics of study. Nonetheless, some vast operating frequency areas cannot be exploited due to the intrinsic and restricted resonance frequency band of most metamaterials.

11.12 Types of Tunable Metamaterials

a) Liquid Crystals Based Tunable Metamaterials
b) The adjustable metamaterials made up of liquid crystals have been the subject of much investigation. Liquid crystals, that have a high optical anisotropy and are incredibly susceptible to outside effects, have been employed to extensively tune the energy difference between two allowable ranges of electron energy of photonic crystals at optical frequencies [65]. Liquid crystals can therefore be utilized to regulate the reversed electromagnetic properties by dynamically controlling the permittivity of the metamaterials. A theoretical investigation of liquid crystals spread throughout nanospheres to create programmable metamaterial was published by [66]. This metamaterial allows tuning of the refractive indices from negative via zero to good values by tuning the permittivity of the liquid crystals. The relative permittivity of liquid crystals isn't resonance dependent, therefore it relies completely on the director axis's inclination angles concerning the optical wave vector. Tunability can therefore be obtained by adding various fields such as magnetic and ac electric, as w, well as additional polarized optical area, or just by changing the sample's temperature. Then Zhao *et al.* theoretically examined the nematic liquid crystals' negative group refraction [67].
c) Its refraction can be altered by temperature or electrical current. The customizable negative turning of nematic liquid crystals was also demonstrated experimentally by measuring the refraction angle of the light beam through the prism-shaped apparatus and adjusting the magnetic field [68]. Conventional metamaterials can be given tunable properties by incorporating liquid crystals because their permittivity is fairly flexible. Werner *et al.* performed a quantitative analysis to assess the adjustable refraction of a near-infrared metamaterial with liquid crystals as the superstrate and substrate [69]. Such metamaterial's refraction was modified from negative to positive by adjusting the liquid crystals' permittivity. A tunable negative permeability metamaterial made of a periodic arrangement of resonators with split rings infused with nematic liquid crystals was also statistically investigated by Zhao *et al.* [70]. An applied electric field that affects the penetrated nematic liquid crystals can ceaselessly and irreversibly change the resonance dip of a negative quantitative measure of the degree of magnetization metamaterial. Several newly invented tunable liquid crystal metamaterials, such as

divided ring, omega-type, and fishnet resonators, have also been mathematically and empirically proved by changing the various parameters (i.e., electric & magnetic fields, and temperature) [71, 72]. These tunable metamaterials are simple to realize. Merely adjusting the parameters will change the tunable qualities. Additionally, there are some drawbacks, such as the limited adjustable range.

a) Tunable Metamaterials that Ferrites related
b) The tunable features are also attainable in ferrites and wire-based metamaterials. This architectural idea was first presented by Dewar [73]. He showed a prototype design that put the wires covered in dielectric materials into the ferrite hosts while hypothetically investigating the functional permeability and permittivity of the composite media. He also discussed ways to reduce losses and improve negative refraction properties. The ferrites produced the negative effective quantified measure of the extent of magnetization when a magnetic bias was added. As a result, it stands to reason that altering the magnetic bias would allow for tuning of the negative permeability band. The improved model was provided by Cai *et al.*, Rachford *et al.*, and our study team, who also numerically illustrated the negative propagation and transmittance properties [74, 75].
c) Using experiments, it was proven that the characteristics of negative transmittance, negative refractions, and transmittance exist [76]. In a wide frequency range, we also looked into favorable refraction properties. One can also add the ferrites to the standard resonator metamaterials to achieve harmonious quality since the tunable properties of the ferrites are easily controlled by altering the applied magnetic bias. The adjustable properties of the ferrites-based negative permeability metamaterials and left-handed metamaterials, accordingly, were examined and empirically proven by [77]. Conversely, adjustable dual-band and multi-band metamaterials, which have been partially examined herein, can also be created with the addition of ferrites to ordinary resonator metamaterials. It is also easy to create these different tunable metamaterials. By adjusting the magnetic bias over a wide frequency range, the tunability was regulated. The large magnetic bias of several thousand is necessary for typical ferrites to have tunability, though. Future research should focus on identifying the unique ferrites that only require a very tiny magnetic bias to achieve tunable negative permeability properties in the GHz and THz ranges.

d) Varactor-loaded Tunable Metamaterials

e) Gil *et al.* and Vélez *et al.* [78] were the first to report adjustable metamaterials packed with varactor. They suggested that microstrip lines connected to varactor-packed (loaded) split-ring resonators may produce metamaterial transmission lines with tuning capabilities, the tunability of which could be managed by incorporating a variable voltage source. The architecture and practical demonstration of a sequence of left-handed metamaterials loaded with varactors and adjustable negative permeability metamaterials. Wang *et al.* proposed a microwave varactor-loaded paired single side with S-ring resonator-based adjustable metamaterial [79]. By analyzing the refractive angle of a prism-like sample with various voltage biases, the adjustable negative refractions were empirically shown. A split ring resonator metamaterial occupied with a capacitor was demonstrated by Aydin *et al.* [80]. The disparity zone between the overlapping circles, the outside fracture area, and the interior division section are all possible locations for the capacitor. By adjusting the capacitor's values under the three situations, the resonant frequencies of the magnet can be adjusted. A split rang resonator metamaterial loaded with a thin film capacitor made of barium strontium titanate has also been demonstrated by Hand *et al.* [81]. By modifying the biases of the voltage supplied to the BST film capacitor, tunability can be achieved. Due to their ease of integration into microwave circuits to accomplish results like filters, antennas, and nonlinear features, these types of adjustable metamaterial have enormous microwave engineering applications [82].

f) Many adjustable metamaterials that are realized by multiple elements have recently emerged. For instance, utilizing the constant temperature superconductor, Ricci *et al.* suggested and practically showed an adjustable superconducting metamaterial [83]. The tunability can be obtained by modifying the dc magnetic field and rf magnetic field. The components of the framework can also be controlled via MEMS technology [84]. By altering the orientations of the magnetic or electric resonators, Tao *et al.* illustrated the THz frequency adjustable metamaterial. The framework of the resonators-based adjustable negative permeability metamaterials is also actually unveiled at microwave and THz frequencies [85]. As the regular deflections of strata grew, the magnetic resonances either were blue-shifted or red-shifted, leading to the tunable attributes.

11.13 Nonlinear Metamaterials

Since the study of metamaterials has advanced so rapidly in the past few years, it is not strange that the concept of nonlinear metamaterials was introduced as soon as the foundations for macroscopic specification were established and close comparison to optical crystals was discovered. The first publications [86], on the emerging field, provide two methods for dealing with nonlinear effects in metamaterials. The two strategies are similar in concept, as we shall see below, but various analytical methodologies were created, each of which proved useful based on the issue at hand. It is evident that nonlinear optics, on the one hand, and microwave engineering with nonlinear components, on the other, have both been successful over more than 50 years. Metamaterials have since made it possible to effectively synthesize the knowledge developed in both research areas. The topic afterward caught the interest of various research groups [87, 88], and we will now go into depth about some of the unique research accomplishments in these and numerous more publications. The many different methodologies used to create metamaterials with nonlinear responses can be conceptually grouped into three different categories, which are schematically presented in Figure 11.12.

The first one stems immediately from the comparison to optical crystals and is architecturally the simplest one. Numerous intrinsic nonlinearities in crystals—mostly at the atomic level—provide a nonlinear response by

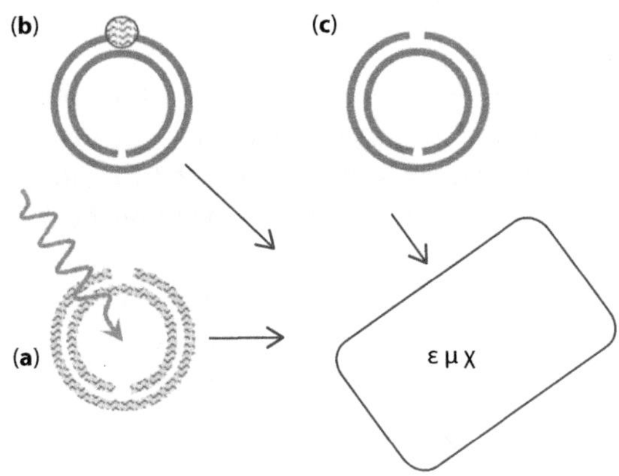

Figure 11.12 Different microscopic metaconstituent perimeters, such as (a) implantation, (b) saturation, and (c) severe concentration, are described theoretically concerning metamacroscopic characteristics.

default at modest levels. the structural aspects of metamaterials. When artificially assembling the modules, the reaction is virtually linear as much as the functional contour has linear parameters. Thus, adding a nonlinear element (such as a diode) to the linear contour, that is, inserting it into the structural element, is a very natural technique to generate nonlinearity, and this is what was proposed when the topic was first introduced. Currently, we refer to this as an insertion approach (Figure 11.12a).

Nevertheless, the fundamental reasoning behind the initial strategy was not particularly novel. Kalinin and Shtykov conceived [89], as an unstructured medium constructed of randomly distributed dipoles, each packed with a diode, long before met materials came into play. The primary goal was to use a wave-only approach for phase coupling at microwave frequencies. However, the scientists discovered that dissipation suppresses the rising third-order nonlinear susceptibility, leaving the effectiveness of that specific design in question at the time. The primary technique, which is rather natural as well, is to combine the host medium's nonlinear qualities with the metamaterial components' linear response. This causes the fields, which are resonantly amplified within the metamaterial elements, to become nonlinearly linked. Now, we refer to that as an immersion method (Figure 11.12b). This different suggestion soon after sparked the quick growth of this study direction.

Since individual elements can be deliberately made and it is feasible to accomplish extremely inhomogeneous electromagnetic field dispersion with the structural unit, it is particularly possible to accomplish this approach's fundamental idea, which goes back to, which essentially implies utilizing favorable characteristics of metamaterials, so there can be a noticeable improvement in the fields. Whenever an outside nonlinear medium is injected into the areas of the augmented field, the outcome is an overarching nonlinear reaction. Despite how dissimilar the two strategies might seem; they have a lot in common. As was just indicated, the nonlinear response of the nonlinear host only happens in a few places where fields are amplified when a metamaterial element is immersed into it.

The nonlinearity in the instance of a ring resonator, for instance, is practically identical to if it were generated by nonlinear equipment placed there in the shape since the electric field can be several orders of magnitude stronger inside the disparity. This similarity continues as long as the wavelength is much larger than the size of the element or as long as the metamaterial hypothesis is valid. Thus, the fundamental phenomenological descriptions for both approaches can be the same, but additional features must be considered for a more detailed study of particular implementations for different nonlinear phenomena. Instead, the area of implementation is where the insertion and immersion roads diverge.

With insertion, any available gadget can be fitted at any point in a very intricate design. The difficulty of this approach rises as the target frequency rises and the size of the pieces must decrease. Small elements are more useful for applications of any form of absorption, like producing nanostructures on a nonlinear substrate. Absorption may be efficient for infrared and optical wavelengths, although insertion is often preferable for the microwave frequency range. We really should briefly discuss the third technique (Figure 11.12c), which is less theoretically advanced due to the numerous challenges posed by the diverse consequences of the different scales and sources at play. According to this reasoning, the inherent nonlinearity of the metal surface, border dispersion events, and even quantum effects may supplement the processes outlined above when the amplitudes and frequencies of the input waves are high enough. In the meantime, the importance of any detailed model in the sense defined above diminishes. In reality, this situation typically indicates that the element size is just marginally smaller than the wavelength. As a consequence, it is challenging to provide a clear theory, certainly not one that is widely appropriate, for such circumstances. Rather, study initiatives use an empirical approach that entails failing and trying again. This area of study has primarily been empirical up to this point, and its roots can be located in the metamaterial environment.

11.14 Absorber of Metamaterial

The emerging area of electromagnetic metamaterials has given rise to novel phenomena like the negative index of refraction, as well as technological innovations like an electromagnetic cloak [50]. The capacity of metamaterials to provide independent, customized electric, and magnetic reactions to incident radiation is key to the realization of such features. Additionally, geometrically scalable, electromagnetic metamaterials enable operation over a sizeable chunk of the electromagnetic spectrum. To date, metamaterials have been demonstrated in every spectral region that is significant to technology, including radio, microwave, near optical, etc. These designer electromagnetic materials offer considerable potential for use in the future and provide an excellent platform for the study of novel spontaneous physical processes. Metamaterials can be identified as an effective medium, by having a multiplicity of electric permittivity $(\tilde{\omega}) = 1 + I2)$ and magnetic measurements of the degree of magnetization $(\tilde{\mu}(\omega) = \mu 1 + i\mu 2)$. However, the frequently disregarded loss elements of the optical constants (2 and μ2) also hold great promise for the development of novel and beneficial materials. They can be altered, for instance, to provide a high absorber. The incident electric and magnetic fields can be absorbed, and the electric and magnetic resonances can be controlled independently. Furthermore, impedance matching and μ can be used to reduce incidence by impedance matching a metamaterial to an unoccupied vacuum.

The cell unit of the absorber consisted of two distinct metallic parts. The electric coupling was given by the electric ring resonator, which was made in a manner reminiscent of that proposed by Padilla *et al.* [56]. This component was made up of two conventional split-ring resonators joined by an inductive ring that ran parallel to the split wire. The limitations of straight wire media [56], led us to employ this design rather than a traditional split-wire design; To increase inductance, new wires can only be added to each unit cell in a split-wire system.

The flux generated by revolving charges had to be precisely vertical to the transmission vector to couple to the incident H-field since magnetic coupling necessitated a more complex configuration. This structure is analogous to the quote-on-quote "fishnet" and coupled nanorod arrangements in that they generate two antiparallel currents in conducting portions, which result in a magnetic response [86], We subsequently turned the magnetic response by tuning the geometry of the severed wire and the spacing between it and the electric resonator. We may independently tune each resonance by decoupling and by adjusting the magnetic coupling without modifying the electric resonator's geometry.

11.15 Acoustic Metamaterials

The primary characteristic that distinguishes a metamaterial from other artificial materials is that it solves long-standing acoustical engineering problems by producing effects that are not present in nature. These are only a few illustrations of the issues that AMM research aims to solve. The seeming rejection of the intuitive laws of physics, which frequently demand bizarre ideas like negative density and negative compressibility, is a recurring motif in the various AMM devices. Acoustic lens designs that are beyond the diffraction limit are made possible by behavior, such as negative refraction, which is caused by negative effective characteristics. Study on acoustic MM was initially spurred by concurrent advancements in electromagnetics, like the case of negative refraction and cloaking. The early researchers who looked into these issues discovered right away that the materials at their disposal lacked the characteristics required for cloaking or negative refraction. The problem may have been solved by simply developing novel substances. Despite challenges, the development of novel materials has been and is the primary motivation for research into acoustic metamaterials. Due to this, innovations for overcoming acoustic metamaterial difficulties that could not have been predicted by electromagnetics have emerged, such as Penta mode materials.

Constant technological advancements, particularly in the areas of computer simulation and additive manufacturing, have greatly eased the difficulty of creating novel materials. Thanks to these technologies and innovative suggestions from the acoustical research community, acoustic metamaterials have rapidly advanced over the last 10 years; a few of these breakthroughs are discussed herewith. Dynamic material characteristics are of interest and might lead to more unusual behavior for acoustical applications. An illustration of this is the concentrating and beam formation from the dolphin's fatty lobe known as the "melon" [90]. The static qualities that we learn about in introductory mechanics classes are not the same as the dynamic properties. Numerous acoustic metamaterial equipment is predicated on negative acoustic compactness and/or deformation. The audible pace, or more particularly the phase speed, is equivalent to the square root of the bulk modulus divided by density. If the other variable is negative, the phase speed is arbitrary, the propagation is stopped, and exponential degradation takes place. When the two acoustic variables are negative, the phase speed, which is once again a real-valued number, denotes vibration. There is a catch, though: the phase and energy velocities are going in separate directions. Acoustic metamaterials can generate "bizarre" occurrences like the below-discussed negative refraction thanks to this reaction. What mechanisms result in undesirable characteristics? A tuned vibration absorber that oscillates in phase or out of phase with the force depending on whether the drive frequency is beneath or higher than the resonance frequency is an example of a device that exhibits negative impedance. Hence, negative inertia can be conceptualized as an out-of-phase time the harmonic motion of a moving mass; for more information, visit http://bit.do/negm.

The first and presumably most well-known acoustic metamaterials are a composite that contained microstructural elements made of heavy masses encircled by a soft rubber annulus and organized sporadically in a three-dimensional solid matrix. It was created to isolate low-frequency sound much more effectively than the classical mass law. Tensioned membranes have been employed as the moving mass in numerous later air-based technologies that utilize negative inertia. Although vibrations frequently have negative inertia, it is less clear how to do so for compaction (inverse of the bulk modulus). Remember that positive pressure causes a reduction in the volume of materials that are present in nature. As a result, positive expansion from the building stress is required for negative compaction to exist. When the hollow volume acts as a vibrant volume source beyond its resonance frequency, this is achievable in immediate contact with the Helmholtz resonator, a well-known acoustical constituent. These resonators can be combined to induce one or both of their undesirable characteristics across a narrow frequency range.

Examine the easiest illustration of an acoustic duct with overlapping spring masses and Helmholtz resonators, as shown in Figure 11.13, to comprehend how a double-negative AMM operates. According to PM = ZMUM and PH = ZHUH, the resonators are described by impedances that connect acoustic pressure (p) with volumetric flow velocity (U). By first taking into account the reflection and propagation of an isolated element at a specific frequency, the effective acoustic characteristics may be computed (Figure 11.13). The 2 × 2 matrix that defines how waves travel through a cycle of the line of propagation is then created by adding the reflection/transmission coefficients.

The second principle that is driving the advancement of acoustic metamaterials is transformation acoustics, which is the basis for systems that use acoustic cloaking and perhaps other sorts of wave bending. A translation, or alteration in dimensions, has an impact on the acoustic equations, and the variables of the new equations correlate to the physical characteristics in the altered region. This is the essential principle of transformation acoustics. For example, a circular acoustic cloak increases the initial core volume by shrinking a large annular region of fluid. If the tighter annulus, designated as the "cloak," accurately replicates the acoustic equations in the

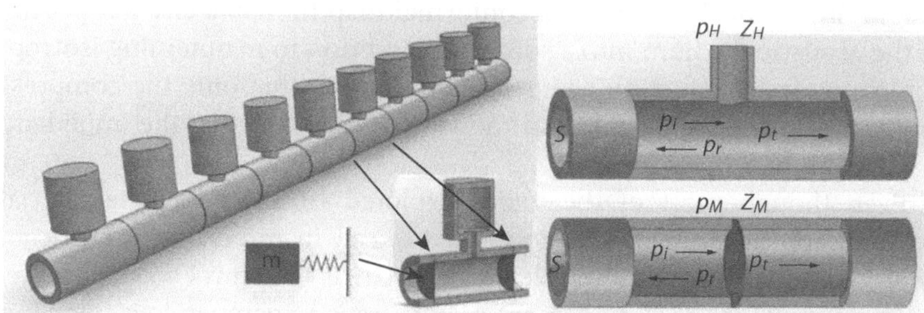

Figure 11.13 An acoustic duct with spontaneous membrane masses and Helmholtz resonators.

original annulus, then anything situated in the enormous center volume, along with the "cloak," distributes sound waves like a much smaller item from the viewpoint of an outside observer.

The most important criterion is that the cloaking material propagates sound more rapidly in the azimuthal path than in the surrounding liquid, while at the same time delaying transmission in the radial direction compared to the background value because of the reduced thickness. The "hidden" object must be surrounded by the acoustic energy, which the cloaking material must direct. As a result, the speed of the sound relies on the direction in which it is propagating at any given place within the cloak. This is not the case with typical fluids. However, systems that consist of layers of several fluids organized like a sandwich are capable of exhibiting acoustic anisotropy. If the fluid densities are very different and the layer thickness is subwavelength, significant inertial anisotropy can be generated. The spherical cloak is an extreme case since it calls for degrees of acoustic anisotropy that are not now feasible. The carpet cloak, also known as the acoustic ground cloak, can be produced with only a modest amount of anisotropy. To obtain the ideal characteristics, Zigoneanu *et al.* [91].

Even though cloaking is the most intriguing example of transformation acoustics, the fundamental concept of coordinate constant also enables us to see traditional effects like focussing in a new way. The short wavelength estimates of ray acoustics, for instance, are the foundation of the convex lens. If the lens itself modifies the wave equation following transformation acoustics so that frequency has no bearing on focusing, a better image is guaranteed. As a result, transformation acoustics creates new opportunities for passive acoustic devices to operate more effectively. Using a monopole cylindrical source, as an illustration, the circle-to-square lens transforms the source's center radiation into a fourfold plane-wave radiation pattern. A plane-wave incident from one of the four directions, on the other hand, will concentrate at the center. transformation acoustics accomplishes this by utilizing a conformal map to map a circular portion of the acoustic medium into a square. In addition to maintaining isotropy, conformal transformation acoustics is unique in that only the compressibility is changed (one could assume via ray acoustics that the impedance is constant, but this is incorrect).

Even though most of the topics explored in acoustic metamaterials research appear strange, it's important to keep in mind that acousticians have traditionally used subwavelength constructs to control acoustic waves. Examples include the development of bass traps to dampen low-frequency sound in concert halls, Helmholtz resonators to control the spread of acoustic waves in ducts, and contrast agents to enhance the capabilities of ultrasonic imaging. The human cochlea, a subwavelength spatial filter, comes to mind as one of nature's beautiful models. Even though these examples are rarely used in acoustic metamaterials, they easily qualify as acoustic metamaterials precursors since they use the same physical principles as individual acoustic metamaterials elements. It is safe to say that many acoustic metamaterial concepts have been waiting for us to come upon them by accident. We are lucky to have at last understood and realized the benefits that correctly constructed subwavelength structures may provide to the acoustics community.

References

1. Pendry, J.B., Holden, A.J., Stewart, W.J., Youngs, I., Extremely low frequency plasmons in metallic mesostructures. *Phys. Rev. Lett.*, 76, 4773, 1996.
2. Ziolkowski, R.W. and Engheta, N., *Metamaterials: Physics and engineering explorations*, pp. 211–221, IEEE Press, Piscataway, New Jersey, 2006.
3. Belov, P.A. and Hao, Y., Subwavelength imaging at optical frequencies using a transmission device formed by a periodic layered metal-dielectric structure operating in the canalization regime. *Phys. Rev. B*, 73, 113110, 2006.
5. Slyusar, V.I., Metamaterials on antenna solutions (PDF), in: *7th International Conference on Antenna Theory and Techniques ICATT*, Lviv, Ukraine, pp. 19–24, 2009.
6. Jacob, Z., Smolyaninov, I.I., Narimanov, E.E., Broadband purcell effect: Radiative decay engineering with metamaterials. *Appl. Phys. Lett.*, 100, 181105, 2012.
7. Pianelli, A., Kowerdziej, R., Dudek, M., Sielezin, K., Parka, J., Graphene-based hyperbolic metamaterial as a switchable reflection modulator. *Opt. Express*, 28, 6708, 2020.
8. Radino, G., Ultramicroscopic observations on the elastic tissue of the human aorta. *Minerva Cardioangiol.*, 7, 395, 1959.
9. Rill, M.S., Kriegler, C.E., Thiel, M., von Freymann, G., Linden, S., Wegener, M., Negative-index bianisotropic photonic metamaterial fabricated by direct laser writing and silver shadow evaporation. *Opt. Lett.*, 34, 1, 2009.
10. Bilotti, F. and Vegni, L., Design of metamaterial-based resonant microwave absorbers with reduced thickness and absence of a metallic backing, in: *Metamaterials and Plasmonics: Fundamentals, Modelling, Applications*, S. Zhouhdi, A. Sihvola, A.P. Vinogradov (Eds.), pp. 165–174, Springer Netherlands, Dordrecht, 2008.
11. Lakhtakia, A., Weiglhofer, W.S., Hodgkinson, I.J., *Complex mediums II: Beyond linear isotropic dielectrics*, SPIE Press, Bellingham, WA, 2001.
12. Zouhdi, S., Sihvola, S., Arsalane, A., *Advances in electromagnetics of complex media and metamaterials*, Kluwer Academic Publisher, Dordrecht, The Netherlands, 2002.

13. Martin, F., Special issue on microwave metamaterials: Theory, fabrication and applications. *EuMA Proc.*, 2, 101–115, 2006.
14. Luo, S.G., Johnson, J.D., Joannopolous, J.B., Negative refraction without negative index in metallic photonic crystals. *Opt. Express*, 11, 746–754, 2003.
15. Sudhakaran, S., Hao, Y., Parini, C.G., An enhanced prediction of negative refraction from EBG-like structures. *Microw. Opt. Technol. Lett.*, 41, 258–261, 2004.
16. Tretyakov, S. and Maslovski, S., Veselago materials: What is possible and impossible about the dispersion of the constitutive parameters. *IEEE Antennas Propag. Mag.*, 49, 37–43, 2007.
17. Landau, L.D. and Lifshitz, E.M., *Electrodynamics of continuous media (course of theoretical physics)*, Butterworth-Heinemann, Burlington, USA, 1984.
18. Shelby, R.A., Smith, D.R., Schultz, S., Experimental verification of a negative index of refraction. *Science*, 292, 77–79, 2001.
19. Fang, N., Sub-diffraction limited optical imaging with a silver superlens. *Science*, 308, 534, 2005.
20. Belov, P., Hao, Y., Sudhakaran, S., Subwavelength microwave imaging using an array of parallel conducting wires as a lens. *Phys. Rev. B*, 73, 033108, 2006.
21. Zhang, S., Yin, L., Fang, N., Focusing ultrasound with an acoustic metamaterial network. *Phys. Rev. Lett.*, 102, 194301, 2009.
22. Ziolkowski, R.W. and Engheta, N., *Metamaterials: Physics and engineering explorations*, John Wiley and Sons, Piscataway, New Jersey, 2006.
23. Erentok, R., Metamaterial-inspired efficient electrically small antennas. *IEEE Trans. Antennas Propag.*, 56, 691–707, 2008.
24. Ziolkowski, R.W. and Kipple, A.D., Application of double negative materials to increase the power radiated by electrically small antennas. *IEEE Trans. Antennas Propag.*, 52, 2626–2640, 2003.
25. Bilotti, F., Anomalous properties of scattering from cavities partially loaded with double-negative or single-negative metamaterials. *Prog. Electromagn. Res.*, 51, 49–63, 2005.
26. Shelby, R.A., Smith, D.R., Schultz, S., Experimental verification of a negative index of refraction. *Science*, 292, 77–79, 2001.
27. Landy, N.I., Perfect metamaterial absorber. *Phys. Rev. Lett.*, 100, 207402, 2008.
28. Pendry, J.B., Schurig, D., Smith, D.R., Controlling electromagnetic fields. *Science*, 312, 1780, 2006.
29. Cai, W. and Shalaev, V., *Optical metamaterials: Fundamentals and applications*, Springer, New York, 2009.
30. Cai, W., Chettiar, U.K., Yuan, H.K., Metamagnetics with rainbow colors. *Opt. Express*, 15, 3333, 2007.
31. Liu, N., Stereometamaterials. *Nat. Photonics*, 3, 157, 2009.
32. Alu, A., Epsilon-near-zero metamaterials and electromagnetic sources: Tailoring the radiation phase pattern. *Phys. Rev. B*, 75, 155410, 2007.
33. Podolskiy, V.A. and Narimanov, E.E., Strongly anisotropic waveguide as a nonmagnetic left-handed system. *Phys. Rev. B*, 71, 1, 2005.

34. Smith, D.R., Kolinko, P., Schurig, D., Negative refraction in indefinite media. *J. Opt. Soc. Am. B*, 21, 1032, 2004.
35. Belov, P.A. and Hao, Y., Subwavelength imaging at optical frequencies using a transmission device formed by a periodic layered metal-dielectric structure operating in the canalization regime. *Phys. Rev. B*, 73, 113110, 2006.
36. Govyadinov, A.A. and Podolskiy, V.A., Metamaterial photonic funnels for subdiffraction, light compression and propagation. *Phys. Rev. B*, 73, 1, 2006.
37. Yao, J., Three-dimensional nanometer-scale optical cavities of indefinite medium. *Proc. Natl. Acad. Sci. U.S.A.*, 108, 11327, 2011.
38. Jacob, Z., Smolyaninov, I.I., Narimanov, E.E., Broadband purcell effect: Radiative decay engineering with metamaterials. *Appl. Phys. Lett.*, 100, 181105, 2012.
39. Smolyaninov, I.I. and Narimanov, E.E., Metric signature transitions in optical metamaterials. *Phys. Rev. Lett.*, 105, 067402, 2010.
40. Kidwai, O., Zhukovsky, S.V., Sipe, J.E., Dipole radiation near hyperbolic metamaterials: Applicability of effective medium approximation. *Opt. Lett.*, 36, 2530, 2011.
41. Cortes, C., Spontaneous radiation of a finite-size dipole emitter in hyperbolic media. *Phys. Rev. A*, 84, 023807, 2011.
42. Cortes, C., Quantum nanophotonics using hyperbolic metamaterials. *J. Opt.*, 14, 129501, 2012.
43. Noginov, M.A., Bulk photonic metamaterial with hyperbolic dispersion. *Appl. Phys. Lett.*, 94, 151105, 2009.
44. Yao, J., Three-dimensional nanometer-scale optical cavities of indefinite medium. *Proc. Natl. Acad. Sci. U.S.A.*, 108, 11327, 2011.
45. Xiong, Y., Two-dimensional imaging by far-field superlens at visible wavelengths. *Nano Lett.*, 7, 3360, 2007.
46. Naik, G.V., Kim, J., Boltasseva, A., Oxides and nitrides as alternative plasmonic materials in the optical range. *Opt. Mater. Express*, 1, 1090, 2011.
47. Korobkin, D., Measurements of the negative refractive index of sub-diffraction waves propagating in an indefinite permittivity medium. *Opt. Express*, 18, 22734, 2010.
48. Lapine, M. and Tretyakov, S., Permittivity medium. *IET Microw. Antennas Propag.*, 18, 3, 2007.
50. Schurig, D., Metamaterial electromagnetic cloak at microwave frequencies. *Science*, 314, 977, 2006.
51. Dolling, G., Simultaneous negative phase and group velocity of light in a metamaterial. *Science*, 312, 892, 2006.
52. Yuan, H.K., A negative permeability material at red light. *Opt. Express*, 15, 1076–1083, 2007.
53. Zhang, S., Fan, W., Panoiu, N.C., Malloy, K.J., Osgood, R.M., Brueck, S.R.J., Optical negative-index bulk metamaterials consisting of 2D perforated metal-dielectric stacks. *Opt. Express*, 14, 6778–6787, 2006.
54. Padilla, W.J., Dynamical electric and magnetic metamaterial response at terahertz frequencies. *Phys. Rev. Lett.*, 96, 107401, 2006.

55. Yen, T.J., Terahertz magnetic response from artificial materials. *Science*, 303, 1494–1496, 2004.
56. Padilla, W.J. and Taylor, A.J., Dynamical electric and magnetic metamaterial response at terahertz frequencies. *Phys. Rev. Lett.*, 96, 107401, 2006.
57. Chen, H.T., Complementary planar terahertz metamaterials. *Opt. Express*, *15*, 1084–1095, 2007.
58. Pendry, J.B., Holden, J., Robbins, D.J., Stewart, W.J., Magnetism from conductors and enhanced nonlinear phenomena. *IEEE Trans. Microw. Theory Tech.*, 47, 2075, 1999.
59. Smith, D.R., Padilla, W.J., Vier, D.C., Nasser, S.C., Schultz, S., Composite medium with simultaneously negative permeability and permittivity. *Phys. Rev. Lett.*, 84, 4184, 2000.
60. Veselago, V.G., The electrodynamics of substances with simultaneously negative values of ε and μ. *Sov. Phys. USP*, 10, 509, 2004.
61. Qiang, D. and Chen, G., Enhancement of evanescent waves in waveguides using metamaterials of negative permittivity and permeability. *Appl. Phys. Lett.*, 84, 669, 2004.
62. Aydin, K. and Ozbay, E., Left-handed metamaterial based superlens for subwavelength imaging of electromagnetic waves. *Appl. Phys. A., Mater. Sci. Process.*, 87, 137, 2007.
63. Tao, H., Landy, N.I., Bingham, C.M., Zhang, X., Averitt, R.D., Padilla, W.J., A metamaterial absorber for the terahertz regime: Design, fabrication and characterization. *Opt. Express*, 16, 181, 2008.
64. Eleftheriades, G.V., Iyer, A.K., Kremer, P.C., Planar negative refractive index media using periodically L-C loaded transmission lines. *IEEE Trans. Microw. Theory Techn.*, 50, 2702, 2002.
65. Li, B., Zhou, J., Li, L., Liu, H.X., Zi, J., Ferroelectric inverse opals with electrically tunable photonic band gap. *Appl. Phys. Lett.*, 83, 4704, 2003.
66. Khoo, I.C., Werner, D.H., Liang, X., Diaz, A., Weiner, B., Nanosphere dispersed liquid crystals for tunable negative-zero-positive index of refraction in the optical and terahertz regimes. *Opt. Lett.*, 31, 2592, 2006.
67. Zhao, Q., Kang, L., Li, B., Zhou, J., Tanh, H., Zhang, B., Tunable negative refraction in nematic liquid crystals. *Appl. Phys. Lett.*, 89, 221918, 2006.
68. Kang, L., Zhao, Q., Li, B., Zhou, J., Zhu, H., Experimental verification of a tunable optical negative refraction in nematic liquid crystals. *Appl. Phys. Lett.*, 90, 181931, 2007.
69. Werner, D.H., Liquid crystal clad near-infrared metamaterials with tunable negative-zero-positive refractive indices. *Opt. Express*, 15, 3342, 2007.
70. Zhao, Q., Kang, L., Du, B., Li, B., Liang, X., Zhang, B., Electrically tunable negative permeability metamaterials based on nematic liquid crystals. *Appl. Phys. Lett.*, 90, 011112, 2007.
71. Zhang, F., Magnetic control of negative permeability metamaterials based on liquid crystals, in: *38th European Microwave Conference*, IEEE, 2008.

72. Kowerdziej, R., Parka, J., Krupka, J., Experimental study of thermally controlled metamaterial containing a liquid crystal layer at microwave frequencies. *Liq. Cryst.*, 38, 743, 2011.
73. Dewar, G., Minimization of losses in a structure having a negative index of refraction. *New J. Phys.*, 7, 161, 2005.
74. Cai, X.B., Zhou, X.M., Hu, G.K., Numerical study on left-handed materials made of ferrite and metallic wires. *Chin. Phys. Lett.*, 23, 348, 2005.
75. Rachford, F., Armstead, D., Harris, V., Vittoria, C., Simulations of ferrite-dielectric-wire composite negative index materials. *Phys. Rev. Lett.*, 99, 057202, 2007.
76. He, Y., He, P., Dae Yoon, S., Harris, V.G., Vittoria, C., Tunable negative index metamaterial using yttrium iron garnet. *J. Magn. Magn. Mater.*, 313, 187, 2007.
77. Kang, L., Zhao, Q., Zhao, H., Zhou, J., Magnetically tunable negative permeability metamaterial composed by split ring resonators and ferrite rods. *Opt. Express*, 16, 8825, 2008.
78. Gil, I., Bonache, J., Garcia, J., Martin, F., Tunable metamaterial transmission lines based on varactor-loaded split-ring resonators. *IEEE Trans. Microw. Theory Tech.*, 54, 2665, 2006.
79. Wang, D., Ran, L., Chen, H., Mu, M., Kong, J.A., Wu, B.I., Active left-handed material collaborated with microwave varactors. *Appl. Phys. Lett.*, 91, 164101, 2007.
80. Aydin, K. and Ozbay, E., Capacitor-loaded split ring resonators as tunable metamaterial components. *J. Appl. Phys.*, 101, 024911, 2007.
81. Hand, T.H. and Cummer, S.A., Frequency tunable electromagnetic metamaterial using ferroelectric loaded split rings. *J. Appl. Phys.*, 103, 066105, 2008.
82. Chen, H., Wu, B.I., Ran, L., Grzegorczyk, T.M., Kong, J.A., Controllable left-handed metamaterial and its application to a steerable antenna. *Appl. Phys. Lett.*, 89, 053509, 2006.
83. Ricci, M.C., Xu, H., Prozorov, R., Zhuravel, A.P., Ustinov, A.V., Anlarge, S.M., Tunability of superconducting metamaterials. *IEEE Trans. Appl. Supercond.*, 17, 918, 2007.
84. Tao, H., MEMS based structurally tunable metamaterials at terahertz frequencies. *J. Infrared Millim. Terahertz Waves*, 32, 580, 2011.
85. Lapine, M., Structural tunability in metamaterials. *Appl. Phys. Lett.*, 95, 084105, 2009.
86. Lapine, M., Gorkunov, M., Ringhofer, K.H., Nonlinearity of a metamaterial arising from diode insertions into resonant conductive elements. *Phys. Rev. E., Stat. Plasma Fluids Relat. Interdiscip. Topics*, *67*, 065601, 2003.
87. Agranovich, V.M., Linear and nonlinear wave propagation in negative refraction metamaterials. *Phys. Rev. B., Condens. Matter. Mater. Phys.*, 69, 165112, 2004.

88. Kockaert, P., Tassin, P., Sande, G., Veretennicoff, I., Tlidi, M., Negative diffraction pattern dynamics in nonlinear cavities with left-handed materials. *Phys. Rev. A*, 74, 033822, 2010.
90. Yamato, M., Characterization of lipids in adipose depots associated with minke and fin whale ears: Comparison with 'acoustic fats' of toothed whales. *Mar. Mamm. Sci.*, 30, 1549, 2014.
91. Zigoneanu, L., Popa, B.-I., Cummer, S.A., Three-dimensional broadband omnidirectional acoustic ground cloak. *Nat. Mater.*, 13, 352, 2014.

12

Plasmonic Metamaterials

M. Rizwan[1]*, A. Ayub[2], M. Sheeza[1] and H. M. Naeem Ullah[3]

[1]School of Physical Sciences, University of the Punjab, Lahore, Pakistan
[2]Department of Physics, University of the Punjab, Lahore, Pakistan
[3]Research Center of Materials Science, Beijing Key Laboratory of Construction Tailorable Advanced Functional Materials and Green Applications, Beijing Institute of Technology, Beijing, P. R. China

Abstract
Plasmonics combined with metamaterials have gained significant consideration during the last decade due to revolutionary influences that were brought both in practical applications as well as in various disciplines of fundamental physics. This analysis represents the new advances, preliminaries, and future perceptions in the evolving discipline of plasmonic metamaterials focusing to unplug novel stimulating openings in the field of nanotechnology, as well as nanoscience.

Keywords: Negative refractive index, plasmonics, surface plasmon polariton, biosensing, negative refraction, localized surfaced plasmon, planar ring resonator

12.1 Introduction

Plasmonics deals with the behavior of light with nanostructures. This interaction of light with nanoscale objects accesses the wide optical plasmonic existence due to the aptitude of sub-wavelength captivity and the improvements of the optical field with nanostructures. Plasmonic optics gave birth to the field of metamaterials with the enhancement of meta-surfaces. Best tunable synthesized materials like low-level material variables and high-level materials variables are distinguished by unusual volume and surface properties. This concept yields when combined electromagnetic conduct of numerous sub-wavelength insertions and constituents as "meta-atoms."

**Corresponding author*: rizwan.spr@pu.edu.pk

Inamuddin and Tariq Altalhi (eds.) Electromagnetic Metamaterials: Properties and Applications, (261–286) © 2023 Scrivener Publishing LLC

Many applications of surface waves provide a path to the broad spectrum of concrete phenomena for the fertile ground [1]. Surface waves are being supported by the expansion of structures having tunable features (i.e., electric and magnetic) by the discovery of metamaterials [2]. Conductors and dielectric interfaces have electromagnetic agitation that is responsible for surface plasmons polaritons (SPPs). Then evaporation is limited in the direction that is perpendicular to them [3–5].

Spoof surface plasmons polaritons are the geometrical induced SPPs that emulate the properties of limited SPPs. The suggested phenomenon operates at fewer frequencies. Spoof surface plasmon is the result of surface structure. The precursor obeys as an ideal prototype for structure surfaces. Meta-surfaces can shape light flow that belongs to a branch of planar metamaterials having exceptional functionality and have recently gained comprehensive attention. The expected phase profile is the major goal of meta-surfaces by scheming sub-wavelength formations at the two usual materials interfaces. A logically designed phase yields the effects of propagating waves that can build completely. It should also be concluded that unusual reflection, as well as refraction, have been proven [6]. Optical devices based on meta surface like "vortex plates, wave plates, and ultra-thin focusing lens" yields different sorts of incident light, such as linearly contrasted light, as well as top beams. This gives rise to promising research for the first time in industry applications [7].

Optical wavefronts can be controlled to differ from the typical propagation techniques instead of imparting on its carefully built interior structure by the discovery of metamaterials. Pendry *et al.* first hypothesized these 20 years ago. By careful formation of constituents in the metamaterials, untypical material properties like refractive index can be attained. By controlling the apparent electromagnetic ε(permittivity) and μ(permeability), the n (refractive index) of a material is:

$$n = \pm \sqrt{e\mu} \quad \text{(Eq. 12.1)}$$

Because of the profound acknowledgment of metallic structures being used, high deprivation and deep dispersive outcome of bulk metamaterials are usually vulnerable. With the use of actual atomic and nanofabrication techniques, the complicated structure formed during 3D metamaterial is difficult. Thus, the evolution of 2D meta-surfaces or metamaterials has been focused on in current studies. These 2D materials tend to build electromagnetic outcomes with fewer fatalities in thin coating formations. The establishment of exteriors gives minimum propagation stage with sub-wavelength thickness. This moves the focus from making materials having a negative (permittivity), as well as μ(permeability) to building exterior formations to modify plane reflection along with transmissions. Electronic band formation gives the difference between the two types of materials in solid-state physics. As metals have overlapping bands (valence and conduction) for the movement of free electrons while insulators have wide band gaps between two bands. Conductors and insulators both are until now able to interrelate with coming electromagnetic fields by various physical techniques. These two materials appeared to be used for the broad future of meta-surfaces [8].

12.2 Negative Refraction and Refractive Indexes

A frequently used electromagnetic magnetic parameter is the refractive index which can be expressed as $n(\omega)^2 = \varepsilon(\omega)\mu(\omega)$. This parameter gives the speed of an electromagnetic wave when it travels in the material. Additionally, it also measures the deviation of the beam of light when it passes through the interface of various materials having a different indexes of refraction. In 1621, Willebrand Snell measured this bending by showing that

$$n_1 \sin\theta_1 = n_2 \sin\theta_2 \qquad \text{(Eq. 12.2)}$$

where n_1, n_2 represents the index of refraction of the first plus second mediums respectively [8]. Also, θ_1 and θ_2 represent angles by the light rays with normal to each other. A ray originating from a source in the medium like free space in action with another material like a slab with $n(\omega) > 0$(refractive index is positive). The rays originating from the point source deviate at the interface between two media, i.e., free space and glass with an angle determined by eq 12.2.

Russian physicist Veselago revealed that $\varepsilon(\omega) < 0$ as well as $\mu(\omega) < 0$, (i.e., electric and magnetic parameters have negative values) and this leads to a negative index of refraction, $n(\omega) < 0$. As Veselago assumed that this property of materials having a negative index of refraction might be formed in intrinsic materials. As metamaterials (artificially structured) have managed electric and magnetic parameters over a wide range of frequencies. So, it is possible to attain conditions (i.e., $\varepsilon < 0$ and $\mu < 0$) in artificially structured MMs. However, Veselago presumed metamaterial can be realized [8].

By Veselago's method, a material having a refractive index ($n < 0$) can be formed by combining such recognized artificial structures that individually give these parameters (ε1 < 0 and μ1 <0). Materials having a negative index of refraction following changes occur:

- For a wave, reverse phenomena happen for its phase velocity.
- Doppler shift is reversed for a source relative to the observer.
- When a charge is in motion, then it moves in a backward direction instead of a forward direction by Cerenkov radiations.
- Radiation tension yields by reversing radiation pressure.
- Diverging lens formation by the converging lens and vice versa.

By these changes in basic electromagnetic phenomena, we get a material having a negative index of refraction. As Veselago's projections were fascinating, at this time, naturally happening materials having negative refractive index were not discovered and his predictions were greatly overlooked [9].

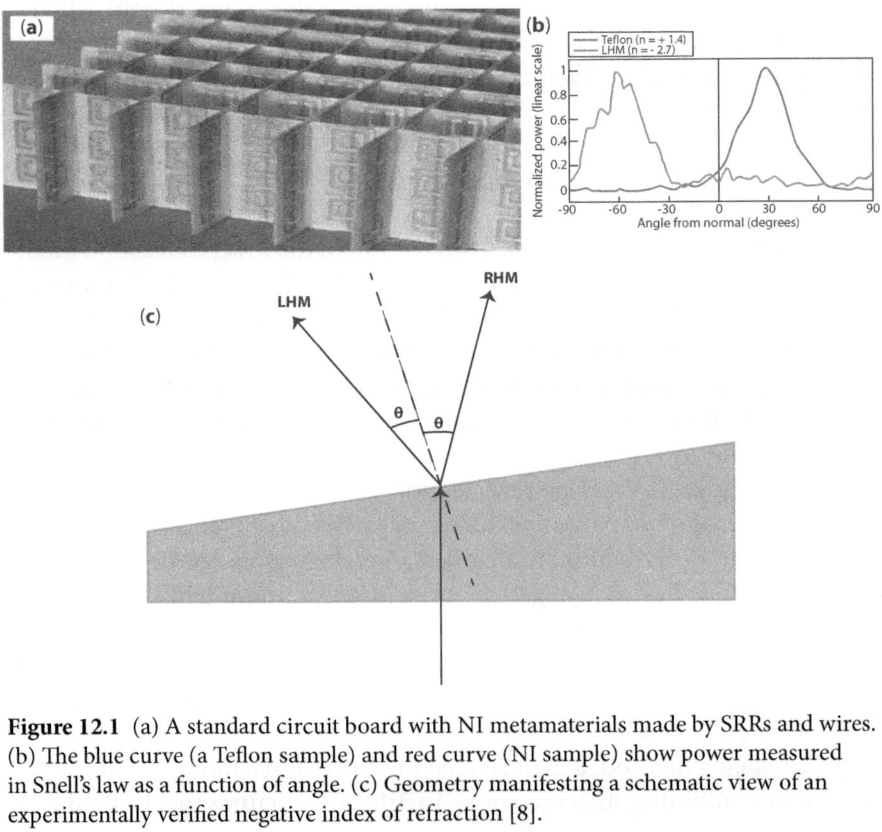

Figure 12.1 (a) A standard circuit board with NI metamaterials made by SRRs and wires. (b) The blue curve (a Teflon sample) and red curve (NI sample) show power measured in Snell's law as a function of angle. (c) Geometry manifesting a schematic view of an experimentally verified negative index of refraction [8].

Smith *et al.* 2000 fabricated negative refractive index material by the use of artificially made metamaterials. These manmade metamaterials with negative refractive index integrated a wire structure having $\varepsilon 1 < 0$ and also an SRR formation having $\mu 1 < 0$ over the identical band of frequencies. That kind of material became a surprise for many scientists [10]. Negative refraction was first performed by Shelby *et al.* in 2004. The negative index of refraction was demonstrated by Snell's law investigation using a metamaterials wedge in a prism-shaped as shown in Figure 12.1. Negative refraction yields by the incidence of microwave radiation on the prism to refract on the surface normal to the opposite side.

For comparison, a Teflon prism having a positive index of refraction is used by the deflection of the same beam. This deflection happens to the opposite side of the surface normally agreeing at an angle to the investigated index of the material. A lot of experimental verification proves the validity of the metamaterials approaches through Veselago's investigations of the negative index of refractions [11].

12.3 Fundamentals of Plasmonics

Plasmonics is an emerging field in the research of nanophotonics as well as nanooptics. Plasmonics deals with the considerations of electron vibration in NS (nanostructures) and NPs (nanoparticles). Surface plasmons possessed optical belongings and also at the nanoscale have the distinctive capacity to enclose light. Also, surface plasmons (SPs) are very liable to the properties possessed by the medium in which they travel and to the neighboring medium. Additionally, the SPs resonances can be managed by modifying the size, periodicity, shape, and materials' nature [12]. By managing these parameters as described above, technological advancement yields new plasmonic mediums through the theoretical, numerical, and computational analysis of the optical properties of plasmonic mediums. With the progress of these optical properties, many applications of plasmonic systems like optical devices, biosensors, and photovoltaic devices are present [13].

12.3.1 Surface Plasmon Polaritons

For each medium and the related BC (boundary condition), by solving Maxwell's equation the electromagnetic field for SPPs on a dielectric metal intersection can be gained. The last shows the persistence of tangential components for both fields (electric as well as magnetic) over the boundary and the finishing of both pitches endlessly beyond the interface. For the introduction of main parameters distinguishing SPPs, let us view a system having dielectric material distinguished by real, isotropic, and ε_1 (dielectric constant with positive value) in the partial space (X_3>0) and a metal distinguished by $\varepsilon(\omega) = \varepsilon_1(\omega) + ¡\varepsilon_2(\omega)$ (complex dielectric function) in half space (X_3<0). First, we examine a transverse magnetic (a p-polarized TM) wave that travels in this medium in x1 direction. The loss of generality in taking this way of travel is because of optical isotropy in the two media. As a result of this polarization, the wave generates have magnetic vector perpendicular to the incidence plane (plane shown by the way of travel and perpendicular to the surface). By solving Maxwell's equations in x1 direction, which are also wavelike, and there is a decrease in amplitude exponentially in every medium by the increase of distance. At the interface ($X_3 = 0$) gives

$$H^{>}(x;t) = (0,\ A,\ 0)e^{ikx_1 - k_3^{(1)}x_3 - i\omega t} \qquad \text{(Eq. 12.3)}$$

$$E^{>}(x;t) = -A\,c\,/\,i\omega\varepsilon_1(k_3^{(1)}, 0, ik)e^{ikxl - k_3^{(1)}x_3 - i\omega t} \qquad \text{(Eq. 12.4)}$$

in the region (x3 > 0),

$$H^{<}(x;t) = (0,\ B,\ 0)e^{ikx_1 - k_3^{(m)}x_3 - i\omega t} \qquad \text{(Eq. 12.5)}$$

$$E^{<}(x;t) = -Bc/i\omega\varepsilon_1(k_3^{(m)}, 0, ik)e^{ikxl - k_3^{(m)}x_3 - i\omega t} \qquad \text{(Eq. 12.6)}$$

in the region (x3 < 0),

Here, $k_3^{(1,m)}$ shows the decrease of the em field results by an increase in the distance taken from the surface.

$$K_3^{(1)} = (k^2 - \varepsilon_1(\omega/c)^2)^{1/2}, \qquad \text{(Eq. 12.7)}$$

$$K_3^{(m)} = (k^2 - \varepsilon(\omega)(\omega/c)^2)^{1/2}, \qquad \text{(Eq. 12.8)}$$

where the $K_3^{(1)}$ and $K_3^{(m)}$ show the real parts necessary to be positive such that Eq 12.7 and Eq 12.8 reflects an em wave at x_3=0 (i.e., localized to the dielectric metal interface). Applying boundary conditions at the interface x_3=0 gives a couple of equations

$$A = -B, Ak_3^{(1)} / \varepsilon_1 = -Bk_3^{(m)} / \varepsilon(\omega). \qquad \text{(Eq. 12.9)}$$

By this, we get dispersion relation by applying the condition of equations having nontrivial solution by linking the ω (frequency) and its k (wavenumber) of the wave which is p-polarized,

$$k_3^{(m)} / k_3^{(1)} = -\varepsilon(\omega) / \varepsilon 1. \qquad \text{(Eq. 12.10)}$$

Let us consider that the metal has a dielectric function that is real ε(ω), then equations (12.3–12.6) reflect a surface EM wave when $k_3^{(1)}$ and $k_3^{(m)}$ are necessary to be real and positive. Surface em wave exists when ε(ω) is necessary to be negative from Eq. (12.10) and this applies to the volume em wave to be pure imaginary for its k (wave vector) under this frequency range. From this, we conclude that when volume em wave cannot travel in metal in the frequency range then surface em wave exists. Equations (12.3 and 12.5) yields field vectors of surface em wave with A=B and their frequency yields from dispersion relation, then Eq (12.10) is known as surface plasmons polaritons (SPPs). It is also known that excitations of electrons are in the conduction band. On squaring both sides of equation (12.10) close to the metal surface there is a coupling of photons.SSPs yield as a function of frequency (ω).

$$K_{sp} = \frac{\omega}{c}\left[\frac{\varepsilon_1 \varepsilon(\omega)}{\varepsilon_1 + \varepsilon(\omega)}\right]^{\frac{1}{2}} \qquad \text{(Eq. 12.11)}$$

This relation is valid when ε(ω) shows even a complex function [14].

12.3.2 Localized Surface Plasmons

Along with SPPs and SPs (surface plasmons) at a planar interface of dielectric metal surface, confined surface em excitations can hold with further configurations like metallic particles as well as abolish different geology. In defined configurations, that kind of surface excitation is known as LSPs (localized surface plasmons) [15]. Their frequency can be obtained in electrostatic (progressive) calculation with adjustable boundary conditions by the solution of Laplace's equation. When a (characteristic size) of a system is comparatively small then λ (wavelength) is comparable to localized surface plasmons frequency, then electrostatic approximation is indifferent to the results of retardation. For instance, in the form of a metallic sphere having radius R and inserted in a medium having dielectric constant ε0, then for electrostatic potential by solving Laplace equation is $\phi^{<}$ (r, Θ, ϕ) at the origin within the sphere is finite [16].

$$\phi^{<}(r, \Theta, \phi) = \sum_{l=0}^{\infty} \sum_{m=-l}^{l} a_{lm} r^{l} Y_{lm}(\Theta, \phi), \quad 0 \le r, \le R,$$

(Eq. 12.12)

Y_{lm} (Θ, ϕ) represents spherical harmonic. In the same way, the electrostatic potential $\phi^{>}$ (r, Θ, ϕ) that vanishes by solving Laplace's equation at the infinity outside the sphere can be in the form of

$$\phi^{>}(r, \Theta, \phi) = \sum_{l=0}^{\infty} \sum_{m=-l}^{\infty} b_{lm} \frac{1}{rl+1} Y_{lm}(\Theta, \phi), \quad r \ge R.$$

(Eq. 12.13)

By application of boundary situations at the top of the sphere, such as the steadiness of $\varepsilon \frac{\partial \phi}{\partial r}$ plus ϕ, determines the frequency of localized surface plasmons,

$$\frac{\epsilon(\omega)}{\epsilon 0} + \frac{l+1}{l} = 0 \qquad \text{(Eq. 12.14)}$$

Free electrons in a metal form the dielectric function, and the solution of Eq. (12.9) is given as

$$\omega_l = \omega_p \left[\frac{l}{\varepsilon 0(l+1)+l} \right]^{1/2}, \quad l = 0,1,2,3, \qquad \text{(Eq. 12.15)}$$

where only l =1 (dipole active excitation) is the main for spheres whose sizes are small. The addition of higher multipoles becomes very important whenever the sphere size increases with the comparison of a big sphere having limit $l \to \infty$. then, the frequency of LSPs reaches SPs of dielectric metal at the planar interface.

$$\omega_\infty = \frac{\omega_p}{(\epsilon_0 + 1)^{\frac{1}{2}}} \qquad \text{(Eq. 12.16)}$$

Now, we have three groups of LSP frequencies for an ellipsoidal particle, and these frequencies are associated with the three axes for an ellipsoidal particle. For small ellipsoidal, these three axes-related frequencies formed dipole-active localized surface plasmons, which are calculated as

$$\frac{\epsilon(\omega)}{\epsilon_0} = 1 - \frac{1}{L_i} \quad \text{(Eq. 12.17)}$$

Here, Li is the supposed depolarization factor that depends merely on the ellipsoidal shape, and also i = 1, 2, 3 shows the coordinate axes associated with ellipsoidal axes. Metal has a negative dielectric function in the frequency range when LSP modes exhibit bulk plasma frequency above the metal is bounded. An SPP has a dispersion relation and is a circulating surface mode. Indifference, LSPs also known as electromagnetic exterior form resonances are restricted to curved metal objects. These are categorized by distinct, complex frequencies that depend on the form and size of the object such that SPs are restricted and also its dielectric function. LSPs could be significantly agitated with the light of suitable frequency plus polarization regardless of k (wave vector) of stimulating light. That is why they also effectually decline with the production of light.

In a metal, the bulk has voids that are equivalent LSP frequencies that can be obtained by changing ε(ω) with 1/ε(ω) in the agreeing Dirichlet problem. A particle and a void of LSP having the same form are linked to each other by

$$\omega_{particle}^2 + \omega_{void}^2 = \omega_p^2 \quad \text{(Eq. 12.18)}$$

Therefore, the resonances linked with SPs (surface plasmons) are constrained to voids determined from known localized surface plasmon frequencies in metal particles [16].

12.3.3 Applications of Plasmonics

Scientists are more ambitious to utilize the advantages that are offered by plasmonic nanoparticles. Current research about plasmonic nanoparticles has explained many potential and beneficial applications like cancer treatment as well as biomedical applications and plasmonic solar cells. Plasmonic nanoparticle research demonstrates the optical properties of metamaterials in range within the visible and near-infrared spectral bands. That band differs in size, structure, shape, and tunable plasmon resonance. Plasmonics (or nanoplasmonics) is a modern topic of research that consists of nanophotonics and nanooptics [17].

Plasmonics treats the exploration of oscillations of electrons in nanoparticles (NPs) and metallic nanostructures. Surface plasmons possess unique optical properties. For example, surface plasmons possess the special capacity, which encloses light at the nanoscales [18]. Surface plasmons are very sensitive to the medium that they contain, as well as the properties of the substances in which they propagate. The surface plasmon resonances can be tuned and adjusted by their size and shape as well as periodicity and nature. However, the technological demonstration allows the researchers to create new plasmonic systems [7]. This can be done by controlling all the factors. Many tools, like theoretical, computational, and numerical, have been introduced which allow for the best understanding regarding the optical properties of plasmonic systems [19]. Thus, the optical properties of plasmonic systems permit a great number of applications like biosensors, optical devices, photovoltaic devices, cancer treatment, color engineering, biomedical, spectroscopy, and thin films.

12.4 Types of Plasmonics Metamaterials

Plasmonic materials have gained great importance because their surfaces can confine incident light, as well as extend the optical path lengths. These include the following types of plasmonic metamaterials.

1. Graphene-based plasmonic metamaterials
2. Nanorod plasmonic metamaterials
3. Plasmonic meta-surfaces
4. Self assembles plasmonic metamaterials
5. Non-linear plasmonic materials
6. 2D plasmonic metamaterials

12.4.1 Graphene-Base Plasmonic Metamaterials

Graphene-based metamaterials had been considered as an organizer for uses such as ideal absorbers. It is used as a light emitter, as well as a modulator and photodetector. It is used in tunable spintronic devices. However, challenges are related to conventional film deposition techniques. These techniques are used to make the complicated metamaterial demanding to produce, which has drastic finite experimental authorization. These metamaterials having optical properties are modulated statically by the manageable process, which is known as laser mediation conversion. The layers of graphene oxide were used as counterparts of graphene. These counterparts contain a lot of degrees of conversion, providing great dormant for devices fabrication and devices. The productive graphene layers show optical conductivity comparable (approximately within 10%) in response to chemical vapors and deposited analogs [20].

Functional photonic devices are led by laser pattering like flat lenses of ultrathin submerged in a device lab-on-chip. It continues the stability and contains sub-wavelength, which focuses on resolution. It is applied within flowing environments without any humiliation, in contrast with the original lens. This type of graphene-based metamaterial introduces a new experimental plan for vast uses like in on-chip integrated photonic and microfluidic devices as well as biomedical. Metamaterial as well as meta-surfaces structures can enhance the presentation of electromagnetic waves, which is done beyond the boundary conditions. Metamaterials are considered unusual materials, made of special micro as well as nanoscale arrays [21]. These arrays permit them to come in contact with electromagnetic waves and as well as other types of energies, which are not present in nature. For example, light contains a negative refractive index, which has the most important effect on metamaterials. Meta-surface is one type of metamaterial structure, made of different parts in order of 2D to overcome the varieties of losses as well as challenges in the fabrication of nanomaterials. Newly generated types of meta-surfaces are taking advantage of graphene. They are introduced as well as used in antenna designing, sensors, and absorbers in the THz band with a cheap cost accomplishment [4].

12.4.2 Nanorod Plasmonic Metamaterials

By the use of LSPPs in the field of biosensing, different nanoparticle formations, such as ellipsoids, spheres, rods, and rings are approved. Among these applicants, nanorods are among those who hold a comparatively high understanding of the index of refraction variations and geometrical integrating, which have been used extensively. On a glass substrate, there is a nanorod layer covered with self-built monolayers. The discrete nanorod is around 15 nm (diameter) and 50 nm (long). For sensing, imprisonment antibodies are pre-linked with the self-build monolayers and definite antigens will fix to them in the analytic solution through measurement [22].

As an executant of detection, an arrangement of nanorods in the plasmonics metamaterials has been presented with new means to overawed the understanding range of sensors with good flexibility based on LSPPs. The additional fundamental advantage of nanorod metamaterials (NMs) based sensors is having porous consistency, which helps greatly to enhance the surface area for the response, as well as benefits to gain a great penetrating depth [23]. Nanorods are precursory restrained with streptavidin particles for operationalization. MMs for biosensing reflect the supremacy of guided means of these materials [24]. Markedly, nanorods or wire formations give rise to the latest technique of nanofabrication for a variety of applications and supplies which permits them to be unified into complex systems. By the lookout of metamaterials (MMs), some nanorod arrangements reflect as a bulk material with rare dispersion relations which are hyperbolic and their dielectric constant tensor have a negative factor laterally the long vertical and positive factor the small vertical of nanorods [25]. Along with this exclusive feature, the complete structure holds a directed photonic mode at NI (near infrared) while the ascertainment of field distribution within the porous film is by the interaction of plasmon facilitation among nanorods. The excited state of this directed mode is the same as the exciting surface plasmon polaritons on a metallic film. For inaccessible nanorods having enough big distance of separation, the near field linking is insignificant and the reaction is dominant by localized surface plasmons [26].

12.4.3 Plasmonic Metal Surfaces

When the light comes in contact with metallic nanostructures via plasmons with resonant stimulating results in combined oscillation of conduction electrons within nanostructures. Nanostructure commonly exists in size in the range of 10 to 10,000 nm. This structure productively behaves like an antenna for light. The current introduction of metamaterials increased the control with help of engineered inner structures of those materials. Pendry *et al.* theorized it about 20 years ago [26]. After this, the progress in the area of artificially fabricated material carries on to stabilize.

Metamaterials possessed a large series of em (electromagnetic) processes which do not occur in nature. These invented materials are formed by the set of complex elements. These elements contain magnitudes that are larger than the of magnitude of molecular unit cells of real materials. Ideally, it permits the MMs to give images about the relation with em waves. This can be represented in terms of parameters which are actual material parameters. MMs depicts as materials that are standardized at the required wavelengths. At optical frequencies, these materials are considered to be a type of invented fabricated materials with unique characteristics [27]. The main property is that the similar permittivity in various directions is different from each other. Thus, it is considered a variant by designing its structures through experimental work [28]. This property explains that metamaterials might act as surface plasmon polaritons (SPPs) in the way of propagation with a more malleable demonstration in contrast to ordinary materials. MMs allow a distinguished multitude of material characteristics. These include artificial magnetism as well as controllable electric permittivity and other many properties involving negative refractive indices. These materials possessed a high index of refraction and strong chirality. These properties remain circulating by the contact of light and artificial materials [12]. By the concept of these great characteristics, light can be operating uniquely. The metamaterials having negative refractive index use split ring resonators and an array of metallic wires to make their unit cells. These unit cells are demonstrated experimentally in the microwave dominion. After that, at optical wavelengths, it was demonstrated as the elemental array is made to the least range to nanoscales [18].

In current years, meta-surface elements are organized periodically, having resonant subwavelength of thin planner arrays. They have gained huge attention and are considered a quickly growing field in research. The reason is that they have extraordinary access to electromagnetic fields. They are modified for impinging waves by using boundary conditions. Usually, meta-surfaces are used to consider remarkably beneficial due to their simplest fabrication procedures as compared to metamaterials having 3D analogs. The modified metamaterials are usually composed of complicated and artificial 3D nanostructures processing rather than em properties [29]. As meta-surfaces are considered a continuously growing field of research, several articles give complete detail about specific reviews [19]. Some present a complete review of the field. Some focus on a certain type of meta-surfaces or specific branches of applications. Some focus on the current progress of GSPMs during the past few years and claim to provide the main objectives for this specific branch of meta-surfaces [30].

12.4.4 Self-Assembled Plasmonic Metamaterials

The combination of nanochemistry and nanooptics has encouraged the exposure of continuous research in the science area which is known as self-assembled metamaterials. In this scientific area, chemical self-assembly techniques are mainly used to fabricate nanostructures needed for applications in optics and photonics. The main purpose of these bottom-up techniques is to fabricate the unit cells to adjust at the nanoscales to demonstrate a desired optical effect range (visible and NI). The team works by Alivisatos and Brousse introduce the probability of arranging the two strong associated nanospheres (known as a dimer) with nanometer precision by using

DNA and as well as molecular linkers. These first arrangements of the production of a unit cell started with self-assembly techniques. This arrangement gave a huge idea to plot the most complex nanostructures by using bottom-up techniques.

MMs fabricated by using bottom-up methods contain many advantages over their top-down techniques. Their offering structures are beneficial like smaller sizes, as well as more ready response to three-dimensional metamaterials. These techniques are mainly quicker and cheaper [31]. Due to this reason, there is a huge importance of this technology for the long last application. Over the last decades, a remarkable number of researchers have gone into the demonstration of plasmonic nanoparticles by using nanochemistry as the main components of metamaterials unit cells. This field is continuously being upgraded and nanoparticles of a huge variety of shapes as well as compositions and sizes can now be produced. This large category of plasmonic materials is demonstrating localized surface plasmons (LSPRs) at optical frequencies. This is becoming of continuous importance to the expanding field of optical metamaterials. These NPs are used to fabricate structures with beneficial optical properties. It is typically necessary to encourage their organization generally through the utilization of particle-substrate and inter-particle relations into specific architectures.

Convenient structures are identified by a high degree of nominal order just on short-range parameters. The précised spatial arrangement is impossible across the larger dimensions in most cases, which leads significantly to amorphous structures. These self-assembled nanostructures need novel analytical means to report their properties, advanced design of functional elements which hold a desired near- and far-field effect, and requires genuine characterization and nanofabrication techniques. Finally, novel applications are to be recognized which are modified to the specifics of the self-assembled nanostructures [32].

12.4.5 Nonlinear Plasmonic Materials

An artificially manufactured material has no properties that are present in nature. In contact with em (electromagnetic) radiation, they are identified by other parameters, such as permittivity, as well as permeability. The refractive index is a result of the product of permeability and permittivity. Metamaterials cannot occupy a positive refractive index like other materials having properties that are found in nature. So, they contain a negative refractive index. Naturally occurring materials cannot show a prominent nonlinear response as compared to artificially manufactured materials [33].

Nonlinear plasmonic material contains periodic as well as nonlinear and transmission mediums. Such materials are considered as a negative refractive index because there is a reason in response to the electromagnetic source. The reason for nonlinearity is the result of electric fields that have large values. Infect in these materials the microscopic electric field is greater than that of the macroscopic electric field of em (electromagnetic) source. Due to this effect, we can enhance the non-linear response of the materials.

A prominent nonlinear behavior can be derived with the help of hysteresis-type dependence in response to light. The response is observed as the magnetic permeability of the incident light within the material. The changing in materials is due to the field intensity which permits them to change themselves from left to right-handed as well as back [34]. For nonlinear optics, there is a need for a nonlinear medium. Most optical materials show weak responses as they show weak responses in response to relatively high changes in the intensity of light. Nonlinear plasmonic materials show control on this limitation, as the average value of the field can be smaller than that of the local field.

A physical process is created when the light is scattered by the materials and frequencies having different wavelengths try to change from its linear trajectory. When the light changes its medium (reflected), it is then scattered. At resonance, metamaterials are made for scattering light by split ring resonators. A similar size of resonant scattering elements is designed within the material. When light propagates through the material the similar size of the resonant element is smallest than that of the wavelength of the frequency of light [35].

12.4.6 2D-Plasmonic Metamaterials

In addition to modern concepts like negative Refractive Index (RI) and photonic crystals in metamaterials are becoming the most popular materials due to their unusual optical properties and can be used in optical instruments. Metals such as Ag, Al, and Au metals were used as metamaterials but their performance was not admirable due to radiative losses. The high amount of energy dissipation, as well as poor tuneability, was included as drawbacks. To solve these problems for systematic plasmonic uses, a group of 2D materials was introduced which presented an important phenomenon of light-matter interaction. That resulted in efficient quantum confinement effects. Many materials including semiconductors and dielectrics are considered plasmonic materials because they contain unusual plasmonic properties. The advanced properties are connected with band gap contrivances as well as the transfer of electrons [36].

Graphene was investigated as the first 2D material with zero band gaps. Graphene has extraordinary conductivity properties because of its higher electronic strength. At the industry level, the realization of graphene is considered for useful applications in the laboratory. Researchers have started investigating more and more 2D materials to artifice their performance for plasmonic applications. More than 140 members of the 2D materials family are playing a pivotal role in advanced and as wells as primary technologies like light-emitting diodes (LEDs), physical catalysis, environmental applications, field-effect transistors (FETs), and sensing applications [37]. Some useful 2D materials are considered hexagonal boron nitride (HBN), black phosphorene, and non-metals having properties of probable plasmonic materials. These 2D materials contain a vast electronic and plasmonic spectrum that covers many properties like high surface area, surface state nature, spin-orbit coupling, and effects like quantum spin Hall effects [22].

12.5 Applications of Plasmonics Metamaterials

Surface plasmon polaritons show distinct characteristics of tight field constraint and strong field improvement. In the locality of metal structures, dielectric media have variations in refractive index that are very delicate. This purpose can be observed by obvious changes of equivalent quantities, like a balance in angle, a change in resonance, and a change in field intensity probable. Thus, surface plasmons polaritons not merely yield label-free, actual contact to investigate biological mechanisms at nanoscales but give rise to very high perseverance. Mutually circulating plus confined surface plasmons polaritons have been functional for biomedical and biochemical applications [38].

12.5.1 Nanochemistry

Nanochemistry is linked with size, defects, form, self-assembly, and bionano such that any new nanoconstruct is linked with all these ideas. In this way, nanoconstruct formations depend on the dependence of size, form, and the surface of self-assembly of constituents into the functional structures, such that they may have defects and also may be useful for bioanalytical, photonic, electronic, and medical issues. Nanochemical techniques can be helpful to form carbon nanomaterials like carbon nanotubes (CNT), fullerenes, and graphene, which have achieved consideration in latest years due to their great electrical and mechanical properties.

Progress in chemistry, as well as nanofabrication, have been in search of the plasmonic nanostructures (PN) of various outlines. These plasmonic nanostructure advances lead to managing the spectral limit of plasmonic resonances plus adjusting the pitch confinement and development beneficial to enhance detection and hot carrier attraction presentation. Another way, plasmonic metamaterials adopt the same manner by exploiting large-scale plasmonic nanostructure arrangements based on gatherings of nanoholes, nanorods, split ring resonators, nanotubes, and further nanostructures. From all these, PNM (plasmonic nanorod metamaterials) is made up of arrangements of plasmonic nanorods which are highly interacting and ranged and mainly beneficial for sensing along with nanochemistry applications. This enables nanochemical reactors in which hot electron effects are completely subjugated and become highly anisotropic metamaterials used in nanophotonics submissions for wave-guiding and imaging proceeds the diffraction limit to improve fluorescence and optical effects in nonlinear forms. Hence, the use of plasmonic nanorods in metamaterials assumed from hot electron prompted followed by organic conversions in metamaterials, which are electrically operated [39].

12.5.2 Biosensing

Although fronting their particular problems, each of them has distinct advantages. A finding sensitivity beyond 10^{-5} refractive index units (RIU) for methods based on circulating surface plasmon polaritons has been attained. But this goal is still stimulating for analysis because of careful nanostyles and also with very minor size. By way of confined surface plasmon polaritons, a kind of device in which by changing chemical components, size, and form of plasmonic structures have attained good tenability and malleability. But it suffers a sensitivity of lower magnitude at least at the range of one.

Biosensing holds a novel promise with unique sensitivity, as well as specificity in the field of plasmonics metamaterials. Firstly, it follows the characteristics of metamaterials such that operating frequency could be broadly adjusted to fix the individual excitations in biomolecules along with higher sensitivity can be formed a very sharp spectral line. Second, adaptability should be high to form the building blocks of plasmonics nanostructures confirming that this device follows the requirements for various applications. New advances in biosensing, which are based on plasmonics metamaterials are as follows;

- meta-surfaces,
- chiral fields,
- nanorod structure,
- fano resonance.

Due to limited space, we cannot discuss more sensing approaches [24].

12.5.3 Filters

Filters like band pass filers (BPFs) are the necessary first component for microwave/RF systems. Recent developments in current wireless, as well as radar communication utilization, demand great performance, and variably managed RF sub-systems. Therefore, automatically manageable microwave filters (MFs) is attaining additional consideration for exploration along with progress. Convectional MF uses outdated microstrip techniques which go through mutual coupling, radiation losses, and cross-talk issues. To overcome these problems, spoof surface plasmons polaritons (SPP)-based RF systems are used.

Natural surface plasmons polaritons are distinctive exterior waves that are exceedingly confined at the boundary of metal-dielectric and concerning optical and near-infrared (NIF) frequencies. It declines exponentially along the plumb track of the boundary and therefore displays sub-wavelength pitch limitations and improvements. Therefore, to perceive the surface plasmons polaritons like features of the locality in metal-dielectric boundary approaches to THz as well as MF (microwave frequencies), plasmonic metamaterial have regularly ridged metallic formations with two-dimensional hole and one-dimensional grooves have been projected and also known as designer or spoof surface plasmon polariton. The cutoff frequency and dispersion curve for that kind of system could be changed by its physical constraint. However, these formations can manipulate and control electromagnetic waves, which can be used to make transmission lines at the sub-wavelength scale for the stimulation of amplifiers, tunable filters, switches, and antennae. In-band stop, manageable behavior exhibits by varying the physical dimension. Now, capacitors with various values show that the low pass frequency has been tuned among the two-unit cell by the relation of dispersion and cutoff frequency. For this, researchers show the pass-band tenability of spoof SPPs. The use of 24 varactor diodes and 25 lumped inductors in various positions to manage the higher and lesser cutoff frequency of the filter of surface plasmon polaritons created a transmission line, which makes these alignments very heavy and complicated for a practical prototype [25].

12.5.4 Planner Ring Resonator

Metamaterials have achieved great attention for different microwave frequencies having a lot of applications because of their extraordinary properties. Such applications highlight the area of filters [40], components having multiple bands, elements for enhancement of bandwidth, design for absorbers, and much more. Metamaterial (MTM) is not found in nature. They are made in artificial manners. Its properties may be manipulated in different ways to distinguish values as required for the application. A metamaterial on an anchor medium is formed by the use of regular arrangements of materials (dielectric plus metallic) [41]. These naturally occurring substances contain positive ε_0 permittivity and μ_0 permeability. These properties may be negative. There is the possibility that any one of them can have a negative value. The two metamaterial classes are double negative (DNG) and single negative (SNG).

Non-planar ring resonator (NRR) contains different beneficial submissions like "ring laser gyroscopes" [42]. Siegman introduced the latest outlook of this NRR. Ring resonator is helpful for the designing of the cavity, improvement of the cavity, and lining of nonpolar resonators. A variety of research is available on the reflector "misalignment-induced optical-axis perturbation" of planar ring resonator (PRR), as well as NRR [43]. Non-planar ring resonator has adjusted disorder or "optical-axis stability." Four equal-sided non-planar rings having optical-axis agitation involve the reactivity of optical-axis dissemination as well as an understanding of optical-axis gradient SG. Optical-axis disseminated over the center depicts the influence to permit the optical axis within the resonator. Optical axis gradient gives the tendency to make the parallel by the optical axis to the lengthwise axis. The obtained results show clarity in response to the simulation. The properties like permeability as well as permittivity plus single cell index of refraction have been obtained by using "CST's resampling module" based on the "Robust retrieval" technique. By the NRW method, the outcomes are tested by a "MATLAB" sequencer. These two methods contain similar belongings through ENG performance as well as near-zero indexes. The unit cell having the equivalent circuit model is designed as well as validated. This is done by using ADS. The performance of the arrays is checked by measurement, which shows the characteristics of the triple band. They contain minor dimensions plus better EMR have projected that ENG metamaterials may be good substrates for applications (S, X-band, and C). It is used in applications like wireless communication [43].

12.5.5 Optical Computing

The word optical computing is derived from which means "electromagnetic radiation." It is not assumed a proper definition. In the field of computer science, a hierarchy continues which explains accurately the "computational complexity." This is consisting of the amount of state as well as how the state is examined. The research for optical computation is told about in 1858 as "Foucalt's knife-edge" test. Over the 65 years of old research, new activities occurred in this field [44]. In 1940 and 1960 in the first decade, holography and lasers respectively were dominated. This results in the form of synthetic aperture radar (SAR) processing as a result of combining lenses as Fourier transforms [45]. In 1970 for the first time, liquid crystals at room temperature give researched with the use of "analog spatial light modulators" (SLMs), which resulted in the form of coupling with a large number of LEDs. That decade gave the first survey of optical resistors and was the first entrance into optical instruments [46]. Micro-mirror technology was first introduced in about 1980. This technology gave a dense method of modulation of light rather than SLMs [47]. This technology also introduced a new platform that was based on an interferometer known as optical transistors [48]. In the 2000s, the research of ring resonators and many fields, such as complicated optics properties, became more popular like optical methods consisting of processing data networks [49]. Manual techniques are preferred in contrast to the optical transistor in most research. Now, the research on optical calculation was inadequate when the fabrication of the instrument would reach 100 nm. As a result, nanophotonic instruments are considered not to be experimental, and the instrument based on electronic devices are considered to have limitations based on scaling [50].

Guiding the significant characteristics of calculation, as well as their type of computation, this was the start of specific sessions of calculations as maintained. They included combinative judgment (CJ) having not any shape as well as limited shape androids (LSA) that have only one shape. It included the cast-down androids (CDA) having one shape. In pivoting machine (PM) has a no of shapes that are developed among somewhat arrangements. The calculation sessions have the boundary of specific calculations. So, LSA does not involve because its static shape could not be enhanced by the use of a directional picture which gives evidence about the random numbers. The PM is the most complicated whose properties describe recognized restrictions of calculation. They gave them the idea of presenting whether it is a limited class of symbols (digitally), as numeric values as presentation of analog.

The changes explain a new aspect of optical computing in the future including specialization of architecture as well as mechanisms and algorithms. This convergence gives a chance to make a communal to find out high capability in the field of optics to enhance novel competencies and changes in the calculation [51].

12.5.6 Photovoltaics

A photovoltaic (PV) is a device that converts solar energy directly into electrical energy. This is done by an effect called the photovoltaic effect. PV cells contain the least efficiency and high cost of production which abstain it to be used on a large scale. A lot of methods and techniques have been adapted to maximize the efficiency of these PV cells. It is tried to apply to lower the price of electricity. For this purpose, to increase the efficiency, various types of photovoltaics are made. In this regard, silicon (Si) based photovoltaic cells might be attached to lower the energy crisis [43]. In the early days of PV cells, a common discernment was that first-generation silicon-based solar cells, in the end, would be altered by a second-generation of lower cost thin-film technology. It was done by involving different types of semiconductors. Thin film polycrystalline has elements containing cadmium sulfide, cadmium telluride, and now silicon as the key value.

The solar cell produces electricity based on the energy they receive from the sun than having the ability to process. Mostly photovoltaic cells just convert 10% to 20% of the energy as they absorb it into electricity. Under perfect conditions in the most efficient cells laboratory, PV cells reach an efficiency round around 45%. The reason is that solar cells can convert electricity which they only store photons from in a specific frequency band. Above this band, all the photons are considered wasted. In the frequency band, some photons have less energy than the required energy to produce electrons. On the other hand, others have a lot of energy and thus the excess is wasted [52].

The efficiency of solar cells is calculated by an arrangement of charge carrier separation efficiency, as well as conductive efficiency. It can be determined by reflectance efficiency and thermodynamic efficiency. Solar cells cannot generate power in darkness. They store only some amount of energy. So, it can be used when the availability of light is not possible. This can be done by a process of charging electrochemical storage batteries, which follows the process of photosynthesis in plants. Photovoltaics are divided into three major types

1. Crystalline silicon cell
2. Thin film solar cell
3. Third-generation solar cells

On the smallest scale, photo voltaic allows us to power without batteries in watches and calculators. Solar power is used in hospitals and health centers in some countries to power water pumps as well as telephone boxes and even refrigeration units. Current developments are working to make a self-cleaning coating for solar panels. It is done to increase their efficiency as well as projects to minimize the waste of materials during fabrication. Thus, solar power will find local uses on a smaller scale in most developing countries with a satisfactory environment. Photovoltaics are also ideal for domestic and as well as small-scale commercial use [53].

12.6 Conclusion

In the field of metamaterial, it is clear that there is a variety of plasmonic materials. These materials can be used as basic components for all applications. The work can be done at all frequencies. Recently, the uses of metamaterial-based devices have limitations to the low-loss areas of available metals. These limitations are far away from the special telecommunication wavelengths. Plasmonic metamaterials are vast and divided into many types based on semiconductors as graphene-based metamaterials, 2D metamaterials, meta-surfaces, and self-assembled metamaterials. Based on their types, they play an important role on domestic as well as economic scales. These metamaterials are used in watches and calculators as low scale. To overcome the crisis of energy resources these metamaterials are used in photovoltaic which converts the sunlight into electricity. Such materials are used in the field of health and bio as used in cancer treatment. Transparent conducting oxides are playing an important role with the help of metamaterials in real life and unravel the interesting physical phenomena in a new generation of metamaterial and transformation-optics devices.

References

1. Yan, H., Li, X., Chandra, B., Tulevski, G., Wu, Y., Freitag, M., Zhu, W., Avouris, P., Xia, F., Tunable infrared plasmonic devices using graphene/insulator stacks. *Nat. Nanotechnol.*, 7, 330–334, 2012.

2. Radkovskaya, A., Tatartschuk, E., Sydoruk, O., Shamonina, E., Stevens, C.J., Edwards, D.J., Solymar, L., Surface waves at an interface of two metamaterial structures with interelement coupling. *Phys. Rev. B*, 82, 045430, 2010.
3. Echtermeyer, T.J., Milana, S., Sassi, U., Eiden, A., Wu, M., Lidorikis, E., Ferrari, A.C., Surface plasmon polariton graphene photodetectors. *Nano. Lett.*, 16, 8–20, 2016.
4. Politano, A. and Chiarello, G., The influence of electron confinement, quantum size effects, and film morphology on the dispersion and the damping of plasmonic modes in Ag and Au thin films. *Prog. Surf. Sci.*, 90, 144–193, 2015.
5. Pendry, J.B., Moreno, L., Vidal, F.J., Mimicking surface plasmons with structured surfaces. *Sci.*, 305, 847–848, 2004.
6. Viti, L., Coquillat, D., Politano, A., Kokh, K.A., Aliev, Z.S., Babanly, M.B., Tereshchenko, O.E., Knap, W., Chulkov, E.V., Vitiello, M.S., Plasma-wave terahertz detection mediated by topological insulators surface states. *Nano Lett.*, 16, 80–87, 2016.
7. Politano, A. and Chiarello, G., Unraveling suitable graphene–metal contacts for graphene-based plasmonic devices. *Nanoscale*, 5, 8215–8220, 2013.
8. Padilla, W.J., Basov, D.N., Smith, D.R., Negative refractive index metamaterials. *Mater. Today*, 9, 28–35, 2006.
9. Burns, G., *Solid state physics*, Academic Press Inc., Cambridge, 1985
10. Pendry, J.B. and Smith, D.R., Comment on `wave refraction in negative-index media: Always positive and very inhomogeneous. *Phys. Rev. Lett.*, 90, 029703, 2003.
11. Houck, A.A., Brock, J.B., Chuang, I.L., Experimental observations of a left-handed material that obeys Snell's law. *Phys. Rev. Lett.*, 90, 137401, 2003.
12. Barbillon, G., Faure, A.C., El Kork, N., Moretti, P., Roux, S., Tillement, O., Ou, M.G., Descamps, A., Perriat, P., Vial, A., Bijeon, J.L., Marquette, C.A., Jacquier, B., How nanoparticles encapsulating fluorophores allow a double detection of biomolecules by localized surface plasmon resonance and luminescence. *Nanotechnol.*, 19, 035705, 2007.
13. Shahbazyan, T.V. and Stockman, M.I., *Plasmonics: Theory and applications*, Springer, Dordrecht: Netherlands, 2013.
14. Ritchie, R.H., Plasma losses by fast electrons in thin films. *Phys. Rev.*, 106, 874–881, 1957.
15. Boardman, A.D., *Electromagnetic surface modes*, John Wiley, Chichester, New York, 1982.
16. Kreibig, U. and Vollmer, M., Theoretical considerations, in: *Optical Properties of Metal Clusters*, U. Kreibig and M. Vollmer (Eds.), pp. 13–201, Springer, Berlin, Heidelberg, 1995.
17. Zhu, Y., Sun, Z., Yan, Z., Jin, Z., Tour, J., Rational design of hybrid graphene films for high-performance transparent electrodes. *ACS Nano*, 5, 6472–6479, 2011.
18. Neto, A.H., Guinea, F., Peres, N.M.R., Novoselov, K.S., Geim, A.K., The electronic properties of graphene. *Rev. Mod. Phys.*, 81, 109–162, 2009.

19. Pospischil, A., Furchi, M.M., Mueller, T., Solar-energy conversion and light emission in an atomic monolayer p-n diode. *Nat. Nanotechnol.*, 9, 257–261, 2014.
20. Neto, A.C., Guinea, F., Peres, N.M., Novoselov, K.S., Geim, A.K., The electronic properties of graphene. *Rev. Mod. Phys.*, 81, 109, 2009.
21. Boltasseva, A. and Shalaev, V.M.J.M., Fabrication of optical negative-index metamaterials: Recent advances and outlook. *Metamaterials*, 2, 1–17, 2008.
22. Sutton, A.P., Metamaterials, in: *Concepts of Materials Science*, A.P. Sutton (Ed.), pp. 114–124, Oxford University Press, Oxford, England, 2021.
23. Martin, C.R., Nanomaterials: A membrane-based synthetic approach. *Science*, 266, 1961–1966, 1994.
24. Wu, C., Khanikaev, A.B., Adato, R., Arju, N., Yanik, A.A., Altug, H., Shvets, G., Fano-resonant asymmetric metamaterials for ultrasensitive spectroscopy and identification of molecular monolayers. *Nat. Mater.*, 11, 69–75, 2011.
25. Jaiswal, R.K., Pandit, N., Pathak, N.P., Spoof surface plasmon polaritons based reconfigurable band-pass filter. *IEEE*, 31, 218–221, 2019.
26. Link, S. and El-Sayed, M.A., Optical properties and ultrafast dynamics of metallic nanocrystals. *Annu. Rev. Phys. Chem.*, 54, 331–366, 2003.
27. Maier, S.A., *Plasmonics: Fundamentals and applications*, Springer, New York, 2007.
28. Barbillon, G., Ivanov, A., Sarychev, A.K., Hybrid Au/Si disk-shaped nanoresonators on gold film for amplified SERS chemical sensing. *Nanomater.*, 9, 1588, 2019.
29. Late, D.J., Huang, Y.K., Liu, B., Acharya, J., Shirodkar, S.N., Luo, J., Yan, A., Charles, D., Waghmare, U.V., Dravid, V.P., Rao, C.N., Sensing behavior of atomically thin-layered MoS_2 transistors. *ACS Nano*, 7, 4879–4891, 2013.
30. Zhang, Y., Chang, T.R., Zhou, B., Cui, Y.T., Yan, H., Liu, Z., Schmitt, F., Lee, J., Moore, R., Chen, Y., Lin, H., Jeng, H.T., Mo, S.K., Hussain, Z., Bansil, A., Shen, Z.X., Direct observation of the transition from indirect to direct bandgap in atomically thin epitaxial $MoSe_2$. *Nat. Nanotechnol.*, 9, 111–115, 2014.
31. Nair, R.R., Blake, P., Grigorenko, A.N., Novoselov, K.S., Booth, T.J., Stauber, T., Peres, N.M., Geim, A.K., Fine structure constant defines visual transparency of graphene. *Sci.*, 320, 1308–1308, 2008.
32. Simovski, C.R., Belov, P.A., Atrashchenko, A.V., Kivshar, Y.S., Wire metamaterials: Physics and applications. *Adv. Mater.*, 24, 4229–4248, 2012.
33. Wurtz, G.A., Dickson, W., O'Connor, D., Atkinson, R., Hendren, W., Evans, P., Pollard, R., Zayats, A.V., Guided plasmonic modes in nanorod assemblies: Strong electromagnetic coupling regime. *Opt. Express*, 16, 7460–7470, 2008.
34. Nasir, M., Peruch, S., Vasilantonakis, N., Wardley, W., Dickson, W., Wurtz, G., Zayats, A., Tuning the effective plasma frequency of nanorod metamaterials from visible to telecom wavelengths. *Appl. Phys. Lett.*, 107, 121110, 2015.
35. Ma, X., Hartmann, N.F., Baldwin, J.K., Doorn, S.K., Htoon, H., Room-temperature single-photon generation from solitary dopants of carbon nanotubes. *Nat. Nanotechnol.*, 10, 671–675, 2015.

36. Pendry, J.B., Holden, A.J., Robbins, D.J., Stewart, W.J., Magnetism from conductors and enhanced nonlinear phenomena. *IEEE Trans. Microw. Theory Tech.*, 47, 2075–2084, 1999.
37. Shekhar, P., Atkinson, J., Jacob, Z., Hyperbolic metamaterials: Fundamentals and applications. *Nano Converg.*, 1, 1–17, 2014.
38. Barbillon, G.J.M., *Plasmonics and its applications*, MDPI, Basel, Switzerland, 2019.
39. Martin, C.R., Nanomaterials: A membrane-based synthetic approach. *Science*, 266, 1961–1966, 1994.
40. Liu, Y. and Zhang, X., Metamaterials: A new frontier of science and technology. *Chem. Soc. Rev.*, 40, 2494–2507, 2011.
41. Soukoulis, C.M. and Wegener, M., Past achievements and future challenges in the development of three-dimensional photonic metamaterials. *Nat. Photonics*, 5, 523–530, 2011.
42. Khoo, E.H., Li, E.P., Crozier, K.B., Plasmonic wave plate based on subwavelength nanoslits. *Opt. Lett.*, 36, 2498–2500, 2011.
43. Li, Y. and Hong, M., Diffractive efficiency optimization in metasurface design via electromagnetic coupling compensation. *Mater. (Basel)*, 12, 1005, 2019.
44. Barbillon, G., Plasmonics and its applications. *Mater. (Basel)*, 12, 1502, 2019.
45. Lawrence, A., The ring laser gyro, in: *Modern Inertial Technology: Navigation, Guidance, and Control*, A. Lawrence (Ed.), pp. 208–224, Springer, New York, 1998.
46. Siegman, A.E., Laser beams and resonators: Beyond the 1960s. *IEEE*, 6, 1389–1399, 2000.
47. Arnaud, J.A., Degenerate optical cavities. II: Effect of misalignments. *Appl. Opt.*, 8, 1909–1917, 1969.
48. Williams, E.D., Ayres, R.U., Heller, M., The 1.7 kilogram microchip: Energy and material use in the production of semiconductor devices. *Environ. Sci. Technol.*, 36, 5504–5510, 2002.
49. Williams, E.J.T.F. and Change, S., Forecasting material and economic flows in the global production chain for silicon. *Technol. Forecast. Soc. Change*, 70, 341–357, 2003.
50. Woditsch, P. and Koch, W., Solar grade silicon feedstock supply for the PV industry. *Sol. Energy Mater. Sol. Cells*, 72, 11–26, 2002.
51. Hajjiah, A., Kandas, I., Shehata, N., Efficiency enhancement of perovskite solar cells with plasmonic nanoparticles: A simulation study. *MDPI*, 11, 1626, 2018.
52. Lu, G., Xu, J., Wen, T., Zhang, W., Zhao, J., Hu, A., Barbillon, G., Gong, Q., Hybrid metal-dielectric nano-aperture antenna for surface enhanced fluorescence. *Mater. (Basel)*, 11, 1435, 2018.
53. Magno, G., Bélier, B., Barbillon, G., Al/Si nanopillars as very sensitive SERS substrates. *Mater. (Basel)*, 11, 1534, 2018.

13

Nonlinear Metamaterials

M. Rizwan[1*], H. Hameed[1], T. Hashmi[1] and A. Ayub[2]

[1]*School of Physical Sciences, University of the Punjab, Lahore, Pakistan*
[2]*Department of Physics, University of the Punjab, Lahore, Pakistan*

Abstract
Nonlinear metamaterials are an exciting class of metamaterials. There have been various efforts to incorporate nonlinear optical effects in metamaterials through various manipulations. Nonlinear metamaterials can have a positive or negative and even a zero refractive index depending on different characteristics. Nonlinear metamaterials are essential in taking full advantage of electromagnetic radiation. Negative index metamaterials are of great significance. Nonlinear effects have a vast domain of applications, such as tunable acoustic metamaterials, reconfigurable refractive index, SRR microwave nonlinear electric metamaterials. Various nonlinear metamaterials such as liquid crystal, varactor, and ferrite-based nonlinear metamaterials are deliberated along with the general design of nonlinear metamaterials and nonlinear effects in these materials. Electric, magnetic, plasmonic, and dielectric nonlinear metamaterials are also some astonishing results of artificially made nonlinear materials. Applications of various nonlinear metamaterials are also explored briefly.

Keywords: Nonlinear metamaterials, negative refractive index, nonlinear effects, plasmonic metamaterials, split ring resonator, negative permittivity

13.1 Introduction

For a long period, physicists have been finding ways of fabricating materials whose nonlinear responses are speedy and broad. The frequency range of microwaves has displayed some advancement in this direction. Nonlinear optics can be utilized in ultra-soft, solid, low-power applications, i.e., at

**Corresponding author*: rizwan.sps@pu.edu.pk

Inamuddin and Tariq Altalhi (eds.) Electromagnetic Metamaterials: Properties and Applications, (287–308) © 2023 Scrivener Publishing LLC

higher frequencies, designing, and manufacturing of these materials would transform nonlinear optics. Advancements in different materials have been accelerated by the invention of metamaterials (MMs). Compound electromagnetic materials should be developed practically such that those materials have characteristics that make them significantly different from their parent components and are appropriate for the implementation of novel functionalism [1]. Over the last decade, many unusual linear optical characteristics have been demonstrated by optical metamaterials. For instance, the refractive index may be engineered to be positive, negative, or even zero by altering the size, periodicity, and other attributes of unit cells of MMs termed meta-atoms at any specified frequency [2]. Similarly, it is reasonable to infer that the same method may be used to change the nonlinear characteristics of MMs. Figure 13.1 provides the cylindrical and cubical view of a simple assembly configuration of nonlinear metamaterials.

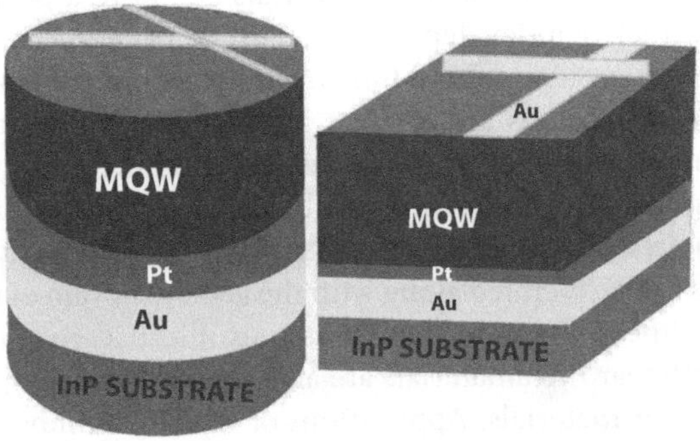

Figure 13.1 Simple basic structure of nonlinear Mms (cylindrical view of nonlinear MMs left, cubical view of MMs right).

The nonlinear optical effect is generated by the cumulative response of the fundamental constituents of MMs, which can be restricted by the characteristics of ordinary atoms or molecules. Some restrictions can be kept in line to maximize specific interactions, while others are rudimentary, dependent on the peculiarities of energy levels or transition elements that create nonlinearity. For many years, scientists have been concerned with the issue of a material's "biggest potential nonlinear optical response." The intriguing functions and theory of nonlinear events play an important role in the study of nonlinear material characteristics. Material is characterized by four parameters for efficient optical interactions and applications. The parameters are nonlinearity, optical transparency, suitable phase-matching conditions, and appropriate resistance to optical damage triggered by strong light radiations. Nonlinear terms are produced by methods for example electronic, atomic (including molecular movements), electrostatics, and thermal process to enable optical switching. When processing logic is required, the electronic method can assist in the quickest response using the Fs-Ps time scale.

A new and rousing gateway to electromagnetic theory and practice has been opened by the discovery of Metamaterials. Meta comes from the Greek word that has meanings of "beyond" or "more than." An artificially manufactured material that displays properties that do not exist in nature is called a nonlinear metamaterial. Its permittivity and material permeability specify its response to electromagnetic radiation. The Refractive index is the result of the product of permittivity and material permeability.

In contrast to naturally occurring materials, nonlinear metamaterials yield a negative refractive index. Moreover, their nonlinear response is more evident and notable. Nonlinear metamaterials belong to the class of negative refractive index metamaterials as the nonlinear behavior is because the macroscopic electric field part of electromagnetic waves could be smaller in comparison to the microscopic electric field part of inclusions. This turns out to be a great implement in intensifying the nonlinear behavior of metamaterials [3].

Metamaterials can be described by homogenous, constitutive parameters in exact analogy to natural materials. In contrast with their atomic counterparts, these patterns and particles show different behavior. These dissimilarities are beneficial as they can achieve properties that are either limited or unachievable by naturally occurring metamaterials. For an instance, a medium consisting of split ring resonators (SRR) can couple to incident magnetic fields to produce a section of negative magnetic permeability that can be produced by a medium composed of SSRs coupled to an incident magnetic field which cannot be achieved by naturally occurring materials.

For nonlinear optics nonlinear media is necessary. However, the nonlinear response is relatively weak for many optical materials such that a great change in electromagnetic field intensity leads to a small change in their properties. The discovery of Nonlinear MMs overcomes this constraint since their local fields are substantially greater than the average value of fields. Nonlinear metamaterials are being developed to better use electromagnetic radiation. Some characteristics of natural materials such as optical response and electromagnetic parameters can be tuned by specific engineering of the unit cell geometry of metamaterials. Here the unit cell indicates the fraction of wavelength of radiated electromagnetic waves that are geometrically and dimensionally arranged [4]. Moreover, permittivity and permeability can be tuned by changing the size and the structure of the unit cell. These two elements can influence the transmission of electromagnetic waves in a substance, hence extending the range of attainable optical and electromagnetic effects.

Optical qualities are not limited to mirrors, lenses, or other common materials. Victor Veselago introduced one of the greatest researched effects, which is the negative refractive index. As the name implies, these materials have the opposite optical characteristics of air, glass, and other common materials. Electromagnetic waves are refracted by negative index materials at high frequencies, in descriptive ways, to zero index materials. Furthermore, as energy transmits in a different direction, compensating mechanisms emerge [5].

A negative refractive index, an intentionally generated macroscopic feature, is of particular importance in nonlinear metamaterials developed via Negative Index metamaterials (NIMs). Nonlinear processes such as parametric amplification and second-harmonic production can cause anomalous behavior in NIMs. This happens due to the transmission of wave vector and Poynting vector of wave in NIM can transform the phase matching conditions for incident waves, which results in backward transmitting waves and spatial distributions of the interacting field intensity [6].

13.2 Nonlinear Effects in Metamaterials

The physical phenomena inspired by nonlinear effects may be examined both theoretically and empirically in an acoustic metamaterial having double negativity containing two kinds of scatterers, such that side holes and tissue positioned beside a pipe. This leads to the results that input acoustic intensities vary with the pass and forbidden bands associated with single and double negativity effects in metamaterials. The reason behind this variation is the nonlinear behavior of both kinds of scatterers [7]. The specific features linked with these characteristics of frequency bands of metamaterials can be interrupted by the nonlinearities; nevertheless, the nonlinear effects have the vast domain of applications, such as tunable acoustic metamaterials, reconfigurable refractive index, SRR microwave nonlinear electric metamaterials, and so on. Sugimoto 1992, first studied the effects of HRs using nonlinear wave equations with losses represented by fractional derivates terms on nonlinear acoustic waves, for acoustic metamaterial of single negative modulus, propagating in a long tunnel. They discovered that the exposure of shockwaves in a tunnel can be effectively stopped by suitably designed resonators [8].

Nonlinear effects are thought to be crucial for the progress of metamaterials, which is why the nonlinear behavior of electromagnetic parameters and optical response characteristics of metamaterials have been extensively explored, leading to the controlled synthesis of second harmonic and actively tunable artificial materials. The development of the second harmonic wave in acoustic MMS was investigated by correlations of optical and electromagnetic characteristics of metamaterials. As a result, nonlinearities in pipe-based metamaterials must be widely recognized, however, research on acoustics metamaterials nonlinearities lags far behind in the linear zone. Before the concept of metamaterials, the arrangement of Helmholtz resonators was employed to control shockwaves in a pipe [9]. Excitations of harmonic waves were investigated by nonlinear Bloch theory in a waveguide with a kind of scatterer with a solo side branch. This theory was used to examine harmonic waves within a metamaterial having a negative modulus. Furthermore, the consequence of nonlinearity on a cutoff frequency parallel to local resonance is investigated. The nonlinear response in the negative index medium explains the first example of direct phase matching of second harmonic waves in opposite directions to the fundamental wave. These nonlinear metamaterials gain a broad spectrum of this combination by merely disrupting the existing patterns. Exotic features of resonant components can be found in a restricted frequency region. However, when dealing with terahertz and optical frequencies, it causes loss, spatial dispersions, and several manufacturing issues [10]. Many forbidden gaps may exist in double negative metamaterials such as low-frequency gaps started at 0 Hz, single negativity double negativity as well as positivity forbidden gaps caused by discrete processes. Nonlinear effects in double-negative metamaterials are more sophisticated in comparison with single-negativity metamaterials due to structural complications and their distinctive frequency bands. These nonlinear effects can alter the distinctive frequency bands, disrupting the unique properties resulting from single and double negative effects for example acoustic isolation and negative refractive index, although nonlinearities may give opportunities to broaden the application sectors of metamaterials [11].

For the evolution of new metamaterials, nonlinear effects play an important role. This effect mostly pays attention to electromagnetic and optical waves yielding the actively harmonious artificial materials with many conspiring nonlinear phenomena, like a parametric down conversation (PDC) [12]. Second or third harmonic generation, and parametric shielding of electromagnetic fields. Metamaterials based on cylindrical pipes have sought more attention due to the high intensities of acoustic fields and resonance as we cannot neglect the nonlinear properties of acoustic. This is why we should thoroughly investigate the nonlinear effect of various sorts of the metamaterial. However, as previously noted, due to fewer data, nonlinear tubular metamaterials fell behind in the linear field [12].

13.3 Design of Nonlinear Metamaterials

The three primary classifications of left-hand materials are negative permittivity, permeability, and negative index materials, one of which belongs to the class of innovative materials structures. Because of their captivating applications in many areas of the field like waveguides and resonators, frequency modulators invisible cloaks, and so on, left-hand metamaterials have attained considerable attention in recent years. The trial and error approach is a standard process to make an optimum design for nonlinear optical that does not lead to a theoretical result and is incompetent but one can get enough knowledge about the non-linear geometric and devices as this process can be automated by deep learning. To optimize the second-order nonlinear material, an ideal plasmonic design needs to be created to form plasmonic configurations for a nonlinear metamaterial using a deep learning framework [5]. Smit *et al.* and Shelby *et al.* announced the first successful design, which is a 2-D periodic structural unit cell of SRRs along with copper bands. Following that, various novel designs and structures of left-hand materials, such as S-shaped resonators, were described. Because of their wire resonance behavior, the frequency bands in most SRR-based left-hand metamaterials are generally limited, although the permittivity bands are usually large, and the narrow DNG Frequency band is the major constraint that the researchers deal with. Designing new unit cells is still an important topic for researching innovative left-hand materials and expanding their uses. Various approaches for investigating and inquiring about the design or arrangement of left-hand materials and their possible uses have been proposed [13].

The form and geometry of the unit cell serve as interpretive factors for designing effective left hand metamaterials. Wires, spheres, and hollow slabs can also be used as unit cells. The geometry of the left-hand material should be simply permissible to acquire the appropriate characteristics of the negative index following possible use since it regulates the geometric parameters. For example, in a split ring resonator, the negative index of the resonator may be modified by altering the spacing between the inner rings and outer rings. This geometry of the resonators makes them more useful [14].

Since the publication of Smith *et al.* first's work demonstrating the occurrence of nonlinear metamaterials, these media have gained prominence. Researchers are currently experimenting with various design methodologies, electromagnetic characteristics, and applications. Metamaterials cannot be utilized in broad frequency areas. The reason is their narrow and intrinsic resonance frequency bands; though with the help of additional materials the creation of tunable metamaterials is possible, for example, LCs, Superconductors, Ferrites, MEMS, varactors, and some capacitors. Several of these are described below.

13.3.1 Liquid Crystal-Based Nonlinear Metamaterials

The main purpose of using LCs is that at optical frequencies, LCs can be utilized to change the band gaps (E_g) of photonic crystals as they have high optical anisotropy exceedingly sensitive to the exterior field [15].

Thus, LCs are extremely useful for controlling the permittivity of metamaterials and tuning reversed electromagnetic properties. Because LCs permittivity can be easily modified, they may be added to traditional metamaterials to create adjustable qualities [4]. For example, the refractive index of nanospheres distributed liquid crystals may be readily adjustable from a negative value to a positive value by adjusting the comparative permittivity of LCs. The permittivity of LCs is affected by the angle of the director axis. Thus an applied magnetic field or electric field or a different polarized optical field or temperature variation can control These properties but it also has some drawbacks for an instance the finite range of tunable materials [16].

13.3.2 Ferrite-Based Tunable Metamaterials

The nonlinear and tunable properties can also be achieved by metamaterials composed of ferrites and wires Dewar first reported this design method [17]. His model was made up of dielectric-clad wires inserted into ferrite hosts. He mathematically examined the permittivity, as well as the permeability of his module, and how to grasp negative index qualities using magnetic bias. The negative permeability band may be easily modified as a result of this biassing. Ferrites can also be added to standard resonator metamaterials since their characteristics are easily adjusted by changing the magnetic bias. This ferrite addition may also result in tunable dual bands or multiband metamaterials. Their creation is extremely simple [18]. Their tunability could be regulated by regulating the magnetic bias over a wide frequency range. Conventional ferrites, on the other hand, require a strong magnetic bias. However, there is a good likelihood that shortly, researchers may discover some specific ferrites that will function with extremely little magnetic bias to accomplish this adjustable negative magnetic permeability in the Giga Hz and Tera Hz zones.

13.3.3 Varactor/Capacitor-Loaded Tunable Metamaterials

Gil and colleagues Velez *et al.* were the first to disclose tunable metamaterials stacked with varactors, which are "diodes whose junction capacitance may be changed by an applied electric field" [19]. They demonstrated that tuning capability in metamaterial transmission lines may be seen using varactor-stacked split ring-ring resonators connected to microstrip lines and that the addition of a variable voltage source can regulate the tunability [20]. Left-hand metamaterials and series adjustable negative permeability metamaterials layered with varactors have previously been created and tested. Experimentally, the refractive index of the prism-like samples with different voltage biases illustrates the tunable negative refraction. The capacitor could be positioned in space among the inner and outer rings of the SRR. To tune the magnetic resonance frequency the capacitance of the capacitor is changed. These metamaterials have fascinating engineering applications such as filter antennas and nonlinear components [21].

13.3.4 Other Tunable Metamaterials

Lately, many different metamaterials accomplished by several distinct constituents are proposed and fabricated, for an instance, a tunable superconducting metamaterial was designed and experimentally proposed by using a fixed temperature superconductor it was reported and demonstrated by Ricci *et al.* Tunability may be accomplished by varying the direct current magnetic field and the Rf magnetic field. MEMS is a technique for controlling the composition and structure of unit cells [22]. THz frequency tunable metamaterials may be shown by altering the direction of a magnetic resonator or an electric resonator. Negative magnetic permeability μ metamaterials established on resonator structure shifting have recently been described at microwave and THz frequencies [23].

13.4 Nonlinear Properties of Metamaterials

In the past, only linear properties of left-hand materials or negative index MMs were studied during wave propagation, which shows that material permeability and permittivity are independent of the electromagnetic field intensity. But on the other hand, tunable structure designing requires the know-how of nonlinear properties where both the permittivity and permeability vary with the intensity of electromagnetic fields. The range of stop band spectra or transmission spectra is affected by this change, which results from the change in permeability by the intensity of the macroscopic magnetic field. By adjusting the field intensity swapping among positive and negative values is possible. As a consequence, left-handed materials can be switched to right-handed ones and vice versa. Left-hand materials cannot be created if the magnetic response is not correct [24].

Typical metals can be utilized as electromagnetic shielding materials because they reflect electromagnetic fields at particular frequencies. Linear left-hand metamaterials cannot be employed for this purpose, but when the nonlinearity of magnetic response is considered, a controlled shielding effect with a parametric reflection is found [25].

For different arrangements of constituent atoms and molecules, some materials display phase transition in their optical properties. The increment of the effect of phase transition on total optical properties and confining external stimuli, such as heat or light within phase transition material can be done by the hybridization of such materials with metamaterials [26].

For nonlinear materials, the linear proportionality with incident fields does not define the electromagnetic response, i.e., the materials' polarization and magnetization in the existence of electric and magnetic fields. All materials have a fast nonlinear response, for enough strong incident fields such that they respond at frequencies of the same order as the applied light. This response is a unique property of nonlinear mechanism only and can help in ultrafast switching from right-handed material to left-handed materials and can even induce changes in the incident wave frequency [27].

In the early stages, the novel properties exhibited by metamaterials urge research into a wide range of newly attainable phenomena which attracted many researchers. Rather than application-oriented the results were quite fundamental. For materials that are linear in the limit of weak incident fields, polarization and magnetization can be extended in power series, such that

$$P(t) = P^{(1)}(t) + P^{(2)}(t) + P^{(3)}(t) + \ldots \quad \text{(Eq. 13.1)}$$

and

$$M(t) = M^{(1)}(t) + M^{(2)}(t) + M^{(3)}(t) + \ldots \quad \text{(Eq. 13.2)}$$

where the superscript denotes the linear, quadratic and cubic field dependence. Here we are dealing with two types of nonlinear responses, namely second-and-third order harmonics and their associated phenomena. Second-order responses, i.e., $P^{(2)}(t)$ and $M^{(2)}(t)$ are related to non-centrosymmetric materials or materials which show no inversion symmetry Third-order responses i.e., $P^{(3)}(t)$ and $M^{(3)}(t)$ are associated with both centrosymmetric and non-centrosymmetric materials [28]. The properties of negative refraction can be demonstrated by the novel types of microstructure materials both theoretically and experimentally. The composite material in left-hand materials is generated using wire arrays and split ring resonators. For microwaves, left-hand materials made of wire arrays and split rings have a negative actual component of permeability and permittivity. Magnetic nonlinear behavior is discovered to be substantially greater in comparison with dielectric nonlinearity because of field amplification in the SRRs. This reliance allows it to alternate between positive and negative values. Such events can be accounted as a second phase shift initiated by outward electromagnetic fields. Veselago analyzed their properties long before, but recently they are experimentally validated [29].

Veselago observed that these Left-handed materials had negative refraction for inverse light pressure, reverse Doppler effect, and interface scattering. Left-hand metamaterials have been explored in the linear regime of wave transmission, where electromagnetic parameters are assumed to be independent of field strength. Nevertheless, understanding such materials' nonlinear characteristics is essential to build adjustable metamaterials in which the field strength modifies the transmission properties of left-handed materials, which is highly unique. For example, the effective permeability of a unit cell comprising SRRs and wires having nonlinear dielectric relies on the intensity of the macroscopic magnetic field in a nilpotent fashion that permits swapping among left- and right-hand materials by adjusting the field strength [30].

Structural losses are indicated by the imaginary part of effective permeability and might be controlled efficiently by the particular choice of the strength of the external high-frequency magnetic field. These are very important for future projections. The magnetic nonlinear behavior of the said structure is considerably greater than the corresponding electric nonlinear behavior due to resonant interactions of the EM field with the lattice of SRRs and the high value of the electric field in the gap of SRR. As a result, in left-handed materials, magnetic nonlinearity should predominate. Furthermore, the SRR gap can only be filled by nonlinear components that can be easily tuned by applying an external field. As electromagnetic waves transmitting in these kinds of materials always have restricted amplitude, the prospect of greatly amplified nonlinear behavior in left-hand metamaterials may need a fundamental rethink of linear theory-based notions. Creating such highly nonlinear compound materials would enable a variety of novel microwave domains such as limiters, beam spatial spectrum transformers, frequency multipliers, switchers, and so on [2].

13.5 Types of Nonlinear Metamaterials

Nonlinear metamaterials are engineered in such a way that they exhibit properties not found in nature, they are artificial materials with many desired properties, nonlinear behavior of metamaterials can be achieved by changing the natural parameters of electromagnetic radiations such as permittivity and permeability. These parameters are made negative so that the index of refraction can be negative which is the ultimate requirement of nonlinear materials. The index of refraction is tuned on specific frequencies to categorize types of metamaterials. These materials can be categorized into several types, in this article few types have been discussed such as.

13.5.1 Nonlinear Electric Materials

Metamaterials with negative refractive index commonly consist of compound structures, those structures show negative magnetic and electric responses on specific frequency bands simultaneously. Negative permittivity is usually created by arrangements of long metallic wires whereas negative permeability or negative magnetic responses are commonly achieved by split ring resonators. The long metallic wire approach has an advantage over SRRs which is wires provide a broadband electric negative response that can overlap the narrow band negative response of split ring resonators easily [29]. But in many applications for example transform optics which necessitate unit cells having nonperiodic arrangements as well as local distinctions in parameters of unit cells, long metallic wires are not suitable. To overcome this problem, electric resonators can be used which are assembled by two resonant loops consisting of the fundamental mode. In this way, the total magnetic dipole moment will be canceled which leaves a nonvanished electric dipole moment. Microwave and Terahertz bands show such structures. A nonlinear magnetic response that is very strong is a result of nonlinear resonant shift for SRRs; a similar approach has been taken by David A. Powell *et al.* in their work to design nonlinear electric resonators for a purpose to attain strong nonlinear electric response [31].

Figure 13.2 shows, the isotropic response is created by the introduction of two sets of boards that are perpendicular. Inside each resonator, to place varactor diodes extra gaps are introduced. The resonant frequency is in tune with the addition of capacitance in series. The geometry of the setup is such that copper-clad FR4 having 8 mm of height and width is used to fabricate resonators and 11 mm is the lattice period, with outer gaps having 0.4 mm separations, 1 mm track width, and 2.4 mm length [32].

The current is flowing in two rings of a resonator in opposite directions, which results in magnetic dipoles equal in magnitude and opposite in orientation; this gives net zero magnetic moment and electric moment with a dominant effect. Impotently net current is passing significantly through

Figure 13.2 Slab of nonlinear electric metamaterials [31].

the chief conductor having a varactor diode. Hence resonant frequency is modified strongly via incident power [31]. Figure 13.3 provides the low frequency and high frequency measurement of non linear metamaterial slab explained in Figure 13.2.

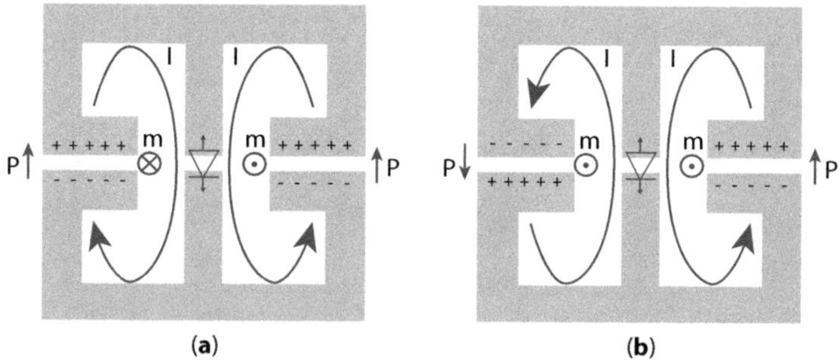

Figure 13.3 (a) Lower-frequency resonance. (b) Higher-frequency resonance [31].

13.5.2 Nonlinear Magnetic Metamaterials

Metamaterials are highly responsive to efficiently modifying magnetic fields. Magnetic metamaterials can be fabricated by changing the properties of SRRs (key building blocks for compound metamaterials more specifically with negative refractive index), and varactor diodes are introduced in each element of split Ring Resonators of compound structure, such that change in the amplitude of transmitting electromagnetic waves makes the entire structure tunable dynamically [6]. Split Ring Resonators are fabricated by introducing the photosensitive semiconductor or a varactor diode inside the resonator gap. The elements in the SRR can be tuned by high-power signals or by applying direct current voltage. Magnetic metamaterials show the Power dependent transmission properties, such as with the change in the resonant frequency transparent structure can be made opaque for higher incident powers. The reverse of the phenomenon can be observed by the nonlinearity of the structure [33].

Metamaterials are capable to manipulate magnetic fields which makes them a strong candidate for enhancement of the M-O effect, Magnetic field confinement, High-quality sensing, Wireless power transfer, Plasmonic perfect absorption, and MRI [34].

Several articles have been published following the tunability of magnetic metamaterials such that ferroelectric films are induced to SRRs substrate, with changing temperature electric permittivity is controlled and magnetic tenability is achieved.

Magnetic responses are tuned with photodoping of low-doped semiconductors inside split ring resonators by changing the semiconductor's conductivity with the help of an external light source [35].

Ferrite rods are added to the still unit cell of SRR and with external magnetic field inductance of ferrite rods is tuned magnetically which modulated the magnetic resonance. In a comparison of these methods, varactors are more practicable in the application of microwaves [35]. Figure 13.4 explains schematic representation of SRRs with their components inner radius (r), spacing (d), ring width (c), a typical SRR and its probing method.

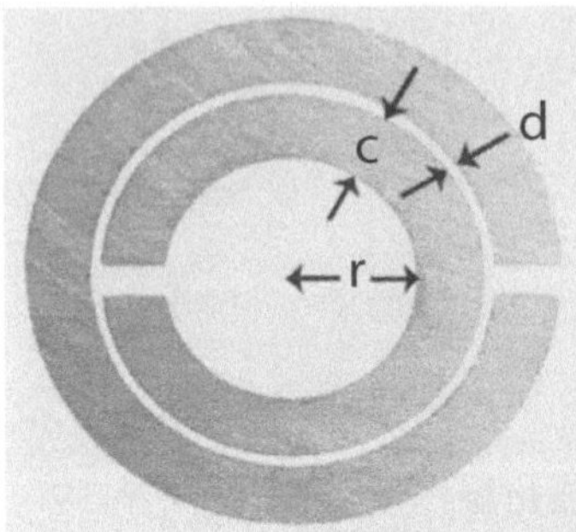

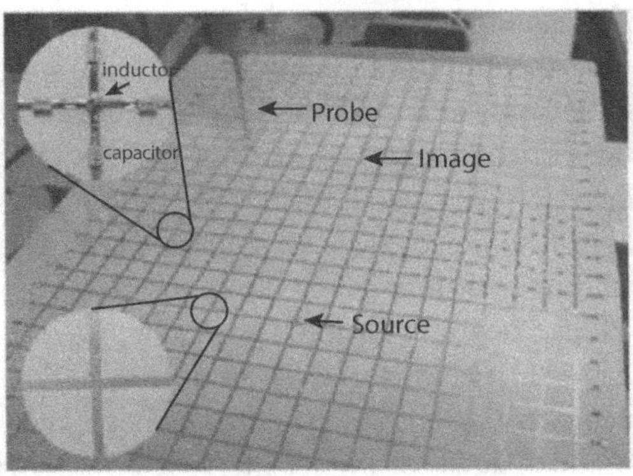

Figure 13.4 SSRs for nonlinear magnetic metamaterials [36].

Rogers R4003 having a thickness of 0.5 mm printed circuit board having 3.4 nominal dielectric constants is used to fabricate the metamaterial sample. Dielectric boards are made by allocation of an appropriate slot having copper nonlinear SRRs with the tin coat. Variable capacitor diode (Skyworks SMV-1405) is contained by each SRR, through which nonlinear current-voltage dependence is introduced, this gives each SRR a nonlinear magnetic dipole moment. SRR is arranged by a 2-D square lattice having $29 \times 4 \times 1$ unit-cell of 10.5 mm in size [33].

13.5.3 Plasmonic Nonlinear Metamaterials

A promising platform is provided by plasmonic metamaterials to explore new applications and complex optical effects, such as Weak processes that can be utilized and enhanced effectively by strong light-matter interaction which is the result of the enhancement of large local field. Miniaturized devices are constructed by ultra-soft optical processes and tight field confinements, which is working as bridges for all-optical photonic circuits. Plasmonic metamaterials spectra can be tuned by shifting the response of every single component and their couplings. It is expected that non-linear metamaterials have a much broader horizon, Such as non-linear optics in PMM. The large local fields are responsible for the strong enhancement of non-linear processes if non-linear materials are embedded into plasmonic nanostructures [37]. Optical properties of metamaterials can be obtained by plasmonic resonances of elements for example nanorods or SRR along with their electromagnetic couplings. Enhancement in the non-linear response is the result of the modification in the plasmonic resonance and their interactions, which is a consequence of refractive index modification of a substrate or inserted dielectric. In comparison with plasmonic crystals, the size and separation among the elements in metamaterials are much lesser than the incident wavelength [38].

13.5.4 Dielectric Nonlinear Metamaterials

For the characterization studies of materials having different shapes and sizes, the dielectric constant of materials needs to be determined precisely, as well as for scientific, industrial, and medical applications relating to the microwave locality. Dielectric parameter gives us information about electromagnetic wave interactions with matter and helps us to identify these interactions, this is very important for the development of various sensor and instrumentation studies in the food and biochemical field. Several methods have been suggested for the determination of dielectric constant in microwave regions, such as the transmission-reflection method, near field sensor method, free space method, and resonant method [39].

The resonant method is commonly preferred for the measurement of permittivity in comparison to other methods because of its sensitivity and high precision. Two resonant methods have been suggested the resonant perturbation method and the other is resonator method. Chakyar *et al.* have aided the resonant perturbation method for complex permittivity dimensions by using metamaterial SRRs for low dielectric values and small loss. SRRs are the structures of metamaterials that give negative permeability and permittivity, many methods have been reported regarding this shift in SRRs. SRRs are very efficient for the study of electrometric of liquids and solids because of their tunable resonant frequency with variations in environmental parameters. SRRs have very vast applications in magneto-inductive devices, sensors, and material characterizations [40]. Figure 13.5 provides setup of a sample probe for the detection of dielectric constant of different materials using resonant perturbation methods for low frequency split ring resonators (SRRs).

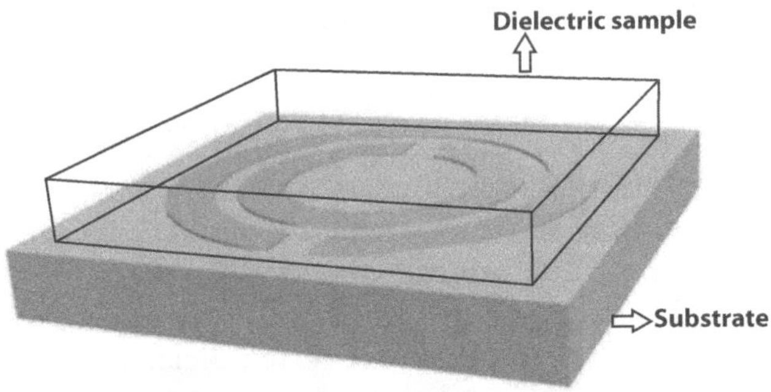

Figure 13.5 Dielectric constant and SRRs test prob [41].

The above picture is the sample prob for the determination of dielectric constant for different materials using resonant perturbation methods for low-frequency SRRs [41].

13.6 Applications

Researchers and material scientists have been working on metamaterials for the past few days as they have shown exotic properties, such properties are not found in nature. Non-linear metamaterials more specifically facilitate us to tune desired properties by tuning the negative index on particular frequencies, which allows non-linear metamaterials to be utilized in many applications such as.

13.6.1 Tunable Split-Ring Resonators for Nonlinear Negative-Index Metamaterials

In recent years, the study of microstructure materials that are engineered carefully has gathered the huge interest of scientists in these materials or metamaterials. These materials have shown great fascinating properties for transmitting electromagnetic waves more specifically negative refractive index [42].

Electromagnetic properties of metamaterials are highly dependent on the macroscopic and homogenous parameters, which are permittivity and permeability. Split ring resonators and short wires are structures used to make these parameters negative by careful engineering. To make the index of refraction negative in the narrow band frequency, negative magnetic permeability μ is given by an array of SRRs in the range of narrow frequency band just above resonance frequency, whereas negative permittivity is given by an array of short metallic wires in a wide range of frequency below effective plasma frequency. Linear metamaterials have been studied extensively where electromagnetic responses are not affected by an external field, efforts toward non-linear metamaterials have been made more specifically non-linear tuneability of split ring resonators [43]. Split ring resonators are LC resonators and the geometry of the ring is important to determine the resonant frequency. The single ring has a diameter of 7, 6 mm is the inner diameter and 0.7 mm is the width of the slit, and fabrication made on the 0.8-mm-thick PC board substrate is the SRR's geometry. The resonance frequency is 4 GHz. The resonance frequency can be taken down to 0.9 GHz when a high Q-capacitor having 2 pF capacitance is attached to the slit. The turns ratio between the SRR size and wavelength large to 50 is beneficial in miniaturizing the microwave device size [44].

Extra materials or extra components have to be introduced in SRRs to tune magnetic responses. Variable capacitance diodes are merged into the SRRs to produce a dynamic tunable system.

13.6.2 SRR Microwave Nonlinear Tunable Metamaterials

Microwave metamaterials are non-natural materials having a negative refractive index, such materials have a waste range of applications, i,e., super lenses and cloaking. Usually, SRRs use array units that provide strong magnetic coupling for tunable metamaterials. Recently, SRR-based nonlinear metamaterials have gathered huge interest; the reason is an enhancement in the strong local electromagnetic field in the unit cell of sub-wavelength [45].

It has already been discussed that the size and structures of metamaterials are much smaller than the incident wavelengths, hence electromagnetic properties of such material can be described by macroscopic and homogeneous parameters such as magnetic permeability and electric permittivity.

Both parameters can be made negative by careful engineering of the photonic atoms to attain the negative refractive index from metamaterials. The composite of SRRs and short metallic wires are the most extensively used structures for this type of engineering [44].

Nonlinear SRRs can be fabricated by inserting the varactor diode in the resonator's gap. This diode permits the elements in the SRR to be tuned by high input power or by DC-applied voltage. Properties of nonlinear metamaterials can be extensively tuned and desired properties can be obtained by changing the applied high input power [46].

13.7 Overview of Nonlinear Metamaterials

Nonlinear metamaterials are specially engineered metamaterials, which are developed in such a way that electromagnetic radiations can be manipulated in novel ways. Electromagnetic and optical properties of regular materials are commonly manipulated by using chemistry, but in metamaterials, these properties (electromagnetic and optical) are engineered with the unit cell geometry of materials. Unit cell materials are those in which the geometry is arranged with dimensions that are fractions of the wavelength of the incident wave [47].

By having control over the adjustment of size and configuration of unit cell many effects can be tuned which gives us control over permeability and permittivity. These are the parameters of determination of propagation of electromagnetic radiations through matter. Hence, the possible optical and electromagnetic effects can prolong [48].

Optical properties of materials are not limited to mirrors, lenses, and conventional materials, but can be prolonged beyond these capabilities. The negative index of refraction is the most studied effect, Victor Veselago proposed this in 1967. Optical properties exhibited by the negative index materials are opposite to air, glass, and other predictable materials. Electromagnetic waves are refracted by negative index materials at appropriate frequencies to negative or zero index materials. Moreover, energy propagation can take place in the opposite direction which may outcome in a compensation mechanism along with many other possibilities [49].

13.8 Conclusion

Nonlinear metamaterials also known as left-handed metamaterials are artificially manufactured and have properties opposite to those of linear or natural materials. Nonlinear properties depend on the intensity of electromagnetic fields. These materials are capable of working at high frequencies on which natural materials failed to respond. The intensity of electromagnetic radiation alters the permeability and permittivity of these materials. Metamaterials cannot be operated in some wide frequency areas due to their intrinsic nature and their narrow resonance frequency bands, but some tunable or nonlinear metamaterials have been created using different materials, such as varactors, capacitors, liquid crystals, superconductors, etc to operate in such wide frequency bands. Nonlinear metamaterials are capable of supporting combinations of both linear and nonlinear properties that are weak or unavailable in conventional nonlinear metamaterials.

The first illustration of direct phase matching of second harmonic waves in opposite directions to fundamental wave is explained by the nonlinear response in negative index medium A large spectrum of these combinations is obtained by these nonlinear metamaterials by simply disturbing the established patterns. Such exotic properties can be found in the narrow frequency band of resonant elements but when dealing with terahertz and optical frequency, it results in loss, spatial dispersions, and many fabrication problems. Metamaterial engineering of nonlinear optics allows us to make more modest adjustments to the existing nonlinear materials. This adjustment enhances the linear and nonlinear properties of metamaterials. Metamaterials have been utilized in many applications due to their fascinating properties, electric and magnetic non-linear metamaterials adopt a similar approach to solve different problems. These materials work on different frequency ranges, and their geometry is quite different from each other. Indeed, nonlinear metamaterials will be the most fascinating and attractive field for future researchers. With the help of these materials, many barriers in nonlinear optics can be overcome and a new gateway to research will be open.

References

1. Padilla, W.J., Basov, D.N., Smith, D.R., Negative refractive index metamaterials. *Mater. Today*, 9, 28–35, 2006.
2. Zharov, A.A., Shadrivov, I.V., Kivshar, Y.S., Nonlinear properties of left-handed metamaterials. *Phys. Rev. Lett.*, 91, 037401, 2003.
3. Engheta, N., Metamaterials with high degrees of freedom: Space, time, and more. *Nanophotonics*, 10, 639–642, 2021.
4. Werner, D.H., Kwon, D.-H., Khoo, I.-C., Kildishev, A.V., Shalaev, V.M., Liquid crystal clad near-infrared metamaterials with tunable negative-zero-positive refractive indices. *Opt. Express*, 15, 3342–3347, 2007.

5. Kshetrimayum, R.S., A brief intro to metamaterials. *IEEE*, 23, 44–46, 2004.
6. Boltasseva, A. and Shalaev, V.M., Fabrication of optical negative-index meta-materials: Recent advances and outlook. Elsevier, *Metamaterials*, 2, 1–17, 2008.
7. Fan, L., Chen, Z., Deng, Y.-C., Ding, J., Ge, H., Zhang, S.-Y., Yang, Y.-T., Zhang, H.J., Nonlinear effects in a metamaterial with double negativity. *Appl. Phys. Lett.*, 105, 041904, 2014.
8. Sugimoto, N.J., Propagation of nonlinear acoustic waves in a tunnel with an array of Helmholtz resonators. *J. Fluid Mech.*, 244, 55–78, 1992.
9. Bradley, C.E., Time-harmonic acoustic Bloch wave propagation in periodic waveguides. Part III. Nonlinear effects. *J. Acoust. Soc. Am.*, 98, 2735–2744, 1995.
10. Bradley, C.E., Time harmonic acoustic Bloch wave propagation in periodic waveguides. Part I. Theory. *J. Acoust. Soc. Am.*, 96, 1844–1853, 1994.
11. Wegener, M.J.S., Metamaterials beyond optics. *Science*, 342, 939–940, 2013.
12. Li, Y., Lan, J., Li, B., Liu, X., Zhang, J., Nonlinear effects in an acoustic metamaterial with simultaneous negative modulus and density. *J. Appl. Phys.*, 120, 145105, 2016.
13. Rose, A., Huang, D., Smith, D., Demonstration of nonlinear magnetoelectric coupling in metamaterials. *Appl. Phys. Lett.*, 101, 051103, 2012.
14. Rizwan, M., Mahmood, T., Rafique, H.M., Tanveer, M., Haider, S.F., Design of a negative refractive index material based on numerical simulation. *Chin. J. Phys.*, 54, 587–591, 2016.
15. Khoo, I.C., Werner, D.H., Liang, X., Diaz, A., Weiner, B., Nanosphere dispersed liquid crystals for tunable negative-zero-positive index of refraction in the optical and terahertz regimes. *Opt. Lett.*, 31, 2592–2594, 2006.
16. Zhang, F., Kang, L., Zhao, Q., Zhou, J., Lippens, D., Magnetic and electric coupling effects of dielectric metamaterial. *New. J. Phys.*, 14, 033031, 2012.
17. Dewar, G., A thin wire array and magnetic host structure with $n<0$. *J. Appl. Phys.*, 97, 10Q101, 2005.
18. Bing, C., Ming, Z., Kai, H., Numerical study on left-handed materials made of ferrite and metallic wires. *Chin. Phys. Lett.*, 23, 348–351, 2006.
19. Gil, I., Bonache, J., Garcia, J., Martin, F., Tunable metamaterial transmission lines based on varactor-loaded split-ring resonators. *IEEE*, 54, 2665–2674, 2006.
20. Aydin, K. and Ozbay, E., Capacitor-loaded split ring resonators as tunable metamaterial components. *J. Appl. Phys.*, 101, 024911, 2007.
21. Li, S., Elsherbeni, A.Z., Yu, W., Li, W., Mao, Y., A novel tunable triple-band left-handed metamaterial. *Int. J. Antennas Propag.*, 2017, 7583736, 2017.
22. Tao, H., Strikwerda, A.C., Fan, K., Padilla, W.J., Zhang, X., Averitt, D.R., MEMS based structurally tunable metamaterials at terahertz frequencies. *J. Infrared Millim. Terahertz Waves*, 32, 580–595, 2011.

23. Zhong, J., Huang, Y., Wen, G., Sun, H., Zhu, W., The design and applications of tunable metamaterials. *Proc. Eng.*, 29, 802–807, 2012.
24. Bayindir, M., Aydin, K., Ozbay, E., Markoš, P., Soukoulis, C.M., Transmission properties of composite metamaterials in free space. *Appl. Phys. Lett.*, 81, 120–122, 2002.
25. Feng, S. and Halterman, K., Parametrically shielding electromagnetic fields by nonlinear metamaterials. *Phys. Rev. Lett.*, 100, 063901, 2008.
26. Goldflam, M., Driscoll, T., Chapler, B., Khatib, O., Jokerst, N., Palit, S., Smith, D., Kim, B.-J., Seo, G., Kim, H.-T., Reconfigurable gradient index using VO_2 memory metamaterials. *Appl. Phys. Lett.*, 99, 044103, 2011.
27. Pendry, J.B., Time reversal and negative refraction. *Science*, 322, 71–73, 2008.
28. O'Brien, K., Suchowski, H., Rho, J., Salandrino, A., Kante, B., Yin, X., Zhang, X., Predicting nonlinear properties of metamaterials from the linear response. *Nat. Mater.*, 14, 379–383, 2015.
29. Veselago, V.G., The electrodynamics of substances with simultaneously negative values of ε And μ. *Sov. Phys. Usp.*, 10, 509–514, 1968.
30. Luo, C., Johnson, S.G., Joannopoulos, J.D., Pendry, J.B., All-angle negative refraction without negative effective index. *Phys. Rev. B*, 65, 201104, 2002.
31. Powell, D.A., Shadrivov, I.V., Kivshar, Y.S., Nonlinear electric metamaterials. *Appl. Phys. Lett.*, 95, 084102, 2009.
32. Schurig, D., Mock, J.J., Justice, B.J., Cummer, S.A., Pendry, J.B., Starr, A.F., Smith, D.R., Metamaterial electromagnetic cloak at microwave frequencies. *Science*, 314, 977–980, 2006.
33. Shadrivov, I.V., Kozyrev, A.B., Weide, D.V.D., Kivshar, Y.S., Nonlinear magnetic metamaterials. *Opt. Express*, 16, 20266–20271, 2008.
34. Zhao, X., Duan, G., Wu, K., Anderson, S.W., Zhang, X., Intelligent metamaterials based on nonlinearity for magnetic resonance imaging. *Adv. Mater.*, 31, 1905461, 2019.
35. Zharov, A.A., Shadrivov, I.V., Kivshar, Y.S., Nonlinear properties of left-handed metamaterials. *Phys. Rev. Lett.*, 91, 037401, 2003.
36. Porfyrakis, P. and Tsitsas, N.L., Nonlinear electromagnetic metamaterials: Aspects on mathematical modeling and physical phenomena. *Microelectron. Eng.*, 216, 111028, 2019.
37. Yao, K. and Liu, Y., Plasmonic metamaterials. *Nanotechnol. Rev.*, 3, 177–210, 2014.
38. Boardman, A.D. and Zayats, A.V., Nonlinear plasmonics, in: *Handbook of Surface Science*, N.V. Richardson and S. Holloway (Eds.), pp. 329–347, Elsevier, North-Holland, 2014.
39. Kivshar, Y.J., All-dielectric meta-optics and non-linear nanophotonics. *Natl. Sci. Rev.*, 5, 144–158, 2018.
40. Qin, M., Pan, C., Chen, Y., Ma, Q., Liu, S., Wu, E., Wu, B., Electromagnetically induced transparency in all-dielectric U-shaped silicon metamaterials. *MDPI*, 8, 1799, 2018.

41. Chakyar, S.P., Simon, S.K., Bindu, C., Andrews, J., Joseph, V.P., Complex permittivity measurement using metamaterial split ring resonators. *J. Appl. Phys.*, 121, 054101, 2017.
42. Shadrivov, I.V., Morrison, S.K., Kivshar, Y., Tunable split-ring resonators for nonlinear negative-index metamaterials. *Appl. Phys. Lett.*, 14, 9344–9349, 2006.
43. Saeed, K., Shafique, M.F., Byrne, M.B., Hunter, I.C., Planar microwave sensors for complex permittivity characterization of materials and their applications, in: *Applied Meaurement Systems*, M.Z. Haq (Ed.), pp. 319–350, InTech, Croatia, 2012.
44. Wang, B., Zhou, J., Koschny, T., Soukoulis, C.M., Nonlinear properties of split-ring resonators. *Opt. Express*, 16, 16058–16063, 2008.
45. Li, D., Zhu, J., Wu, S., Xiong, X., Liu, Y., Liu, Q.H., Field-circuit co-simulation for microwave metamaterials with nonlinear components, in: *Proceedings of the International Symposium on Antennas & Propagation*, pp. 1131–1133, 2013.
46. Shadrivov, I.V., Kozyrev, A.B., van der Weide, D.W., Kivshar, Y.S., Tunable transmission and harmonic generation in nonlinear metamaterials. *Appl. Phys. Lett.*, 93, 161903, 2008.
47. Turco, E., Barchiesi, E., Giorgio, I., Dell'Isola, F., A Lagrangian Hencky-type non-linear model suitable for metamaterials design of shearable and extensible slender deformable bodies alternative to Timoshenko theory. *Int. J. Nonlinear Mech.*, 123, 103481, 2020.
48. Sihvola, A.J., Metamaterials in electromagnetics. *Metamaterials*, 1, 2–11, 2007.
49. Gardezi, S.M., Pirie, H., Carr, S., Dorrell, W., Hoffman, J.E., Simulating twistronics in acoustic metamaterials. *2D Mater.*, 8, 031002, 2021.

14

Promising Future of Tunable Metamaterials

Tanveer Ahmad Wani[1]* and A. Geetha Bhavani[2]

[1]*Department of Physics, Noida International University Greater Noida, UP, India*
[2]*Department of Chemistry, Noida International University Greater Noida, UP, India*

Abstract

There are certain non-natural electromagnetic properties of materials like negative index of refraction exhibited by a new class of materials, known as metamaterials. The electromagnetic waves interact with the inclusions, inducing electric and magnetic moments, which in turn affect the macroscopic effective permittivity and permeability of the bulk composite "medium." Effective permittivity and effective permeability are basic engineering parameters of metamaterials. During the past decade, there had been substantial development in their fabrication and design. This has led to a new excitement in the scholarly ranks. Metamaterials are expected to have an impact across the entire range of technologies where electromagnetic radiation are used and will provide a flexible platform for technological advancement. Nowadays, a large number of products are produced and commercialized. Starting from longer wavelengths in the microwave spectral domain, the commercialization efforts are moving to higher frequencies. It is expected that further developments will lead to the applications of tunable metamaterials at higher frequency ranges, including terahertz (THz), infrared and visible. The future belongs to the metamaterials and the incredible revolution because of them.

Keywords: Metamaterials, electromagnetism, permittivity, permeability, tunable

**Corresponding author*: tanveer.ahmad@niu.edu.in

Inamuddin and Tariq Altalhi (eds.) Electromagnetic Metamaterials: Properties and Applications, (309–326) © 2023 Scrivener Publishing LLC

14.1 Introduction

A metamaterial is made up of an assembly of unit cells known as meta-atoms. Each meta-atom has a size much smaller than its interacting wavelength. These are made microscopically from dielectrics and metals—the conventional materials. But, the shape, size, and geometry of meta-materials can disturb light macroscopically (Figure 14.1). This may result in resonances and uncommon values for permittivity and permeability (ε & μ).

Metamaterial can be characterized by using Maxwell equations. Transformation of Maxwell equations have a prominent role in describing metamaterial, which is given below:

$$\nabla \times \vec{E} = -j\omega\mu\vec{H};\ \nabla \times \vec{D} = \rho$$
$$\nabla \times \vec{H} = \vec{J} + j\omega\epsilon\vec{E};\ \nabla \times \overrightarrow{B} = 0$$

For the plane waves, these equations can be reduced to

$$\vec{k} \times \vec{E} = \omega\mu\vec{H};\ \vec{k} \times \vec{H} = -\omega\epsilon\vec{E}$$

Therefore, for positive ε and μ, $\vec{E}, \vec{H}, \vec{k}$ form a right-handed orthogonal system. When ε and μ are negative the above equation changes to

$$\vec{k} \times \vec{E} = -\omega\mu\vec{H};\ \vec{k} \times \vec{H} = \omega\epsilon\vec{E}$$

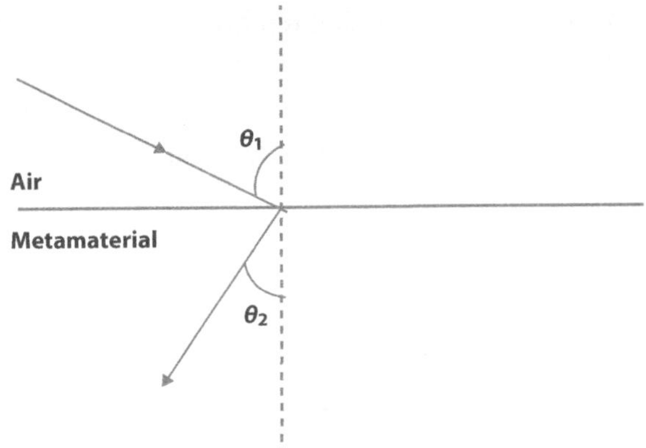

Figure 14.1 Light incident from air to metamaterial.

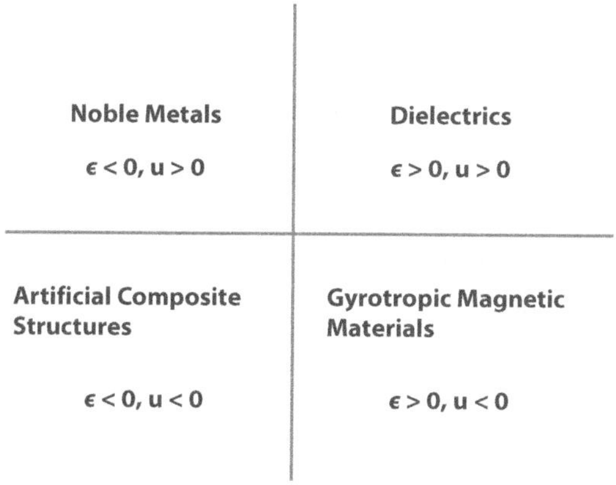

Figure 14.2 Classification based on ε and μ.

The above case shows left-handed materials and their opposite direction and left-hand triplet of $\vec{E}$, $\vec{H}$, $\vec{k}$.

It is well known that the response of a system to the presence of an electromagnetic field is determined to a large extent by the properties of the materials involved. We describe these properties by defining the macroscopic parameters permittivity ε and permeability μ of these materials. The electric permittivity and permeability can be used for classifying the meta-materials and grouped into four combinations as given below (Figure 14.2):

a) $\varepsilon > 0$ and $\mu > 0$
c) $\varepsilon > 0$ and $\mu < 0$
d) $\varepsilon < 0$ and $\mu > 0$
e) $\varepsilon < 0$ and $\mu < 0$

A group of metamaterials known as left-handed metamaterials, exhibit both negative ε and μ [1]. The term originated since the wave vector is anti-parallel as compared with the cross product of electric and magnetic fields in right-handed metamaterials. These negative values result in negative index of refraction. This offers a boundless opportunity in this class of materials and in turn has garnered attention among research community.

14.1.1 Examples of Metamaterials

Meta-materials have subwavelength nature. Due to this, the metamaterials operating at microwave frequency have a unit cell of the size in millimeters.

The unit cell of metamaterial bears a size of nanometers in case of visible spectrum. Another important property of metamaterials is that they can absorb light in a fine frequency range and in turn absorb/block a specific color. The electromagnetic properties of metamaterials are understood from the composition of nanoscale objects, in contrary, to the conventional materials. The electromagnetic properties in case of conventional materials are understood at atomic/molecular level. This difference leads to the study of permittivity and permeability as easily tuned.

14.1.1.1 Electromagnetic Metamaterials

A metamaterial affects lesser on electromagnetic waves as compared to wavelength of electromagnetic radiation.

i. Single negative metamaterials (SNG): Single negative metamaterials have either negative permittivity or negative permeability.
ii. Double negative metamaterials (DNG): Double negative metamaterials are the metamaterials that have both permittivity and permeability as negative with negative index of refraction.
iii. Electromagnetic band gap metamaterials (EBG): Electromagnetic band gap metamaterials control the propagation of light.
iv. Bi-isotropic and bianisotropic metamaterials: In many examples of electromagnetic metamaterials, the electric field causes magnetic polarization, and the magnetic field induces an electrical polarization, i.e., magneto electric coupling.

14.1.1.2 Chiral Metamaterials

Chiral metamaterials consist of arrays of dielectric gammadions or planar metallic on a substrate. When a linearly polarized light is incident on the array, it becomes elliptically polarized upon interaction with the gammadions with the same handedness as the gammadion itself.

14.1.1.3 Terahertz Metamaterials

Terahertz metamaterials are the combination of artificial materials that interact at terahertz (THz) frequencies. Terahertz waves lie just before the start of the microwave band to far end of the infrared band.

14.1.1.4 Photonic Metamaterials

Photonic metamaterials are the type of electromagnetic metamaterials that are designed to interact with optical frequencies. Photonic metamaterials radiate the source at optical wavelengths.

14.1.1.5 Tunable Metamaterials

These are the metamaterials that have the ability to randomly change the frequency of a refractive index. An incident electromagnetic wave gives variable response with these metamaterials. This includes how an incident electromagnetic wave interacts with a metamaterial in remote controlling.

14.1.1.6 Frequency Selective Surface Based-Metamaterials (FSS)

FSS-based metamaterials are the substitute to the fixed frequency metamaterials with static geometry and spacing in the unit cells used to find out the frequency response of a given metamaterials.

14.1.1.7 Nonlinear Metamaterials

Nonlinear metamaterials are artificial materials in which the nonlinearity exists. This is due to less macroscopic electric field of the electromagnetic source than the microscopic electric field of the inclusions.

14.2 Tuning Methods

Since the existence of the negative index type of media [1–3], researchers across the globe have shown a special interest in the study of metamaterials. Few studies are mentioned below:

a) The studies by Parimi *et al.* [2] and Smith *et al.* [4] have explained the methods of design.
b) Aydin, K. *et al.* [5] Qiang, D *et al.* [6] have focused on the electromagnetic properties.
c) The applications of metamaterials have been demonstrated in the studies of Tao, H. [7] Pendry, J. B *et al.* [8].

As discussed, the metamaterials can absorb light in a fine frequency range, thereby, cannot be used in a wide range of frequencies. The tunable metamaterials have been proposed are given as:

14.2.1 Tuning by Additional Materials

Tuning by additional material includes the following:

a) Liquid crystals [9–21]- A thermodynamic stable phase characterized by anisotropy of properties without the existence of a three-dimensional crystal lattice, generally lying in the temperature range between the solid and isotropic liquid phase, hence the term mesophase.
b) Ferrites [22–34]- Ferrites are explained as any of a group of nonmetallic, ceramic-like, usually ferromagnetic compounds of ferric oxide with other oxides, especially a compound characterized by extremely high electrical resistivity.
c) Varactors [35–37]- A varactor (or a tuning diode) is a diode designed to be used as a voltage-controlled capacitor.
d) Capacitors [38, 39]- The capacitor is a component which has the ability or "capacity" to store energy in the form of an electrical charge producing a potential difference (Static Voltage) across its plates, much like a small rechargeable battery.
e) Superconductors [40]- A superconductor is a material that can conduct electricity or transport electrons from one atom to another with no resistance.
f) MEMS [41]- MEMS is an acronym that stands for microelectromechanical systems. It describes a manufacturing technology used to create microscale integrated devices or systems that combine mechanical and electrical components. These devices and systems have the ability to sense, control and actuate on the micro scale, and generate effects on the macro scale.
g) Tunable structures [42–44] and applications [45–50]- Active materials are capable of converting free energy into directional motion, giving rise to notable dynamical phenomena. Developing a general understanding of their structure in relation to the underlying nonequilibrium physics would provide a route toward control of their dynamic behavior and pave the way for potential applications.

14.2.2 Tuning by Changing the Structural Geometry

This is achieved by mechanically deforming the shape of the constituent elements or their mutual arrangement in the metamaterial, which affects the overall properties due to mutual coupling between these elements, e.g., elastic deformation of the structure.

14.2.3 Tuning by Changing the Constituent Materials

This is achieved by changing the properties of the materials composing the individual metaatoms, e.g., actively changing the conductivity of semiconductors by injecting free electrons into them.

14.2.4 Tuning by Changing of the Surrounding Environment

Achieved by immersing the metamaterial in an environment, which properties can change, e.g., by a liquid crystal (LC).

14.3 Types of Tunable Metamaterials

14.3.1 Thermally Tunable Metamaterials

There are certain properties that vary with the changing temperatures. The common materials for thermally tunable metamaterials are Vanadium Dioxide and Germanium Antimony Telluride. At a transition temperature of 68°C [51], the vanadium dioxide changes from metal to dielectric or vice versa. The surface Plasmon resonance in Au-VO2 [52–55] has been attained thermally. Similarly, Germanium Antimony Telluride experiences a phase transition from amorphous to crystalline [56]. The thermal tuning of metamaterials provides compatibility and enough tuning range in integrated circuits [57]. The shortcoming of thermal tuning of metamaterials is slow modulation speed. This is primarily because of required time in heating/cooling. Further, the thermal tuning scope is outside the visible spectrum.

14.3.1.1 Optically Driven Tunable Metamaterials

The disadvantages found in thermally tunable metamaterials can be eradicated by changing the optical sources. Thus, optically driven tunable metamaterials have fast modulation speed and also tuning scope is within visible spectrum. However, these tunable metamaterials have a narrow range and cannot be accustomed to each cell. The application includes high-capacity communications and real time holograms. The quick response makes them in adaptive optics.

14.3.2 Structurally Deformable Metamaterials

The structurally deformable metamaterials are having effective exceptional properties as a result of arrangement of structures. These properties can be tuned by external excitations. These external excitations can distort the shape and structure.

14.3.3 Electrically Tunable Metamaterials

The electromagnetic properties of metamaterials are changed by electromagnetic tuning. This is achieved by controlling the number of electrical carriers and adjusting the band gap. Another method of obtaining the tunability is by designing two dimensional structures and liquid crystals. The electrical carriers are increased by applying the voltage, which in turn increases carrier density; thereby changing its optical properties. Indium Tin Oxide is an example of electrically tunable metamaterial.

14.4 Significant Developments

The solutions provided by metamaterials offer a considerable opportunity in the area of science and engineering (Figure 14.3).

Few technologies are listed below:

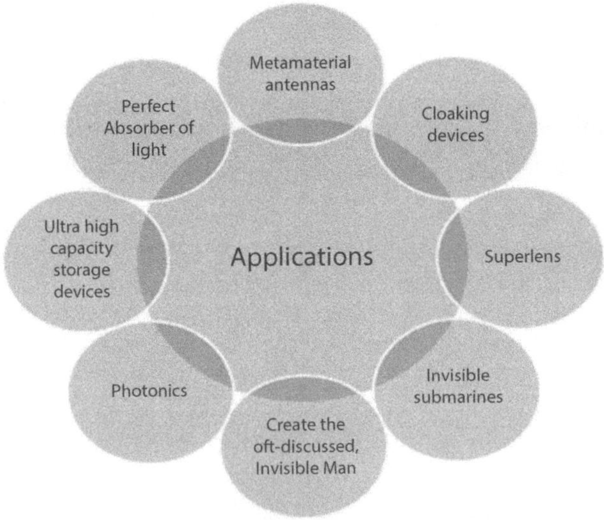

Figure 14.3 Applications of metamaterials.

14.4.1 Vehicles with Mobile Broadband

The vehicles are built with flat panels, which provide them mobile broadband. These vehicles are light and slim.

14.4.2 Transportation Security Administration

Evolv Edge has developed a visitor screening machine with great automation. The solution is handling approximately 900 walkthroughs per hour. It also provides explosion detection, firearm and face recognition solutions.

14.4.3 Tracking Planes, Trains, and Automobiles

An Echodyne's system claims to track a plane, phantom drone and a palm sized drone at 3 Km, 750 m and 200 m, respectively. The system also tracks down ground objects with vehicles as well as pedestrians at two miles and one mile, respectively.

14.4.4 Holographic Something

The technology uses holographic beamforming by wireless service providers. The advantage of this technology is its capacity to increase speed and reuse same band of spectrum. It aims to provide fifty times cheap solutions as of available today.

14.4.5 Wireless Charging with Metamaterials

Metaboards technology aims at developing wireless charging. This provides us charging various devices without plugging.

14.4.6 Seeing Around Corners with Radar

A radar system in amalgamation with artificial intelligence and metamaterials is capable of seeing around the corners.

14.4.7 Manipulating Light

Metamaterial Technologies Inc has come up with a variety of products like:

a) Invisible meta-mesh for touch screen equipments

b) Method for making light emitting diodes effective
c) Solar film
d) Laser-glare protection solutions.

14.4.8 Sound-Proof 'Invisible Window'

This is a type of invisible window that blocks sound and easily lets air in. The technology can be used to reduce noise pollution in buildings, pub gardens and traffic

14.4.9 Terahertz Instruments

In order to characterize materials, tetra hertz instruments offer a solution. The technology aims to work in defect and fault detection areas in semi-conductor industry.

14.5 Future

Much work is being undertaken (Table 14.1) in the recent present and significant progress is being made to create the protective shields for radiation and protective shields against seismic activity. The walls built from metamaterials are supposed to rebound seismic waves. It is very much possible that metamaterials be used in construction industry in the near future in order to provide defense against natural disasters. These could also tap off some energy and recycle for some other uses.

Metamaterials focus on greener technology. The advancement in solar batteries and radio noise harvesting has garnered interest in researchers to design metamaterials that could tap off the energy.

The magneto inductive waves can lead the power and data transfer fields to an unimaginable level. The wirelesss charging for devices without any alignment shall pay a way for modern data connectivity.

Metamaterials provide solutions for technologies related to housing and transportation. The issues of safety and security, as well as the environmental impacts could be addressed. The high-quality imaging in health technologies would pay a way for improved personal data security.

Table 14.1 Recent research in the field of metamaterials.

S. no.	Recent research in metamaterials
1	Semiconductor nanowire metamaterial for broadband near-unity absorption [58]
2	Terahertz thermal curve analysis for label-free identification of pathogens [59]
3	Integrated photonic metasystem for image classifications at telecommunication wavelength [60]
4	Reaching the highest efficiency of spin Hall effect of light in the near-infrared using all-dielectric metasurfaces [61]
5	The visual appearances of disordered optical metasurfaces [62]
6	Pixel-level Bayer-type color router based on metasurfaces [63]
7	All-optical switching based on plasmon-induced Enhancement of Index of Refraction [64]
8	Light absorption enhancement in thin film GaAs solar cells using dielectric nanoparticles [65]
9	High-efficiency broadband achromatic metalens for near-IR biological imaging window [66]
10	Ultra-compact snapshot spectral light-field imaging [67]
11	High-precision digital terahertz phase manipulation within a multichannel field perturbation coding chip [68]
12	Full-color enhanced second harmonic generation using rainbow trapping in ultrathin hyperbolic metamaterials [69]
13	Cavity-enhanced linear dichroism in a van der Waals antiferromagnet [70]
14	Perovskite metasurfaces with large superstructural chirality [71]
15	On-chip zero-index metamaterials [72]
16	Near-infrared to ultra-violet frequency conversion in chalcogenide metasurfaces [73]
17	Dynamic recognition and mirage using neuro-metamaterials [74]

(*Continued*)

Table 14.1 Recent research in the field of metamaterials. (*Continued*)

S. no.	Recent research in metamaterials
18	High-capacity topological coding based on nested vortex knots and links [75]
19	Electromagnetic wave-based extreme deep learning with nonlinear time-Floquet entanglement [76]
20	Heat transfer control using a thermal analogue of coherent perfect absorption [77]
21	Inverse design enables large-scale high-performance meta-optics reshaping virtual reality [78]
22	On-chip nanophotonic topological rainbow [79]
23	An ultrastrongly coupled single terahertz meta-atom [80]
24	An Archimedes' screw for light [81]
25	A broadband achromatic metalens in the visible [82]
26	Nonlinear metamaterials for holography [83]
27	Digital metamaterials [84]

14.6 Conclusion

The metamaterial market was valued at USD 305 million in 2021 and is projected to reach USD 1457 million by 2026. The use of metamaterials in solar power systems, metamaterial-based radar for drones, and advancements in 5G are the key factors driving the growth of metamaterial market.

Increasing demand for metamaterials for solar power systems and metamaterial-based radar for drones, as well as advancements in telecommunication technologies, is creating opportunities for the market. The factors driving the growth of the meta material market are the variety in design functionalities, anti-laser coating application, and development in radar and LiDAR for autonomous vehicles. The metamaterial market in North America is projected to reach 834.8 million by 2026; it is expected to grow at a CAGR of 32%. The metamaterial market growth for North America can be attributed to the increasing demand from the aerospace and defense vertical, Government agencies, such as US DOD, DARPA, and NASA, and extensively funded research universities and industry players for the development of metamaterial-based antennas.

Metamaterial-based products are still in the R&D phase. Hence, for designing techniques, industry players rely on research centers or universities. As a result, metamaterials are synthesized in very small quantities. Sometimes, these techniques are not suitable for the mass production of metamaterials. Due to the unavailability of a suitable technology for mass production, the cost of metamaterials is high. Finding low-cost manufacturing techniques for the mass production of metamaterials is the biggest challenge for the market.

To sum up, metamaterials will represent a multi-billion market in the coming decade with vast improvements in radar technology and LiDAR for vehicles, antennas for telecommunication, 5G networks, wireless charging, damping, noise reduction and more. This is a whole new area of science and engineering, which is advancing at such an amazing speed that we cannot imagine what greater changes it will bring about in the near future.

References

1. Veselago, V.G., The electrodynamics of substances with simultaneously negative values of ε and μ. *Sov. Phys. Usp.*, 10, 509–514, 1968.
2. Smith, D.R., Padilla, W.J., Vier, D.C., Nasser, S.C., Schultz, S., Composite medium with simultaneously negative permeability and permittivity. *Phys. Rev. Lett.*, 84, 4184–4187, 2000.
3. Eleftheriades, G.V., Iyer, A.K., Kremer, P.C., Planar negative refractive index media using periodically L-C loaded transmission lines. *IEEE Trans. MTT*, 50, 2702–2712, 2002.
4. Parimi, P.V., Lu, W.T., Derov, J.S., Sokoloff, J., Derov, J.S., Sridhar, S., Negative refraction and left-handed electromagnetism in microwave photonic crystals. *Phys. Rev. Lett.*, 92, 127401–127414, 2004.
5. Qiang, D. and Chen, G., Enhancement of evanescent waves in waveguides using metamaterials of negative permittivity and permeability. *Appl. Phys. Lett.*, 84, 669–671, 2004.
6. Aydin, K. and Ozbay, E., Left-handed metamaterial based superlens for subwavelength imaging of electromagnetic waves. *Appl. Phys. A*, 87, 137–141, 2007.
7. Pendry, J.B., Negative refraction makes a perfect lens. *Phys. Rev. Lett.*, 85, 3966–3969, 2000.
8. Tao, H., Landy, N.I., Bingham, C.M., Zhang, X., Averitt, R.D., Padilla, W.J., A metamaterial absorber for the terahertz regime: Design, fabrication and characterization. *Opt. Express*, 16, 7181–7188, 2008.

9. Li, B., Zhou, J., Li, L., Wang, X.J., Liu, X.H., Zi, J., Ferroelectric inverse opals with electrically tunable photonic band gap. *Appl. Phys. Lett.*, 83, 4704–4706, 2003.
10. Haurylau, M., Anderson, S.P., Marshall, K.L., Fauchet, P.M., Electrical modulation of silicon-based two-dimensional photonic bandgap structures. *Appl. Phys. Lett.*, 88, 061103-1-3, 2006.
11. Khoo, I.C., Werner, D.H., Liang, X., Diaz, A., Nanosphere dispersed liquid crystals for tunable negative–zero–positive index of refraction in the optical and terahertz regimes. *Opt. Lett.*, 31, 2592–2594, 2006.
12. Zhao, Q., Kang, L., Li, B., Zhou, J., Tang, H., Zhang, B., Tunable negative refraction in nematic liquid crystals. *Appl. Phys. Lett.*, 89, 221918, 2006.
13. Kang, L., Zhao, Q., Li, B., Zhou, J., Zhu, H., Experimental verification of a tunable optical negative refraction in nematic liquid crystals. *Appl. Phys. Lett.*, 90, 181931, 2007.
14. Werner, D.H., Kwon, D.H., Khoo, I.C., Kildishev, A.V., Shalaev, V.M., Liquid crystal clad near-infrared metamaterials with tunable negative-zero-positive refractive indices. *Opt. Express*, 15, 3342–3347, 2007.
15. Zhao, Q., Kang, L., Du, B., Li, B., Zhou, J., Tang, H., Liang, X., Zhang, B., Electrically tunable negative permeability metamaterials based on nematic liquid crystals. *Appl. Phys. Lett.*, 90, 011112, 2007.
16. Zhang, F., Zhao, Q., Kang, L., Gaillot, D.P., Zhao, X., Zhou, J., Lippens, D., Magnetic control of negative permeability metamaterials based on liquid crystals. *Appl. Phys. Lett.*, 90, 193104, 2008.
17. Zhang, F., Kang, L., Zhao, Q., Zhou, J., Zhao, X., Lippens, D., Magnetically tunable left-handed metamaterials by liquid crystal orientation. *Opt. Express*, 17, 4360–4366, 2009.
18. Minovich, A., Neshev, D.N., Powell, D.A., Shadrivov, I.V., Kivshar, Y.S., Tunable fishnet metamaterials infiltrated by liquid crystals. *Appl. Phys. Lett.*, 96, 193103, 2010.
19. Zhang, F., Zhang, W., Zhao, Q., Sun, J., Qiu, K., Zhou, J., Lippens, D., Electrically controllable fishnet metamaterial based on nematic liquid crystal. *Opt. Express*, 19, 1563–1568, 2011.
20. Kowerdziej, R., Parka, J., Nyga, P., Salski, B., Simulation of a tunable metamaterial with nematic liquid crystal layers. *Liq. Cryst.*, 38, 377–379, 2011.
21. Kowerdziej, R., Parka, J., Krupka, J., Experimental study of thermally controlled metamaterial containing a liquid crystal layer at microwave frequencies. *Liq. Cryst.*, 38, 743–747, 2011.
22. Dewar, G., A thin wire array and magnetic host structure with $n < 0$. *J. Appl. Phys.*, 97, 10101, 2005.
23. Dewar, G., Minimization of losses in a structure having a negative index of refraction. *New J. Phys.*, 7, 161, 2005.
24. Cai, X.B., Zhou, X.M., Hu, G.K., Numerical study on left-handed materials made of ferrite and metallic wires. *Chin. Phys. Lett.*, 23, 348–351, 2005.

25. Rachford, F.J., Armstead, D.N., Harris, V.G., Vittoria, C., Simulations of ferrite-dielectric-wire composite negative index materials. *Phys. Rev. Lett.*, 99, 057202, 2007.
26. Cao, Y.J., Wen, G.J., Wu, K.M., Xu, X.H., A novel approach to design microwave medium of negative refractive index and simulation verification. *Chin. Sci. Bull.*, 52, 433–439, 2007.
27. He, Y., He, P., Yoon, S.D., Parimi, P.V., Rachford, F.J., Harris, V.G., Vittoria, C., Tunable negative index metamaterial using yttrium iron garnet. *J. Magn. Magn. Mater.*, 313, 187–191, 2007.
28. Zhao, H., Zhou, J., Zhao, Q., Li, B., Kang, L., Bai, Y., Magnetotunable left-handed material consisting of Yttrium iron garnet slab and metallic wires. *Appl. Phys. Lett.*, 91, 131107, 2007.
29. Zhao, H., Kang, L., Zhou, J., Zhao, Q., Li, L., Peng, L., Bai, Y., Experimental demonstration of tunable negative phase velocity and negative refraction in a ferromagnetic/ferroelectric composite metamaterial. *Appl. Phys. Lett.*, 93, 201106, 2007.
30. Huang, Y.J., Wen, G.J., Li, T.Q., Xie, K., Low-loss, broadband and tunable negative refractive index metamaterial. *J. Electromagn. Anal. Appl.*, 2, 104–110, 2010.
31. Huang, Y.J., Wen, G.J., Li, T.Q., Xie, K., Positive-negative-positive metamaterial consisting of ferromagnetic host and wire array. *Appl. Comput. Electromagn. Soc. J.*, 25, 696–702, 2010.
32. Kang, L., Zhao, Q., Zhao, H., Zhou, J., Magnetically tunable negative permeability metamaterial composed by split ring resonators and ferrite rods. *Opt. Express*, 16, 8825–8834, 2008.
33. Kang, L., Zhao, Q., Zhao, H., Zhou, J., Ferrite-based magnetically tunable left-handed metamaterial composed of SRRs and wires. *Opt. Express*, 16, 17269–17275, 2008.
34. Huang, Y.J., Wen, G.J., Yang, Y.J., Xie, K., Tunable dual-band ferrite-based metamaterials with dual negative refractions. *Appl. Phys. A*, 106, 79–86, 2012.
35. Gil, I., Bonache, J., García, J., Martín, F., Tunable metamaterial transmission lines based on varactor-loaded split-ring resonators. *IEEE Trans. MTT*, 54, 2665–2674, 2006.
36. Vélez, A., Bonache, J., Martin, F., Varactor-loaded complementary split ring resonators (VLCSRR) and their application to tunable metamaterial transmission lines. *IEEE Microw. Wirel. Compon. Lett.*, 18, 28–30, 2008.
37. Wang, D., Ran, L., Chen, H., Mu, M., Kong, J.A., Wu, B.I., Active left-handed material collaborated with microwave varactors. *Appl. Phys. Lett.*, 91, 164101, 2007.
38. Aydin, K. and Ozbay, E., Capacitor-loaded split ring resonators as tunable metamaterial components. *J. Appl. Phys.*, 101, 024911, 2007.
39. Hand, T.H. and Cummer, S.A., Frequency tunable electromagnetic metamaterial using ferroelectric loaded split rings. *J. Appl. Phys.*, 103, 066105, 2008.

40. Ricci, M.C., Xu, H., Prozorov, R., Zhuravel, A.P., Ustionov, A.V., Anlage, S.M., Tunability of superconducting metamaterials. *IEEE Trans. Appl. Supercond.*, 17, 918–921, 2007.
41. Tao, H., Strikwerda, A.C., Fan, K., Padilla, W.J., Zhang, X., Averitt, R.D., MEMS based structurally tunable metamaterials at terahertz frequencies. *J. Infrared Millim. Terahertz Waves*, 32, 580–595, 2011.
42. Lapine, M., Powell, D., Gorkunov, M., Shadrivov, I., Marques, R., Kivshar, Y., Structural tunability in metamaterials. *Appl. Phys. Lett.*, 95, 084105, 2009.
43. Ekmekci, E., Strikwerda, A.C., Fan, K., Keiser, G., Zhang, X., Sayan, G., Averitt, R.D., Frequency tunable terahertz metamaterials using broadside coupled split-ring resonators. *Phys. Rev. B*, 83, 193103, 2011.
44. Wang, Y., Du, C., Dong, X., *Creating a simplified frequency-tunable metamaterial*, vol. 29, SPIE Newsroom, pp. 802–907, 2011.
45. Chen, H., Wu, B.I., Ran., L., Grzegorczyk, T.M., Kong, J.A., Controllable left-handed metamaterial and its application to a steerable antenna. *Appl. Phys. Lett.*, 89, 053509, 2006.
46. Shadrivov, I.V., Morrison, S.K., Kivshar, Y.S., Tunable split-ring resonators for nonlinear negative-index metamaterials. *Opt. Express*, 14, 9344–9349, 2006.
47. He, P., Parimi, P.V., He, Y., Harris, V.G., Vittoria, C., Tunable negative refractive index metamaterial phase shifter. *Electron. Lett.*, 43, 1440–1441, 2007.
48. Huang, Y.J., Wen, G.J., Li, T.Q., Xie, K., Left-handed metamaterial with $\varepsilon = -\varepsilon 0$ and $\mu = -\mu 0$ and some applications. *International Conference on Applied Superconductivity and Electromagnetic Devices*, pp. 119–122, 2009.
49. Zhu, B., Feng, Y., Zhao, J., Huang, C., Wang, Z., Jiang, T., Polarization modulation by tunable electromagnetic metamaterial reflector/absorber. *Opt. Express*, 18, 23196–23203, 2010.
50. Abadi, M.S.H., Shafai, C., Shafai, L., A reconfigurable frequency selective surface using switched slots in the ground plane. *14th International Symposium on Antenna Technology and Applied Electromagnetics & the American Electromagnetics Conference*, pp. 1–4, 2010.
51. Naorem, R., Dayal, G., Ramakrishna, S., Rajeswaran, B., Umarji, A.M., Thermally switchable metamaterial absorber with a VO_2 ground plane. *Opt. Commun.*, 346, 154–157, 2015.
52. Cortie, M.B., Dowd, A., Harris, N., Ford, M.J., Core-shell nanoparticles with self-regulating plasmonic functionality. *Phys. Rev. B*, 75, 113405, 2007.
53. Jin, P., Tazawa, M., Xu, G., Reversible tuning of surface plasmon resonance of silver nanoparticles using a thermochromic matrix. *J. Appl. Phys.*, 99, 096106, 2006.
54. Xu, G., Chen, Y., Tazawa, M., Jin, P., Surface plasmon resonance of silver nanoparticles on vanadium dioxide. *J. Phys. Chem. B*, 110, 2051–2056, 2006.
55. Maaza, M., Nemraoui, O., Sella, C., Beye, A.C., Barak, B., Thermal induced tunability of surface plasmon resonance in Au–VO_2 nano-photonics. *Opt. Commun.*, 254, 188–195, 2005.

56. Makarov, S.V., Zalogina, A.S., Tajik, M., Zuev, D.A., Rybin, M.V., Kuchmizhak, A.A., Juodkazis, S., Kivshar, Y., Light-induced tuning and reconfiguration of nanophotonic structures. *Laser Photonics Rev.*, 11, 1700108, 2017.
57. Hosseini, N., Lim, S., Hwang, H., Rho, J., Reliable $Ge_2Sb_2Te_5$-integrated high-density nanoscale conductive bridge random access memory using facile nitrogen-doping strategy. *Adv. Electron. Mater.*, 11, 1800360, 2018.
58. Tekcan, B., van Kasteren, B., Grayli, S.V., Shen, D., Tam, M.C., Ban, D., Wasilewski, Z., Tsen, A.W., Reimer, M.E., Semiconductor nanowire metamaterial for broadband near-unity absorption. *Sci. Rep.*, 12, 9663, 2022.
59. Jun, S.W. and Ahn, Y.H., Terahertz thermal curve analysis for label-free identification of pathogens. *Nat. Commun.*, 13, 3470, 2022.
60. Wang, Z., Chang, L., Wang, F., Li, T., Ju, T., Integrated photonic metasystem for image classifications at telecommunication wavelength. *Nat. Commun.*, 13, 2131, 2022.
61. Kim, M., Lee, D., Yang, Y., Kim, Y., Rho, J., Reaching the highest efficiency of spin hall effect of light in the near-infrared using all-dielectric metasurfaces. *Nat. Commun.*, 13, 2036, 2022.
62. Vynck, K., Pacanowski, R., Agreda, A., Dufay, A., Granier, X., Lalanne, P., The visual appearances of disordered optical metasurfaces. *Nat. Mater.*, 21, 1035–1041, 2022.
63. Zou, X., Zhang, Y., Lin, R., Gong, G., Wang, S., Zhu, S., Wang, Z., Pixel-level Bayer-type colour router based on metasurfaces. *Nat. Commun.*, 13, 3288, 2022.
64. Dhama, R., Panahpour, A., Pihlava, T., Ghindani, D., Caglayan, H., All-optical switching based on plasmon-induced enhancement of index of refraction. *Nat. Commun.*, 13, 3114, 2022.
65. Chaudhry, F.A., Escandell, L., Fraguas, E., Light absorption enhancement in thin film GaAs solar cells using dielectric nanoparticles. *Sci. Rep.*, 12, 9240, 2022.
66. Wang, Y., Chen, Q., Yang, W., Ji, Z., Jin, L., Ma, X., Song, Q., Han, J., Shalaev, V.M., Xiao, S., High-efficiency broadband achromatic metalens for near-IR biological imaging window. *Nat. Commun.*, 12, 5560, 2021.
67. Hua, X., Wang, Y., Wang, S., Ultra-compact snapshot spectral light-field imaging. *Nat. Commun.*, 13, 2732, 2022.
68. Zeng, H., Liang, H., Zhang, Y., Wang, L., Liang, S., Gong, S., Li, Z., Yang, Z., Lan, F., Feng, Z., Gong, Y., Yang, Z., Mittleman, D.M., High-precision digital terahertz phase manipulation within a multichannel field perturbation coding chip. *Nat. Photonics*, 15, 751–757, 2021.
69. Li, J., Hu, G., Shi, L., He, N., Li, D., Shang, Q., MZhang, Q., Fu, H., Zhou, L., Xiong, W., Guan, J., Wang, J., He, S., Chen, S., Full-color enhanced second harmonic generation using rainbow trapping in ultrathin hyperbolic metamaterials. *Nat. Commun.*, 12, 6425, 2021.
70. Zhang, H., Ni, Z., Stevens, C.E., Cavity-enhanced linear dichroism in a van der Waals antiferromagnet. *Nat. Photonics*, 16, 311–317, 2022.

71. Long, G., Adamo, G., Tian, J., Perovskite metasurfaces with large superstructural chirality. *Nat. Commun.*, 13, 1551, 2022.
72. Li, Y., Kita, S., Muñoz, P., Reshef, O., Vulis, D.I., Yin, M., Loncar, M., Mazur, E., On-chip zero-index metamaterials. *Nat. Photonics*, 9, 738–742, 2015.
73. Gao, J., Vincenti, M.A., Frantz, J., Near-infrared to ultraviolet frequency conversion in chalcogenide metasurfaces. *Nat. Commun.*, 12, 5833, 2021.
74. Qian, C., Wang, Z., Qian, H., Dynamic recognition and mirage using neuro-metamaterials. *Nat. Commun.*, 13, 2694, 2022.
75. Kong, L.J., Zhang, W., Li, P., Guo, X., Zhang, J., Zhang, F., Zhao, J., Zhang, X., High-capacity topological coding based on nested vortex knots and links. *Nat. Commun.*, 13, 2705, 2022.
76. Momeni, A. and Fleury, R., Electromagnetic wave-based extreme deep learning with nonlinear time-Floquet entanglement. *Nat. Commun.*, 13, 2651, 2022.
77. Li, Y., Qi, M., Li, J., Cao, P.C., Wang, D., Zhu, X.F., Qiu, C.W., Chen, H., Heat transfer control using a thermal analogue of coherent perfect absorption. *Nat. Commun.*, 13, 2683, 2022.
78. Li, Z., Pestourie, R., Park, J.S., Huang, Y.W., Johnson, S.G., Capasso, F., Inverse design enables large-scale high-performance meta-optics reshaping virtual reality. *Nat. Commun.*, 13, 2409, 2022.
79. Lu, C., Sun, Y.Z., Wang, C., Zhang, H., Zhao, W., Hu, X., Xiao, M., Ding, W., Liu, Y.C., Chan, C.T., On-chip nanophotonic topological rainbow. *Nat. Commun.*, 13, 2586, 2022.
80. Rajabali, S., Markmann, S., Jöchl, E., Beck, M., Lehner, C.A., Wegscheider, W., Fiast, J., Scalari, G., An ultrastrongly coupled single terahertz meta-atom. *Nat. Commun.*, 13, 2528, 2022.
81. Galiffi, E., Huidobro, P.A., Pendry, J.B., An Archimedes' screw for light. *Nat. Commun.*, 13, 2523, 2022.
82. Wang, S., Wu, P.C., Su, V.C., A broadband achromatic metalens in the visible. *Nat. Nanotechnol.*, 13, 227–232, 2018.
83. Almeida, E., Bitton, O., Prior, Y., Nonlinear metamaterials for holography. *Nat. Commun.*, 7, 12533, 2016.
84. Giovampaola, C. and Engheta, N., Digital metamaterials. *Nat. Mater.*, 13, 1115–1121, 2014.

15

Metamaterials for Sound Filtering

Sneha Kagale[1], Radhika Malkar[1], Manishkumar Tiwari[2] and Pravin D. Patil[3*]

[1]Department of Biotechnology Engineering, Kolhapur Institute of Technology College of Engineering (Autonomous), Kolhapur, India
[2]Department of Chemical Engineering, SVKM's NMIMS Mukesh Patel School of Technology Management & Engineering, Mumbai, Maharashtra, India
[3]Department of Basic Science and Humanities, Mukesh Patel School of Technology Management and Engineering, SVKM's NMIMS University, Mumbai, Maharashtra, India

Abstract
For decades, metamaterials are used for various applications like antennae, absorbers, cloaking devices, insulation material, etc due to their exquisite physical, topological, and geometrical features. Several types of metamaterials are specific in their function based on their fabrication and assembly viz., piezoelectric, photonic, chiral, electromagnetic, etc among which Acoustic metamaterial and Phononic crystals are regarded as important for sound filtering applications. There are various sources of noise type of sound that specifically needs attention for it to reduce/absorb/insulate/attenuate the background sound which will help reduce the public hazard caused by noise from various sources like industries, underwater, construction site, etc. This chapter highlights the introduction to metamaterials and types of metamaterials. Further, summarizes the metamaterials used in sound filtering specifically Acoustic MetaMaterial (AMM) and Phononic Crystals (PC) with their types, methods of fabrication, and assembly.

Keywords: Metamaterials, acoustic metamaterials, phononic crystals, fabrication, assembly, sound, noise filter

15.1 Introduction

Since the dawn of the 21st century, metamaterials have attracted a lot of attention as it is a convergent research field involving researchers from

Corresponding author: dr.pravinpatil.ict@gmail.com

Inamuddin and Tariq Altalhi (eds.) Electromagnetic Metamaterials: Properties and Applications, (327–340) © 2023 Scrivener Publishing LLC

nanotechnology, acoustics, solid-state physics, microwave and optical sciences, materials science, mechanics, and high-performance computing in the various fields due to its properties like increase in the bandwidth, directivity, radiated power and control over the electromagnetic radiation, etc. In Greek, "meta" means beyond or superior. Being said superior materials, they have been used for controlling wave propagation for decades. The definitions of metamaterials have progressively broadened "metamaterials can achieve unusual and advantageous properties of electromagnetic properties" [1].

Vibrations and noise can be the potential source to convert these sources into electrical power. As batteries have a delimited life span, the energy conversion can indeed be used in sensors, homes, devices, workshops, etc instead of batteries. The efficient means of harvesting the energy is by using the transduction mechanism in the device. Piezoelectric devices use this mechanism of energy harvesting that converts mechanical energy into electrical energy. Recently there have been various noise sources like industries with heavy machinery, road traffic, etc. Acoustic barriers have emerged to alleviate the noise. The efficiency of these noise barriers is around 50% only. To overcome this, acoustic metamaterials are widely used as the noise barrier whose efficiency is 1.5 times more than traditional noise barriers used in industries made of vinyl, wood, etc. [2].

15.1.1 Types of Metamaterials

Various types of metamaterials having wide range of applications are classified as shown in Figure 15.1.

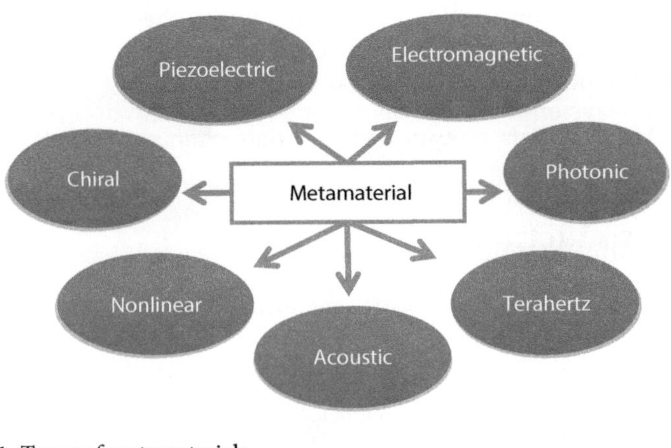

Figure 15.1 Types of metamaterials.

15.1.1.1 *Piezoelectric Metamaterial*

For metamaterials to develop the desired characteristics, piezoelectric structures are essential. They are extensively applied in Micro-ElectroMechanical Systems (MEMS) and sensors. It exhibits archaic mechanical properties and exceptional flexibility. The energy is converted between the energy domains of electrical and mechanical because of the piezoelectric property it also allows for harnessing the energy flow(input) and optimizes the bandgap (both size and location). They are used in sensors, actuators, hydrophones, echocardiograms, and ultrasound imaging [3].

15.1.1.2 *Electromagnetic Metamaterial*

There are several ways to alter the material properties as the field of magnets along with electricity have obtained significant attention. These metamaterials utilize electrical and magnetic energy to manipulate the material properties periodically. They offer advantages such as a wide range of bandwidth and tunable bandgap. The disadvantage of using this metamaterial is that it requires an ample amount of energy. These are widely used in smart materials, MEMS, and sound filtering [3].

15.1.1.3 *Chiral Metamaterial*

A "chiral" is a fragment of material that, after being transmitted or rotated, cannot match its mirror counterpart. Chiral metamaterials behave unusually from their mirror image analog. Due to the unique properties of lightweight mechanical chiral metamaterials, they are used for energy harvesting and vibration attenuation. The unique structure of chiral metamaterial provides flexible topology and lightweight construction making it suitable for biosensors specifically ultra-sensitive. It provides an important attribute that is essential in nanophotonics viz., for controlling the field of light through an external device [3].

15.1.1.4 *Nonlinear Metamaterial*

It is a metamaterial, artificially constructed with the nonlinear geometric properties of a cell. They can be used in high-resolution and intense local field enhancement. There is no need for stimulation from external sources for this type of metamaterial, disadvantages associated with this are narrow bandgap and power dependency [3].

15.1.1.5 Terahertz Metamaterial

These metamaterials work in the range of 0.1 to 10 THz frequency. It is one of the novel metamaterials which has several advantages in controlling radiations in the terahertz range and also has applications in the field of spectroscopy and remote sensing. The disadvantage associated with this type of metamaterial is that there is a limit to propagating terahertz bands through the atmosphere and using them in heavy equipment [3].

15.1.1.6 Acoustic Metamaterial

These materials have received a lot of attention due to their periodic nature which controls, directs, and manipulates sound wave propagation and acoustic cloaks, noise reduction, Non-Destructive Testing (NDT) test, etc. It steers, directs, and manipulates acoustoelastic waves with excellent properties that are very rare or rather unavailable. The advantages and disadvantages of these metamaterials are low-frequency elastic wave propagation, sandwich panels, cost-effectiveness and they lack flexibility, tenability also undesirable loss respectively [3].

15.1.1.7 Photonic Metamaterial

These substances' refractive indices are high, low, or even negative, which allows them to influence optical waves in sub-wavelength dimensions. Its periodic structure is built to control, direct and manipulate the optical wave propagation. It is used in super lens and optical wave manipulation also it is economical and has sandwich panels. It lacks flexibility and tunability adding to undesirable loss [3].

General applications of metamaterials: There are various applications of metamaterials in Table 15.1.

15.2 Acoustic Metamaterials

These materials can control and alter the sound waves which cannot be accomplished by traditional materials. The propagation of waves in periodic structures is traced to the span of the decennary [8]. The latest usage attributed to engineered structures has begun with photonic and phononic crystals [9, 10]. A rubber-like material was used to cover an opaque core during attaining a local unit of resonance [11]. Additionally, they proposed speculation of a local-resonance phononic crystal, that gave rise to an era

Table 15.1 Applications of metamaterials.

Metamaterials applications	Description
Antenna	A metamaterial is utilized to create materials with zero refractive indexes that have strong directivity and negative permeability, better optional frequency, power enhancement, and regulated electromagnetic radiation [4].
Absorber	A metamaterial is utilized to create absorbers that are more effective than typical ones. high electromagnetic radiation is absorbed [4].
Superlens	Metamaterials-based superlens achieves resolution beyond the diffraction limit.
Cloaking device	Metamaterial cloak is based upon coordinate transformation concepts, deflecting microwave beams in a process object appears as if almost nothing was there [5].
Seismic	Metamaterials are used in sensor detection, aerospace, and public safety [6].
Weapons of Mass Destruction Detectors (WMD)	Metamaterials are used in the defense field by the air force and army for exploring biological agents and chemical explosives.
Invisible subs	Submarines are concealed from sonar by the use of metamaterials, which manipulate sound waves by bending them in their direction.
Revolutionary electronics	Electronic applications can be made smaller and faster using photonic equipment based on metamaterials.
Biosensor	Metamaterials are used for label-free biomolecules, and environmental monitoring purposes [7].

of acoustic metamaterials. Also, the mass density of a phononic crystal for the local resonance shall be negative near the local resonance frequency [12]. Recently, a periodic composite structure that was solid-liquid in form (with silicon rubber) was embedded in water [13]. This structure has a negative equivalent mass density and elastic modulus with a frequency range. The introduction of acoustic metamaterials was due to the inspiration taken from electromagnetic metamaterials, for an instance, Helmholtz resonators were designed or arranged periodically with a negative equivalent coefficient of elasticity in the frequency band of resonance [14]. The construction of acoustics with negative equivalent moduli and the mass density was done using the theory of ordinary phononic crystals [15]. Later, a group of researchers designed a 1-dimensional porous silicon-carbide crystal inserted between rubber layers [16]. In essence, the acoustic metamaterials' ability to manipulate sound waves is due to their negative elastic modulus, zero refractive index, and negative effective dynamic mass density.

The acoustics concept has advanced because of several modifications in the materials via experimentation to achieve required acoustic parameters and is referred to as "acoustic metamaterials" [17]. Using experiments to determine the defining characteristics of acoustic metamaterials, a group of scientists were able to obtain a broad-band negative refractive index for spatial structure overlapped with double negative parameters by limiting the use of a local resonance mechanism, that is a substitute for the structure's limited bandwidth [18]. 3 Dimensional double-negative acoustic metamaterials, which have a spatial folding in 3 dimensions, were the subject of later experimentation [19].

A spatial curl structure was also experimented on to reduce the low-frequency noise of aircraft engines [20]. Along with this, hyperbolic anisotropic acoustic metamaterials have double negative equivalent parameters. This also achieves negative refraction from almost all angles as it is the hyperbolic isofrequency curve. There are metamaterials whose structures are designed to be like perforated plates which produce negative refraction, and sub-wavelength resolution profiles at longer stretches [21].

Besides research on the characteristics of acoustic manipulation, there is research on the topological properties too [16, 22]. Programmable metamaterials [23, 24] were developed and the nonlinear programmable acoustic metamaterial structure enabled real-time harmonic and switchable bandgaps [25]. "Metamaterial bricks" that are programmable were utilized to produce diffraction-limited sound fields [26]. This made the creation of sound modulators and the design of acoustic devices simpler. In addition, a programmable metamaterial was based on the gigahertz band that was a triple-origami [27]. Ternary foldable origami program was written in the form of "0," "1," and "2" enabling a huge degree of controlling the elastic and the sound waves. To convert incident surface sound waves into volume shear waves, programmable acoustic metamaterials and piezoelectric regulation were also integrated. This method resulted in the development of practical surface acoustic wave devices [28].

There are several developments, modifications, and manipulations in the acoustics and topological properties of each type of acoustic metamaterial to date. Few historical developments are provided in this section. Further, highlights the types of acoustic metamaterials.

15.2.1 Types and Applications of Acoustic Metamaterials

There are prominent types of acoustic metamaterials are shown in Figure 15.2. The prominent ones involved in acoustic filtering are programmable acoustic metamaterials. There are various others involved in different applications like acoustic barriers, noise control, noise reduction, etc. The types of acoustic materials are listed based on their capacity for sound insulation, and absorption, which here is categorized as Passive and Active acoustic metamaterials further categorization is shown in Figure 15.2. Passive sound insulation and absorption, active sound insulation and absorption are the methods in which acoustics metamaterials work. The applications of acoustic metamaterials are listed in Table 15.2.

15.3 Phononic Crystals

Phonon is a quantum of the acoustic field [29]. The word was originated from Greek (meaning voice). The range of frequencies spans broadly from a few millihertz up to dozens of terahertz. The fundamental features of

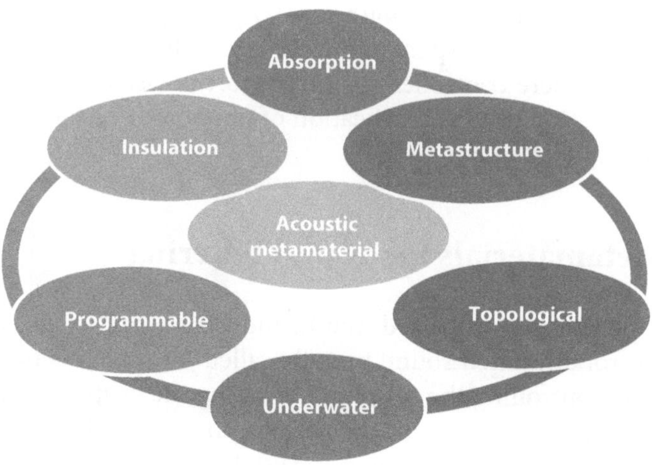

Figure 15.2 General types of acoustic metamaterial.

Table 15.2 Applications of general acoustic metamaterials.

Sr. no.	Types	Applications
1	Metamaterial absorption	Space acoustic absorber, noise control
2	Metamaterial metastructure	Acoustic barrier, industrial noise reduction
3	Metamaterial topological	Acoustic functional design, directional transmission
4	Metamaterial underwater	Underwater noise reduction, acoustic stealth
5	Metamaterial programmable	Acoustic filter, Acoustic Levitation
6	Metamaterial insulation	Acoustic barrier, Industrial noise reduction

phonon include dispersion elation viz., spectral, spatial, and mean free path viz., lifetime, and attenuation [30, 31]. There has been a lot of progress in the structural designs of phononic crystals of the phononic spectrum. The structural design modifications of phononic crystals based on the phononic spectrum are (1) sub-Gigahertz phononic crystals are from infrasonic to ultrasonic designs and tunable and active phononic structures; (2) hypersonic phononic crystals include hypersonic crystals in various dimensions and structures built with low-dimensional nanostructures; (3) optomechanical crystals, which are specifically used in telecommunications are based on the wavelength of GHz photons and electromagnetic radiation; (4) thermal transport phononic crystal helps in thermoelectrics and in devices where controlled heat transport is essential [29]. The pictorial representation of the classification of phononic crystal based on the spectrum is depicted in Figure 15.3.

15.4 Metamaterials for Sound Filtering

Noise is a public hazard caused due to industries, traffic, etc. there is a need to control unwanted sound broadly called "noise." This noise can be controlled by controlling the source, controlling the propagation pathway via isolation, absorption, and attenuation, and protecting the receiver. Acoustic metamaterials and phononic crystals are the materials that play a major role in sound filtering. Acoustic metamaterials and phononic crystals

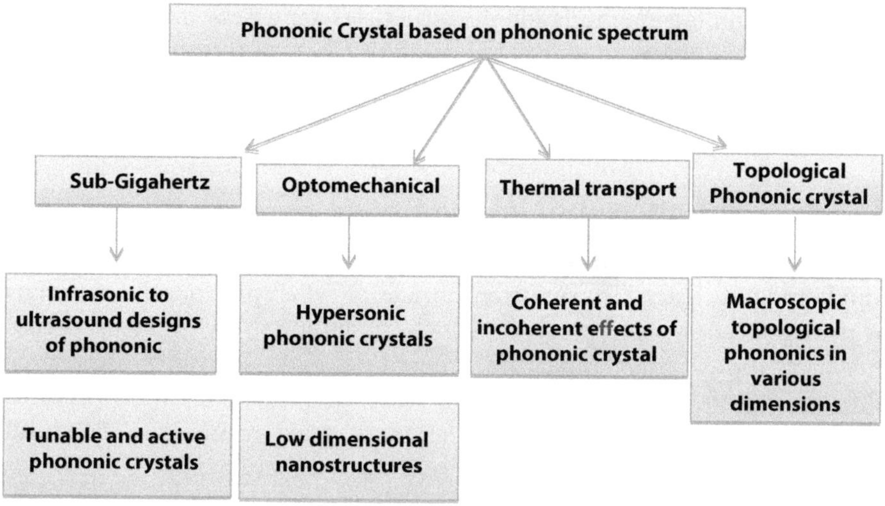

Figure 15.3 Phononic crystal classification based on phononic spectrum.

provide improved manipulation of sound waves via refraction, reflection, transmission, diffraction or attenuation, and absorption. These metamaterials are used for underwater detection, noise cancellation, energy harvesting, and medical imaging [32–34]. These applications will be beneficial in the sectors like environmental sustainability, health care, and well-being. They can be used in the technologies that are next-generation for immersive multisensory experiences [34–37].

15.4.1 Fabrication and Assembly of Metamaterials for Sound Filtering and Attenuation

Constructing an acoustic metamaterial (AMM) or phononic crystals (PC) involves technology and process to manufacture and various strategies for the amalgamation of construction materials. AMMs and PCs shall be fabricated monolithically; however, extensive requirements of design can make it challenging for making prototypes. The properties that AMMs and PCs have, and are governed by their geometric, mechanical, and material characteristics. Therefore, any imperfections in the fabrication can negatively impact their acoustic performance. Incorporating the stimuli-responsive structures and actuation mechanisms in AMM and PC will make them dynamic and increase the building's complexity. Before the fabrication and assembly of AMM and PC, it is essential to know and understand the link between the physical characteristics of the design and its operation.

Once the building methods for the structures are assessed and finalized, then we define their suitability and usability. For an instance, a material bearing high acoustic impedance is preferred to facilitate acoustic transmission or reflection [34].

15.4.2 Fabrication of AMM and PC

The manufacturing methods for AMMs and PCs can be categorized into single and multi-step. The one-step fabrication will reproduce the materials with large accuracy and precision which can be suitable with the structure materials comprising of repeating, ordered elements. The multi-phase production method is utilized for highly heterogeneous, complex structures or that consist of small features at the microscale or nanoscale structural levels [34]. The types of fabrication are shown in Figure 15.4.

One-step fabrication involves a single process excluding the post-processing to produce full AMM/PC or bits of them to further assemble them. Techniques like 3D printing and machining are used for this intent which is perfect for Helmholtz resonators [38, 39], and periodic lattices [40, 41]. These methods are economic, easy to use, and accessible. Hybrid manufacturing uses integrated approaches of various single-step processes for fabrication. This kind of additive or subtractive process will help in reducing the time, and cost, improving and increasing the tolerances and surface quality, respectively. In a study, a robotic arm was utilized with printing and machining heads which are changeable including the heated print bed. This allowed minimal material wastage [42].

Multistep fabrication is essential as the single-step fabrication methods which are not sufficient to manufacture all types of AMM and PCs. Microfabrication can deposit micro-nanometer scale structures that are

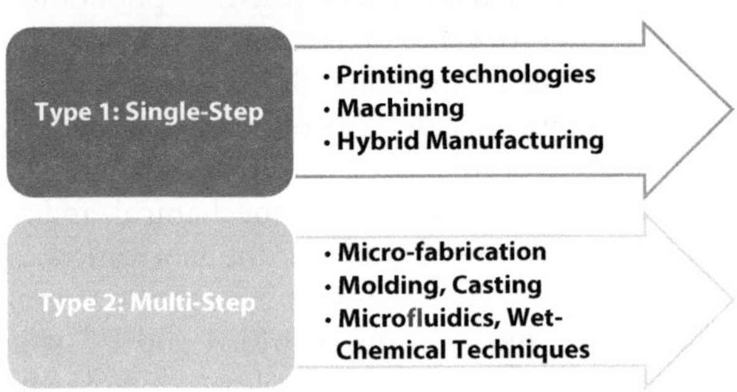

Figure 15.4 Types of AMM and PC fabrication.

easily not possible by even super-resolution 3D printers such as poly jet printers. In microfabrication, the dry plasma etching process can be used for AMM and PC fabrication [43, 44]. Molding and casting is another way of fabrication in a multi-step manner specifically used for soft, deformable structures of AMM and PC. It is a two-step procedure; initially, the mold is manufactured before casting and curing the material [45]. 3-Dimensional molds that are printed using 3D technology can cast elastomeric structures, where mechanical buckling could change the hardness of materials to open/close the bandgaps [46]. Microfluidics and wet-chemical techniques are also methods of fabrication for AMM and PC [34].

15.4.3 Assembly of AMM and PC

Assembly of AMM and PC can be done using adhesives like epoxies and glues to fabricate heterogeneous or homogenous structures [34]. Adhesives can bind two different types of materials. Specifically for the structures that are developed using 3D - technology, the adhesives will assist in decreasing the pressure of post-processing. Bonding is an important step used to reduce the use of support structures. Acoustic levitation is a strategy used to assemble microspheres. The technique of Langmuir-Blodgett involves fabricating molecular layers (mono) which are used for the deposition of spheres. Also, press fitting or interference fitting is used in fastening objects together by using friction. A method of being self-assembled gives the triangular-shaped cell convective technique for AMM [47]. Additionally, smart theoretical designs and strategies involved in assemblies have paved a way for acoustic wave manipulation [34].

15.5 Conclusion

Metamaterials for sound filtering and/or noise filtering are essential as they prevent the public from the hazards to which they get exposed. There are several types of metamaterials whose purpose also varies based on their utility in different industries, among which AMM and PC are essential for sound filtering and/or attenuation. Fabrication and assembly of these metamaterials are important aspects to consider as acoustic wave manipulation which can be possible based on the geometry and topology of the metamaterials. Recapitulating, the invention, design, fabrication, and assembly of AMMs and PCs have led to solving the problems related to wave manipulation in-turn sound filtering.

References

1. Sergei, T., Augustine, U., Nikolay, Z., The century of metamaterials. *J. Opt. (United Kingdom)*, 19, 8, 3–4, 2017.
2. Fariha, M., *Acoustoelastic metamaterial with simultaneous noise filtering and energy harvesting capability from ambient vibrations*, Univ. South. Calif., U.S. State of North Carolina, 2019.
3. Ali, V., Mohammad, K., Mina., R., Amin, Y., Hamid, S., Metamaterials and their applications: An overview. *Proc. Inst. Mech. Eng. Part L J. Mater. Des. Appl.*, 236, 11, 1–40, 2021.
4. Shridhar, M. and Yogeshwar, K., Metamaterial properties and applications. *Int. J. Inf. Technol. Knowl. Manage.*, 4, 1, 85–89, 2011.
5. Jayakrishnan, M. and Liya, M., A survey of electromagnetic waves based metamaterials and applications in various domains. *Proc. 3rd Int. Conf. Smart Syst. Inven. Technol. ICSSIT 2020*, pp. 662–666, 2020.
6. Yahya, S., Metamaterials in antenna applications:Classifications, designs and applications. *2006 IEEE Int. Work.*, vol. 2, pp. 1–4, 2006.
7. Giles, S. and Andreas, M., Chip-based microsystems for genomic and proteomic analysis. *Trends Anal. Chem.*, 19, 6, 364–378, 2000.
8. Leon, B., *Wave propagation in periodic structures electric filters and crystal lattices*, pp. 1–274, Dover Publications, New York and London, 1953.
9. Zhou, X. and Schoepf, T., Detection and formation process of overheated electrical joints due to faulty connections. *IET Conf. Publ.*, 2012, 605, 288–295, 2012.
10. Hussein, M., Leamy, M., Ruzzene, M., Dynamics of phononic materials and structures: Historical origins, recent progress, and future outlook. *Appl. Mech. Rev.*, 66, 4, 1–38, 2014.
11. Liu, Z., Xixiang, Z., Yiwei, M., Zhu, Y., Zhiyu, Y., Chan, C., Ping, S., Locally resonant sonic materials. *Science*, 289, 5485, 1734–1736, 2000.
12. Liu, Z., Chan, C., Sheng, P., Analytic model of phononic crystals with local resonances. *Phys. Rev. B - Condens. Matter Mater. Phys.*, 71, 1, 1–8, 2005.
13. Li, J. and Chan, C., Double-negative acoustic metamaterial. *Phys. Rev. E - Stat. Phys. Plasmas Fluids Relat. Interdiscip. Top.*, 70, 5, 4, 2004.
14. Nicholas, F., Dongjuan, X., Jianyi, X., Muralidhar, A., Werayut, S., Cheng, S., Xiang, Z., Ultrasonic metamaterials with negative modulus. *Nat. Mater.*, 5, 6, 452–456, 2006.
15. Liu, X., Hu, G., Huang, G., Sun, C., An elastic metamaterial with simultaneously negative mass density and bulk modulus. *Appl. Phys. Lett.*, 98, 25, 2012–2015, 2011.
16. Gao, N., Zhang, Z., Deng, J., Guo, X., Cheng, B., Hou, H., Acoustic metamaterials for noise reduction: A review. *Adv. Mater. Technol.*, 7, 6, 2022.
17. Torrent, D. and Sánchez-Dehesa, J., Acoustic cloaking in two dimensions: A feasible approach. *New J. Phys.*, 10, 063015, 2008.

18. Liang, Z. and Li, J., Extreme acoustic metamaterial by coiling up space extreme acoustic metamaterial by coiling up space. *Phys. Rev. Lett.*, 108, 114301, 2012.
19. Garcia-chocano, V. and Sanchez-dehesa, J., Negative refraction and energy funneling by hyperbolic materials: An experimental demonstration in acoustics negative refraction and energy funneling by hyperbolic materials: An experimental demonstration in acoustics. *Phys. Rev. Lett.*, 112, 144301, 2014.
20. Zhao, X., Liu, G., Zhang, C., Xia, D., Lu, Z., Fractal acoustic metamaterials for transformer noise reduction. *Appl. Phys. Lett.*, 113, 7, 2018.
21. Christensen, J. and García De Abajo, F., Anisotropic metamaterials for full control of acoustic waves. *Phys. Rev. Lett.*, 108, 12, 1–5, 2012.
22. Ma, G., Xiao, M., Chan, C., Topological phases in acoustic and mechanical systems. *Nat. Rev. Phys.*, 1, 4, 281–294, 2019.
23. Di Caprio, G., Dardano, P., Coppola, G., Cabrini, S., Mocella, V., Digital holographic microscopy characterization of superdirective beam by metamaterial. *Opt. Lett.*, 37, 7, 1142, 2012.
24. Della Giovampaola, C. and Engheta, N., Digital metamaterials. *Nat. Mater.*, 13, 12, 1115–1121, 2014.
25. Tianzhi, Y., Zhi, S., Eoin, C., Ye, Z., Jia, S., Yi, S., Li, C., Peter, H., A programmable nonlinear acoustic metamaterial. *AIP Adv.*, 7, 9, 2017.
26. Memoli, G., Caleap, M., Asakawa, M., Sahoo, D., Drinkwater, B., Subramanian, S., Metamaterial bricks and quantization of meta-surfaces. *Nat. Commun.*, 8, 1–8, 2017.
27. Le, D. and Lim, S., Four-mode programmable metamaterial using ternary foldable origami. *ACS Appl. Mater. Interfaces*, 11, 31, 28554–28561, 2019.
28. Alan, S., Allam, A., Erturk, A., Programmable mode conversion and bandgap formation for surface acoustic waves using piezoelectric metamaterials. *Appl. Phys. Lett.*, 115, 9, 2019.
29. Vasileiadis, T., Varghese, J., Babacic, V., Gomis-Bresco, J., Navarro Urrios, D., Graczykowski, B., Progress and perspectives on phononic crystals. *J. Appl. Phys.*, 129, 16, 2021.
30. Ziman, J., *Electrons and phonons: The theory of transport phenomena in solids*, pp. 1–33, Oxford University Press, New York, NY, USA, 2001.
31. Ronda, S., Aragón, J., Iglesias, E., de Espinosa, F., The use of phononic crystals to design piezoelectric power transducers. *Sensors (Switzerland)*, 17, 4, 2017.
32. Shen, C., Xie, Y., Sui, N., Wang, W., Cummer, S., Jing, Y., Broadband acoustic hyperbolic metamaterial. *Phys. Rev. Lett.*, 115, 25, 1–5, 2015.
33. Qi, S., Oudich, M., Li, Y., Assouar, B., Acoustic energy harvesting based on a planar acoustic metamaterial. *Appl. Phys. Lett.*, 108, 26, 2016.
34. Choi, C., Bansal, S., Münzenrieder, N., Subramanian, S., Fabricating and assembling acoustic metamaterials and phononic crystals. *Adv. Eng. Mater.*, 23, 2, 2021.

35. Chen, Y., Meng, W., Xin, S., Fu, H., SmartSweep: Context-aware modeling on a single image. *SIGGRAPH Asia 2017 Posters, SA 2017*, pp. 1–2, 2017.
36. Constantin, A., Korte, J., Fails, J., Alexandru, C., Dragomir, M., Pain, H., Good, J., Garzotto, F., Eriksson, E., Waller, A., Expecting the unexpected in participatory design. *Conf. Hum. Factors Comput. Syst. - Proc.*, pp. 7–10, 2019.
37. Vi, C., Ablart, D., Gatti, E., Velasco, C., Obrist, M., Not just seeing, but also feeling art: Mid-air haptic experiences integrated in a multisensory art exhibition. *Int. J. Hum. Comput. Stud.*, 108, 1–14, 2017.
38. Li, Y., Jiang, X., Liang, B., Cheng, J., Zhang, L., Metascreen-based acoustic passive phased array. *Phys. Rev. Appl.*, 4, 2, 024003-01–024003-7, 2015.
39. Zhao, L. and Zhou, S., Compact acoustic rainbow trapping in a bioinspired spiral array of graded locally resonant metamaterials. *Sensors (Switzerland)*, 19, 4, 2019.
40. Fotsing, E., Dubourg, A., Ross, A., Mardjono, J., Acoustic properties of a periodic micro-structures obtained by additive manufacturing. *Appl. Acoust.*, 148, 322–331, 2019.
41. Zhu, R., Liu, X., Hu, G., Sun, C., Huang, G., Negative refraction of elastic waves at the deep-subwavelength scale in a single-phase metamaterial. *Nat. Commun.*, 5, 1–8, 2014.
42. Li, L., Haghighi, A., Yang, Y., A novel 6-axis hybrid additive-subtractive manufacturing process: Design and case studies. *J. Manuf. Process.*, 33, 150–160, 2018.
43. Ryoo, H. and Jeon, W., Dual-frequency sound-absorbing metasurface based on visco-thermal effects with frequency dependence. *J. Appl. Phys.*, 123, 11, 2018.
44. Chiou, M., Lin, Y., Ono, T., Esashi, M., Yeh, S., Wu, T., Focusing and waveguiding of Lamb waves in micro-fabricated piezoelectric phononic plates. *Ultrasonics*, 54, 7, 1984–1990, 2014.
45. Harne, R., Song, Y., Dai, Q., Trapping and attenuating broadband vibroacoustic energy with hyperdamping metamaterials. *Extreme Mech. Lett.*, 12, 41–47, 2017.
46. Wang, P., Casadei, F., Shan, S., Weaver, J., Bertoldi, D., Harnessing buckling to design tunable locally resonant acoustic metamaterials. *Phys. Rev. Lett.*, 113, 1, 1–5, 2014.
47. Otsuka, P., Sylvain, M., Osamu, M., Motonobu, T., Alexei, M., Tian, G., Nicholas, F., Nicholas, B., Vitalyi, G., Oliver, W., Time-domain imaging of gigahertz surface waves on an acoustic metamaterial. *New J. Phys.*, 20, 1, 2018.

16

Radar Cross-Section Reducing Metamaterials

Samson Rwahwire[1,2]* and Ivan Ssebagala[1]

[1]Department of Polymer, Textile and Industrial Engineering, Faculty of Engineering and Technology, Busitema University, Tororo, Uganda
[2]Center of Materials and Energy Engineering, Busitema University, Tororo, Uganda

Abstract
The revolution and development of systems and/or gadgets that utilize radio detection and ranging (RADAR) in their operation set the ball rolling for the invention of metamaterials that are capable of reducing and/or blocking the reflected electromagnetic waves (a technology used to detect, locate, and recognize certain structures, people, or equipment) such that radar systems cannot detect, locate, recognize or even track some sensitive/confidential human beings or equipment. In more recent years, the development of metamaterials for the reduction of radar cross-section (which is a critical parameter in microwave applications) for special equipment/targets has been of great interest. In this chapter, concepts of metamaterial, radar technology, radar cross-section reduction, and different techniques of RCS reduction have been discussed. Finally, the chapter concludes and sights out some future outlooks based on the current progress and advancements in metamaterials for radar cross-section reduction.

Keywords: Metamaterials, radar cross-section (RCS), RCS reduction

16.1 Introduction

Metamaterials can be defined as synthetic composite materials which exhibit supernormal physical properties nonexistent in other natural materials; those possessing a negative refractive index in particular have had a

**Corresponding author*: rbsjunior@gmail.com

Inamuddin and Tariq Altalhi (eds.) Electromagnetic Metamaterials: Properties and Applications, (341–362) © 2023 Scrivener Publishing LLC

major focus including other unique properties of very low density, negative Poisson's ratio, tunable dynamic properties, and high damage resistance [1, 2]. A negative Poisson's ratio is one of the most special/unique properties of metamaterials; if loaded in compression, the strain is perpendicular to the compression direction which is not the case with other natural materials whereby a compression loading causes either lateral strains or none (this signifies a Poisson's ratio very close to zero) [3–5]. Metamaterials are generally artificially engineered but there exist also some natural materials that exhibit negative Poisson's ratios for instance α-cristobalite structure of Silicon dioxide and Citrus maxima (pomelo) peels. These types of materials tend to scatter electromagnetic waves more than natural materials. Metamaterials exhibit unconventional characteristics that primarily rely on micro-architectural structure designs [6]. Metamaterials are generally made from a combination of various composites for instance metals and plastics. The evolution of metamaterials has led to an interest in the electromagnetic wave theory and this has made it easy to control and manipulate electromagnetic waves more effectively and efficiently by blocking, absorbing, enhancing, and/or bending waves [7, 8].

Control and manipulation of electromagnetic waves have captured interest in various fields, especially in the medical and military, this is achieved through designing and developing several functional materials such as liquid crystals, photoactive materials, ferroelectric materials, and phase-shifting materials [9]. Metamaterials can have found applications in healthcare devices, engineering products such as airplanes and satellites, optical filters, sensor detection, radomes, and shielding structures from earthquakes [10]. However, for each field application, the metamaterials have to meet the minimum requirements for the designed function for example for use in the medical field, the metamaterial has to be compatible with the body cells, have remarkable mechanical properties, and not interfere with proper body growth [11].

The reduction of radar cross-section for targets for example naval vessels, military aircraft, spacecraft, ships, and other equipment is one of the leading applications of metamaterials [12]. The concept is that metamaterials tend to reduce or completely block the reflected electromagnetic waves from the target or equipment under consideration from reaching the radar system rendering the system unable to locate, recognize, track or detect the target. The use of radar systems is a complete imitation of the nature of a bat.

16.1.1 The Electromagnetic Radiation and Spectrum

An electromagnetic (EM) spectrum represents the full spreading out of all the electromagnetic radiations following the order of decreasing wavelength and increasing frequency (Figure 16.1). Gamma rays are the strongest EM radiations since they have the highest frequency and the smallest wavelength. The reverse is true for Radio waves since they have the lowest frequency [13, 14].

EM waves are transverse waves because they are composed of both electric and magnetic fields. EM radiations do travel in space and they do not need a medium for transmission.

All EM radiations travel at the speed of light ($3x10^8$ m/s). For any radiation, the frequency and energy are determined by the following formulas;

$$\text{Frequency, } f = c/\lambda$$

$$\text{Energy, } E = hxf = hc/\lambda$$

Where; c is the speed of light in a vacuum
λ is the wavelength of the radiation and
h is the Plank's constant

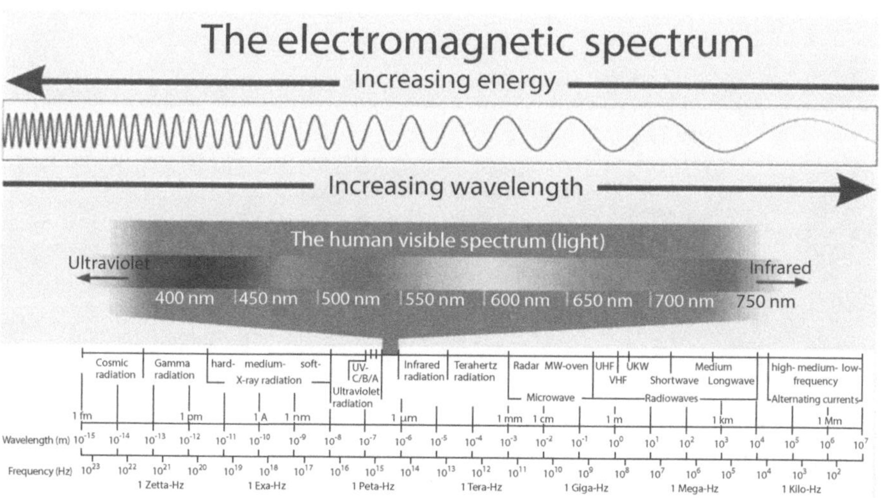

Figure 16.1 An electromagnetic spectrum.

Different waves or radiations are used for different applications for example Infrared waves are used for locating objects in the darkness, ultraviolet light for sterilization, and radiowaves in satellite transmissions, RADAR, and cooking food. More specifically microwaves (frequency range between 1 and 1000 GHz with the respective wavelength of 30 to 0.03 cm) that lie in the radio waves region are used in RADAR applications [15].

16.2 Radiodetection and Ranging

The concept of RADAR mimics nature, for instance, a bat can sense and detect the size of an object or obstacle by sending a sound signal which quickly propagates through the air and when it hits the object, an echo is sent back which is processed by the bat and it can tell the distance and size of the object that sent an echo. Similarly, radar is an electromagnetic sensor that locates, detects, recognizes, and/or tracks objects by transmitting an electromagnetic wave towards an object commonly known as the target and uses the returned echo to perform the above functions (Figure 16.2) [16, 17]. Furthermore, radar can determine the angular position, size, shape, direction, and velocity of the target apart from its location or distance. It normally operates in the microwave part of the electromagnetic spectrum commonly measured in Hertz (Hz).

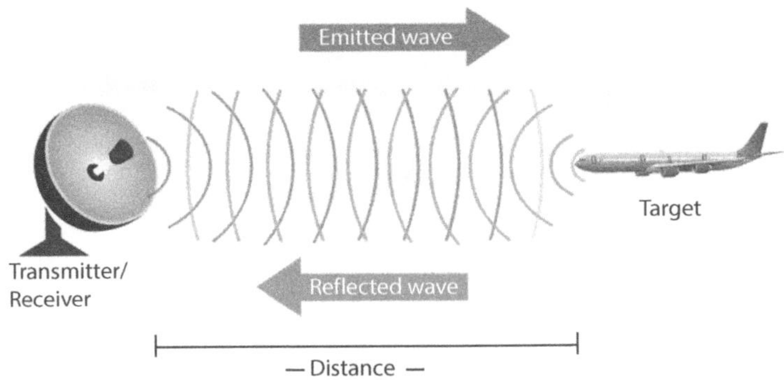

Figure 16.2 The working principle of radar.

16.3 RADAR Cross-Section

RADAR cross-section (RCS) also known as radar signature measures the targets to reflect radar signals to the receiver or the ratio of power density backscattered to the radar receiver to the incident power density on the target (Figure 16.3) [18, 19]. Radar cross-section is one of the most important aspects applied in stealth technology. Target RCS in other words is a comparison between the reflected radar signal strength from the target to the reflected signal strength from a perfectly smooth sphere of 1 m^2 cross-sectional area as illustrated in Figure 16.4 below [20];

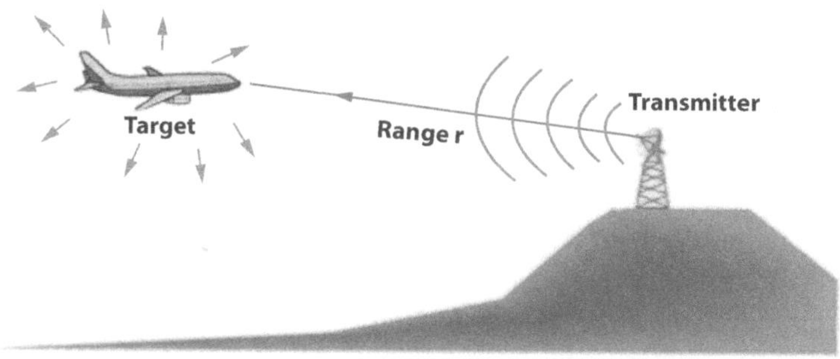

Figure 16.3 The concept of RADAR cross-section.

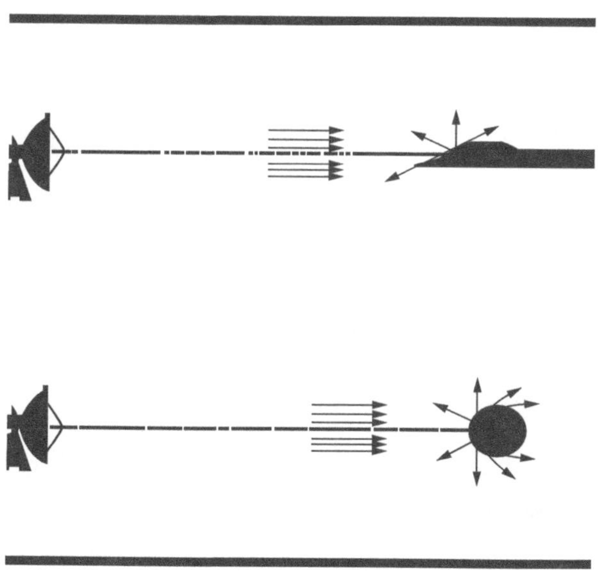

Figure 16.4 The effect of target shape on radar.

RCS is denoted by sigma (σ) and it is measured in m^2 or dBm^2, mathematically it is defined as;

$$RCS\ (\sigma) = 4\pi R^2(Pr/Pi) \tag{16.1}$$

Pi = power intensity of a plane wave striking the target
Pr = power per unit solid angle reflected by the target
R = distance between target and receiver

However, from the conceptual diagram above, it is clear that not all the radiations fall on the target, hence, RCS of the target is viewed as a function of three factors:

σ = Projected cross-section X Reflectivity X Directivity

Reflectivity is the percentage of the intercepted power scattered by the target.

Directivity is the degree to which the emitted radiation concentrates in a single direction.

The range between the target and the radar system can be calculated from; $R = c\Delta T/2$ where c is the speed of light and ΔT is the time elapsed between the electromagnetic wave generation to the return of the reflected waves after hitting the target.

RCS can be either a monostatic RCS (Figure 16.5a) where the transmitter and the receiver antenna are located in the same place or a bistatic RCS (Figure 16.5b) where the transmitter and the receiver antenna are not located in the same place [21].

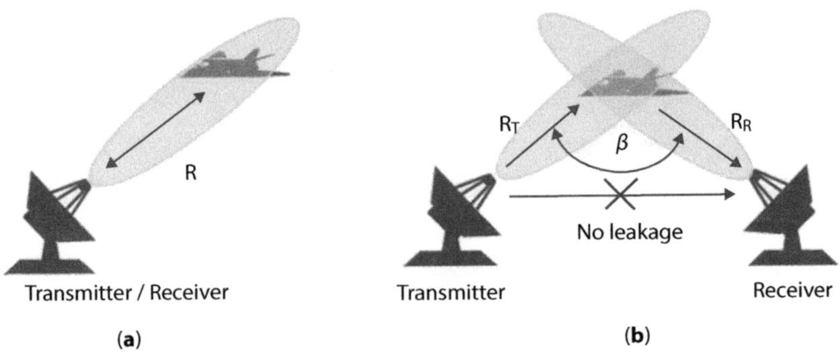

Figure 16.5 Radar types.

A target's Radar cross-section depends on the surface's reflectivity, the target's size, and the directivity of the radar reflection caused by the target's geometric shape [18]. Some factors do influence/affect radar cross-section as discussed below:

Target's Size
The size of the object affects the response; a larger object causes a stronger reflection which eventually gives a bigger RCS. In addition, radars are tailor-made to detect specific bands; for instance, S-band radar cannot detect very small droplets like drizzles but it can detect rain droplets.

Material type or nature
Metals (even a very thin layer of metal) are known for producing very strong reflective radar signals while materials like wood, plastics, fiber-glass, and clothes/less shiny textiles are less reflective or even transparent to radar (close to randoms). Meanwhile, some materials are fabricated radar active in nature like radar antennas which automatically increase the radar cross-section.

Radar absorbent paint
Aircraft like the SR-71 Blackbird is painted with a paint containing tiny metallic coated balls called iron ball paint which tends to convert illuminating radar energy into heat other than reflecting it.

Object Shape and Orientation
Aircraft surfaces for instance the sea shadow US Navy and F-117A are designed to be flat (purpose shaping) so that the radar illuminating falls on the target's surface at a very large angle of incidence which eventually causes the reflected radar to be at respectively a large angle, sufficient scattering of the reflected radar signal happens in other terms. Round shapes produce strong reflection so that's why the edges of stealth aircraft are made sharp to reduce signal reflection. Lastly, orientation plays a critical role in the detection of the target relative to the radar station.

Smooth surfaces
The strength of the emitter and the range between the transmitter and the target do not affect the calculation of RCS but only the reflectivity of the target matters [20, 22]. From equation 16.1 above, a large value of RCS (σ) for a given target implies that the target can easily be located, detected, tracked, or recognized. For purposes of security and confidentiality (more especially developers of weapon systems) of some special equipment such

as military aircraft, ships, vehicles, etc., [19]. There's a need to reduce the RCS for such targets or equipment such that they cannot easily be detected or located. Recently, various attempts have been made to reduce radar cross-section (RCS) for targets by numerous researchers using generally four major [23] different methods/techniques as discussed below:

16.3.1 Use of Radar-Absorbing Materials

Radar-absorbing materials are either used in the original fabrication of the target or incorporated to cover very reflective materials and/or surfaces (Figure 16.6) [24].

Generally, three types of radar-absorbing materials are employed, namely; resonant, non-resonant large volume, and non-resonant magnetic RAMs. Below is the current research progress in line with RAMs:

Liang & Haochuan (2021) showed that Salisbury Screen filled with plasma could change the absorbing resonance frequency by changing the resonance frequency of the plasma, as well as the frequency at which the electrons collide. Additionally, the absorbing frequency of the Salisbury screen was widened compared to the conventional Salisbury screen [25].

Duan *et al.* (2021) 3D printed a metamaterials structure of about 10mm which could achieve a -10dB RL absorption over a frequency range of 5.1-40 GHz plus an absorbing bandwidth of -15dB in 7.7-36.3 GHz [26]. The meta-structure could still possess strong microwave absorption for transverse electric polarization with the striking angle between 0°–55° and 0°–70° for transverse magnetic polarization. This is a very promising strong microwave absorption metastructure for the future.

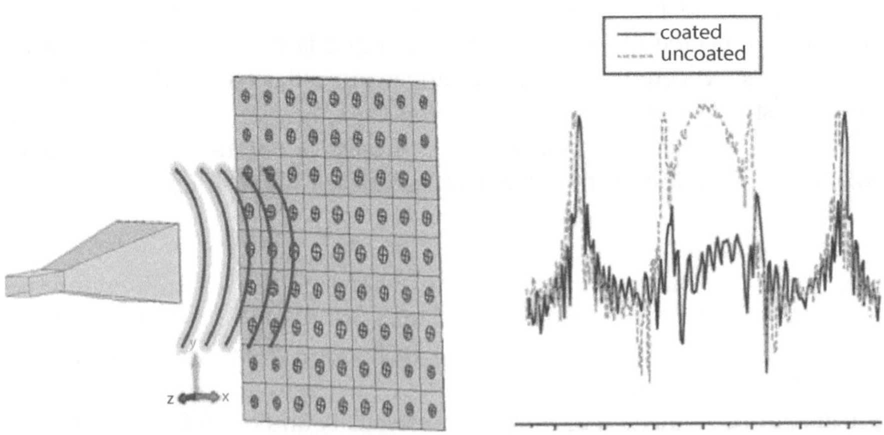

Figure 16.6 The effect of RAM on radar signal.

Kim & Kim (2021) fabricated composites reinforced by Aramid fiber (AF) and coated intrinsically with conducting polymers namely poly(3,4-ethylenedioxythiophene) polystyrene sulfonate (PEDOT-PSS) and protonated polyaniline with camphor sulfonic acid (PANI-CSA) which were tested on a bandwidth of 8-12 GHz plus for mechanical properties. It was observed that the PANI-CSA-coated aramid fiber radar absorbing structures (RAS) did not absorb radar signals owing to the poor integration of PANI-CSA to the composite while PEDOT-PSS coated AF realized high reflection loss in the test bandwidth [27]. Optimum radar absorption for PEDOT-PSS coated AFRP was achieved at a weight ratio of PEDOT equaling to ¼ with a 99% absorption at 10.4 GHz plus a -10dB between 9.1-11.9 GHz. Furthermore, the PEDOT-PSS coating showed future application since it also enhanced the mechanical properties of the composite by 124% for tensile strength and 55% for stiffness.

Most existing multi-tube structures suffer initial peak load stacking and fluctuations. Feng *et al.* (2022) configured and attempted to reduce the initial peak load and the load fluctuations of signal in most combined multi-tube structures by developing a differentiated combined design considering dislocation superposition of peak load in terms of height and diaphragm configuration for individual tubes [28]. Results showed that for a 6-4GT-ΔJ-ΔH 2-Adj tube constructed basing on differentiated configuration design idea with optimal set heights and spaced diaphragm of the tube; both the initial peak load and the load fluctuations were reduced 45% and 77% respectively whilst embracing a high energy absorption capability. This was attributed to the ability of separating fully the multi-tube structures peak loads.

Li *et al.* (2019) fabricated a lightweight broad bandwidth combinatorial foam metamaterial as shown in Figure 16.7 composed of a carbon foam material, metal pattern, and an FR4 dielectric material which exhibited a radiation loss of -10dB over a bandwidth of 14GHz between 4-18GHz [29].

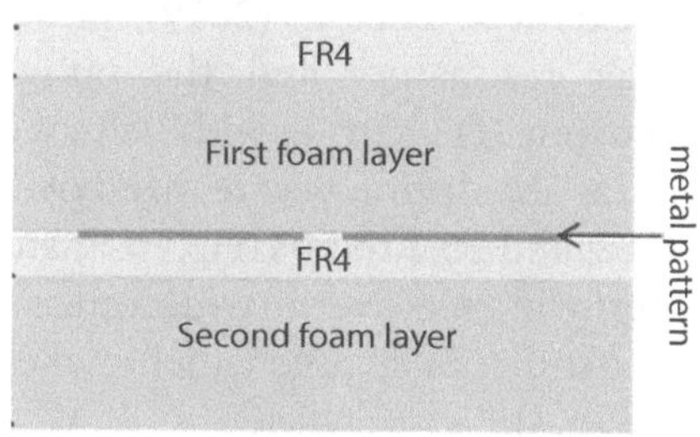

Figure 16.7 A radar-absorbing combinatorial foam metamaterial.

However, the main challenge with lightweight and thin composite structures is the difficulty to achieve high radiation absorption over a wide bandwidth hence this area still requires more research.

Such a high radiation loss is attributed to the adjustment of the metal pattern geometrical function as well as the introduction of a dielectric material.

Li *et al.* (2019) also simulated and experimented on a metamaterial based on silicon carbide foam material where the electromagnetic wave absorption was studied using different methods like power loss density distribution, electromagnetic field distribution, etc., and the reflection loss below −10 dB was over 4 to 18 GHz. This is mainly due to the inherent absorption capability of silicon carbide or carbon foam and the design of the metamaterial plus the significant metamaterial's thickness of 10 mm. Furthermore, the presence of silicon coupled with a big thickness enhanced the thermal stability to up to 400°C [30].

Shen *et al.* (2018) developed a broad bandwidth radar absorbing sandwich structure using two microwave resistive films primarily for wave absorption which gave a wave reflection of less than −10 dB over a frequency bandwidth of 2.6 to 21 GHz (18.4 GHz) which is remarkably incorporated with arrays of fiber columns which contributed to a lesser extent to microwave absorption but significantly designed for enhancing mechanical properties of the sandwich. Microwave absorption was studied using the Equivalent circuit model while the mechanical properties focused on flexural/bending strength [31].

Li *et al.* (2019a) integrated a scattering metasurface with a magnetic absorbing material for broad bandwidth RCS reduction with a thickness of 2mm only which eventually gave a reduction in RCS of 6dB over a frequency bandwidth of 5-34 GHz [32]. Such a good reduction is attributed to the collective destruction/scattering of the reflected echo together with the absorption of the reflected echo. This work greatly tackled and solved the issues associated with thin lightweight radar absorption materials.

Similarly to a lesser extent to Li *et al.* (2019), Zhou *et al.* (2022) introduced cylindrical water resonators and the optimized metamaterial absorber was prepared using 3D printing with help of injected water and a TPU container, 90% radar absorption was realized over a frequency bandwidth of 5.74 to 19.7 GHz and 25.2 to 40 GHz, insensitivity to polarization and produced good performance over a wide range of incidence angles [33]. Additionally, the absorber is flexible which makes it suitable even for curved surfaces (Figure 16.8).

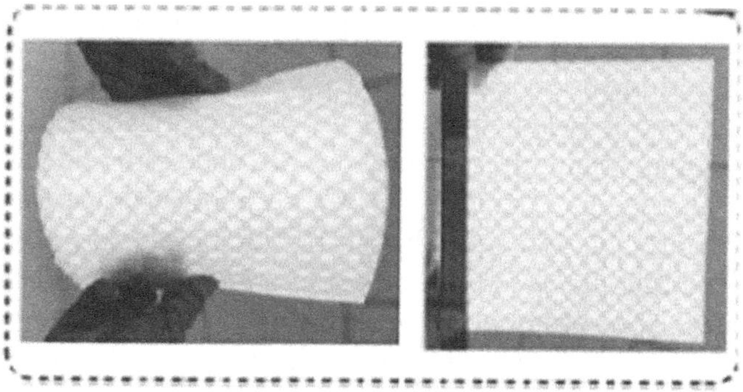

Figure 16.8 Sandwiched cylindrical water-based metamaterial.

Such good properties are attributed to the improved matching of impedance together with sandwiched cylindrical water resonators plus the resonance effect of electromagnetic energy which enhances electromagnetic wave dissipation.

Tau *et al.* (2020) used $Mn_{1-x}Zn_xFe_2O_4$ nanoparticles obtained from natural iron sand to formulate a ferrofluid for both radar absorption and magnetic sensing. From Figure 16.9, it is evident that at maximum Mn, the best reflection loss of -15dB was observed at around 10.9 GHz. It can be concluded that good radar absorption is attained with an increase in Mn ions and a decrease in Zn ions [34].

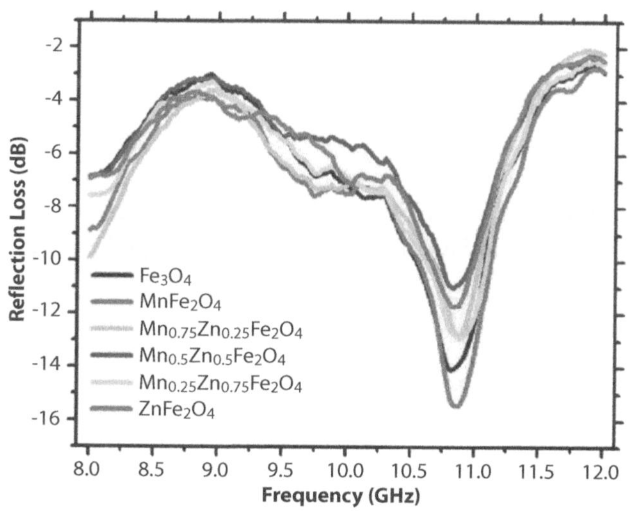

Figure 16.9 Radar absorption by $Mn_{1-x}Zn_xFe_2O_4$ nanoparticles over a frequency of 8–12 GHz.

16.3.2 Polarization of the Impinging/Illuminating Wave

Wave polarization simply expresses how an electric field line in an electromagnetic field is oriented (Figure 16.10). Polarization can either be in one particular orientation throughout or rotate with each wave cycle. It involves placing a substrate/polarizer perpendicular to the direction of waveform propagation such that scattered waves/wave planes can be viewed in a single plane [35]. An electromagnetic (EM) wave is simply an interaction or a combination of an electric field (EF) and a magnetic field (MF) traveling in space. The two fields move in planes or vibrate in directions that are at right angles (perpendicular to one another). When a waveform is vibrating in more than one plane, it's known as an unpolarized waveform, and such examples are microwaves (radio) waves since they are transverse [36].
where E = Electric field (in x-y plane)
B = Magnetic field (in x-z plane), propagation is in the x-direction

Polarization of waves is applied in RCS reduction since it helps to enhance the cut-off wavelength of metasurfaces as well as preventing undesired backscattering or back reflections [37]. Three types of polarization exist, namely (1) Linear polarization where the waveform is limited to one plane toward the propagation direction. (2) Circular polarization comprises two linear components in the wave with equal amplitudes but $\pi/2$ out of phase. (3) Elliptical polarization is one where the wave follows the propagation in an elliptical manner and the amplitude plus phase difference of the two components are not equal.

Sharma *et al.* (2018) designed a novel circular polarized antenna combined with a metasurface where the metasurface was for the reduction of RCS coupled with circular polarization of radar [38]. It is concluded that great performance is achieved by the metasurface but of course, the introduction of the dual SRR helped in matching impedance. A -10dB RCS

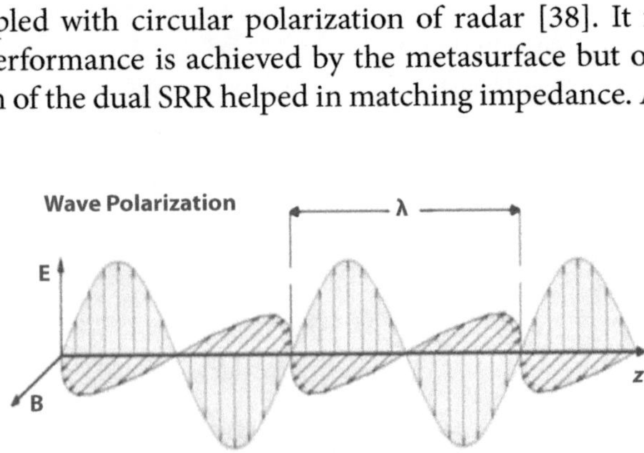

Figure 16.10 A wave polarization pattern.

reduction is achieved over a bandwidth of 41.3% of the bandwidth. Wei *et al.* (2022) designed and fabricated with the major aim of manipulating (polarizing) terahertz waves [39]. The metasurface was able to polarize waves up to 89% over the bandwidth of 2.28-6.75 THz and achieved an RCS reduction of -20dB between 2.1-6.9 THz. This RCS reduction is remarkable and very promising as compared to most RAMs performance. Yang *et al.* (2021) studied the relationship between the phase modulation of metasurfaces and the reduction of RCS and it was found that for the split square-ring module, effective RCS reduction happens at 180° phase gradient reflected horizontal and vertical polarized waves whereas, for the split circular-ring module, effective RCS reduction depended on the difference of phase between u and v as well as amplitudes of x and y polarized waves [40]. J. J. Yang *et al.* (2018) used a matrix-type coding method to code the metasurfaces meant for reducing RCS utilizing a split ring resonator (SRR) structure through microwave polarization [41]. Results show that the polarization conversion rate went up to 90% in a range of 6–15 GHz and the successive RCS reduction was more than 10 dB over a frequency bandwidth of 5.5–15 GHz. J. J. Yang *et al.* (2018) Designed a wave polarization metasurface incorporated with a patch antenna. Both the simulation and experimental results revealed that great RCS reduction of up to -41.5dB over a wide frequency range between 9 and 24 GHz could be achieved [41]. This is attributed to the improvement of the initial peak gain of 1.25 dB, and wide bandwidth of impedance of 13.8–15.2 GHz due to the patch antenna which also improved the radiation characteristics like amplitude, phase angle, etc. Fang *et al.* (2021) integrated a wave phase canceller with RAM characterized by lightweight with a backbone of stereo meta-atoms. The metamaterial can give an RCS reduction of -10dB over 7.65-25 GHz and over -15dB for 8.7-22.8 GHz for both polarized transverse electric and magnetic fields at normal incidence.

It can be observed from the graph above (Figure 16.11) that most of the radiations are absorbed by the metamaterial implying that there is greater RCS reduction since the reflected signal is very small [42]. Ramachandran *et al.* (2022) used passive polarization to develop a metasurface (MS) for RCS reduction at several band resonance frequencies. The MS had a thickness of 1.524 mm composed of Rogers RT6002 substrate with five circular-shaped metamaterials. The MS captured three resonance frequencies of 3.924, 7.254, and 14.832 GHz [8]. The MS exhibited an RCS reduction of −60.218 and −38.875 dBm^2 for the last two above frequencies. It is also evident W. Li, Zhang, *et al.* (2019b) that the integration of a scattering metasurface and a magnetic absorbing metamaterial (MAM) in the same plane could yield great radar cross-section reduction of up to 6 dB

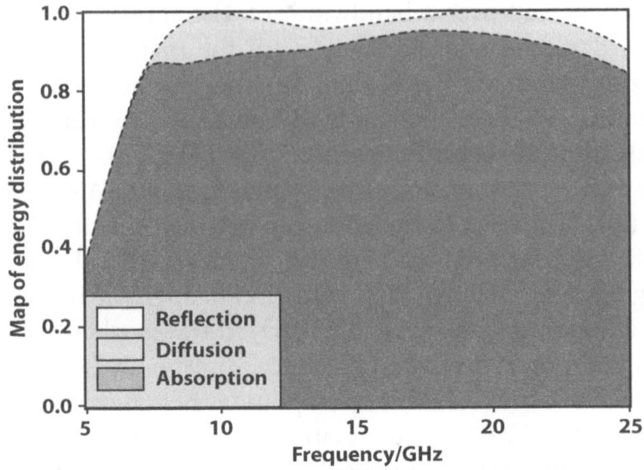

Figure 16.11 The energy distribution for the above-designed structure.

over a bandwidth of 5 to 34 GHz (Figure 16.12) [43]. This integration of a metasurface and a MAM (hybrid formation) has the added advantage of reducing RCS by a bigger margin because they complement each other coupled with having a very small thickness, lightweight, good mechanical properties, and enhanced radar absorption. The figure below shows (a) the electromagnetic and magnetic absorbing metamaterial hybrid, and (b) the RCS reduction magnitude over a wide bandwidth. From graph (b), an RCS reduction of up to 22 dBm^2 at about 13.5 GHz was attained.

Guthi & Damera, (2022) polarized an antenna in a circular form with only one layer substrate and then added stubs in the L-shaped slot on the circular patch antenna. Due to the above-stated integration of equipment, a remarkable bandwidth impedance increase of 20% and a peak gain of

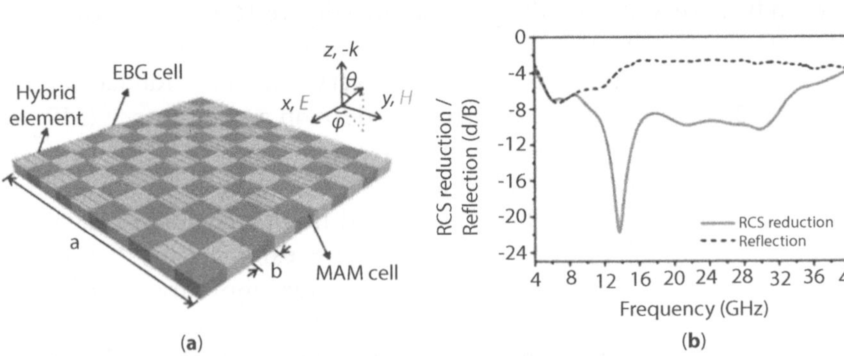

Figure 16.12 A hybrid of EBG and MAM with the respectively achieved RCS reduction.

10.43dB were achieved. An increase in the bandwidth impedance of a given system implies a proportional reduction in the radar cross-section of the system [44]. A multi-dollar observation here is that just a single layer of metasurface was able to produce remarkable and significant results of polarization, implying that the addition of more substrate layers could realize better results.

16.3.3 Active Cancellation of the Scattered Field/Backscatter

Cancellation of backscatter can either be passive or active cancellation, the latter being the most commonly applied class. Passive signal cancellation simply involves an obstacle capable of blocking the backscatter signal from being received by the receiver of the radar (Figure 16.13). Following the concept of noise cancellation by earphones, active backscatter signal cancellation involves the generation of a radar signal but completely out of phase to the anticipated reflection of the incident radar signal by a target hence creating a destructive interference between the generated and the reflected radar signals, eventually reducing the backscatter power [45]. Since the radar cross-section depends on the backscatter power, a reduction in RCS can be achieved. Application of the active cancellation technique requires knowing the incident angle and the waveform characteristics for example amplitude, frequency, wavelength, etc. This technique is somewhat difficult because it is not very easy to establish the nature of the reflected signal relating to the whole aspect of the target [46]. This technique is easy in terms of simulation but a bit difficult in practical engineering applications.

Sengupta *et al.* (2020) proposed an active radar canceller (Microstrip) composed of an amplifier, isolator, and a phase shift connected through a transmission line basically for low frequencies. Results show that RCS

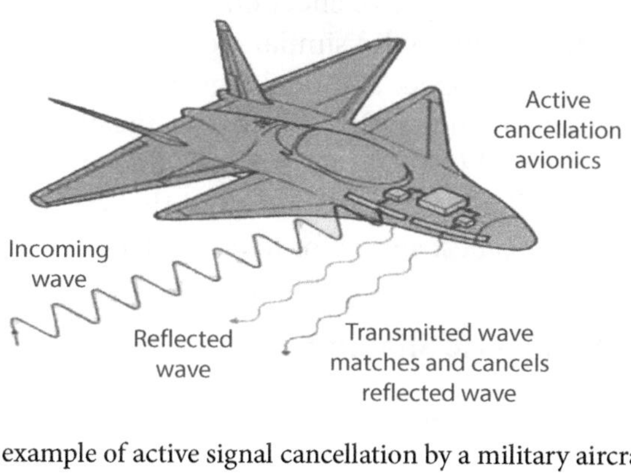

Figure 16.13 An example of active signal cancellation by a military aircraft.

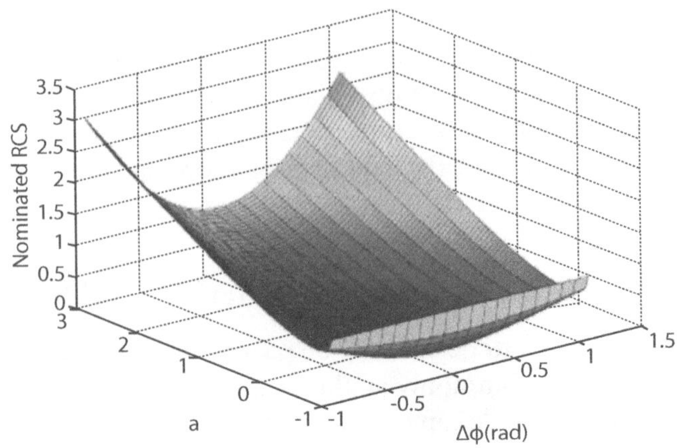

Figure 16.14 The relationship between RCS and error in phase plus amplitude.

reduction of the system increases with an increase in the bandwidth of the patch antenna as well as reducing the delays in the transmission line which could be attributed to many components in the line (in other words due to long signal processing). Also, the angular region of scattering cancellation achieved kept on reducing with the increase in target/object size while curved surfaces showed promising results in the angular region [47].

Xu & Xu, (2014) studied how the error in frequency, phase, and amplitude affect the active radar cancellation stealth system. Below is an optimized graph for the relationship.

It can be observed that amplitude error has a very small impact on RCS, followed by phase error. Additionally, frequency error has a very minimal tolerance implying that frequency is so critical in active cancelation systems (Figure 16.14). So to ensure frequency consistency, the use of the transmission system could be opted for. Lastly, a combination of all three errors could cause a huge negative effect on the RCS of the target [48, 49]. Similarly, Huang & Xu, (2015) did similar work.

Yi *et al.* (2017) developed a model/algorithm based on Nth-order Spectrum Spread and Compression (SSC) specifically for nonlinear frequency modulated signal (NLFM) in trying to solve the problems associated with active cancellation [50]. It was found that the active cancellation technique/technology is best suited for low-frequency stealth difficulties.

16.3.4 Target/Purpose Shaping

This RCS reduction technique is the oldest of all. The primary purpose of target shaping is to orient surfaces or bodies by creating surfaces (using the idea of the angle of incidence about the angle of reflection) that can scatter the illuminating signals or waves at directions very far away from the detectable radar angles of a given radar antenna [17], this RCS reduction method requires that it be considered right from the design of each component of the entire target object/equipment otherwise considering it after the manufacture of a component would compromise with other component design features to meet its primary functions. Literature shows that oval or cylindrical shapes could give a better reduction of RCS by reflecting wave signals far away from the radar antenna receiver as compared to flat surfaces. Such aircraft include the F-117A (Figure 16.15), sea shadow US Navy (Figure 16.16), and B-2-type stealth aircraft. Its major limitation is that a passive radar (multi static radar) can defeat it [45]. This shortfall can be corrected by using a metasurface (MS) capable of redirecting the scattered waves from the radar receiver minus adjusting the target's geometry. Or else, some new advances in technologies have seen the emergence

Figure 16.15 The complicated shaping of F-117A- for purposes of RCS reduction.

Figure 16.16 The sea shadow US Navy.

of morphing techniques where adaptive blades can constantly modify/change their shapes to perform optimally over the entire application. Many metamaterials have been developed based on the folding mechanism that can change their shapes like the self-folding origami lattices that transform from planar to large 3D shapes and they are widely applied in space structures [1, 17].

16.4 Conclusion and Outlook

The progress and advancements in metamaterials for radar cross-section reduction have been discussed. In the previous sections, we discussed the progress of Radar cross-section reduction through the use of metamaterials. Several researchers have attempted to reduce radar cross-section for targets using different methods/techniques whereby a few have been presented in this paper with their respective results (applicability), which are to a high level remarkable and significant. However, there still exist many obstacles in the way to fully controlling and manipulating radar waves that present future challenges in this particular field. A few realized gaps presented per RCS reduction include the following;

a) Most RAMs have been found able to achieve a -10dB reflection loss between a frequency bandwidth of 2GHz to about 40GHz only. This leaves a very wide radar frequency bandwidth from about 40 GHz upwards uncontrolled or not catered for implying that radar waves having frequencies above 40 GHz may not be absorbed by the current RAMs available. Also, there is a need to improve the available RAMs to achieve reflection losses above 10 dB.
b) RCS reduction by the polarization of radar waves by use of metasurfaces (polarizer) produces remarkable of up to -20dB but still over a considerably small frequency bandwidth of 5 to 35 GHz. Researchers should consider integrating the RAM and polarization techniques as a way of achieving larger values of reflection loss over wider frequency bandwidths.
c) The gap existing in the cancellation technique is that it is still only suitable for reducing RCS at very low frequencies plus it is difficult to practically be applied in engineering. Most research about cancellation is still in the simulation stage. Second, it is hard to predict the illuminating waveform

characteristics like amplitude, frequency, etc., which have to be the base to design the canceling signal hence the need for the development of models for this cause.

d) The major setback with target/purpose shaping RCS reduction technique is that it cannot be applied on already made components and/or equipment and still even if it is considered during the design and development of components/surfaces, it would interfere with the components' primary design functionality. So there is a need to research more in the direction of self-folding mechanisms such that equipment components can adapt to different shapes for optimized RCS reduction during the equipment application.

References

1. Ucar, H., Radar cross section reduction. *J. Nav. Sci. Eng.*, *9*, 72–87, 2013.
2. Willis, N.J., *Bistatic radar*, vol. 2, SciTech Publishing, Near LIC building. Ansari Road, Darya Ganj, 2005.
3. Cardoso, O., Borges, P., Velhinho, A., Materials international structural metamaterials with negative mechanical/thermomechanical indices: A review. *Prog. Nat. Sci.*, 31, 801–808, 2021.
4. Chen, S., Liu, W., Li, Z., Cheng, H., J.T., Polarization state manipulation of electromagnetic waves with metamaterials and its applications, in nanophotonics, in: *Metamaterials, Devices and Applications*, A.L. Borja (Ed.), 2017.
5. Course, B. G. E. P., *Spectrum summary notes.*
6. Dorn, C. and Kochmann, D.M., Ray theory for elastic wave propagation in graded metamaterials. *J. Mech. Phys. Solids*, 168, 105049, 2022.
7. Duan, Y., Liang, Q., Yang, Z., Li, Z., Yin, H., Cao, Y., Li, D., A wide-angle broadband electromagnetic absorbing metastructure using 3D printing technology. *Mater. Des.*, *208*, 109900, 2021.
8. Wikipedia, Electromagnetic, History, Classification & Other. *A metamaterial*, pp. 1–18, 2017.
9. Fang, W., Zhou, F., Wang, Y., Chen, P., Broadband, wide-angle, polarization-independent and lightweight low-scattering coding metamaterial based on stereo meta-atoms. *Results Phys.*, *20*, 103687, 2021.
10. Feng, Z., Xie, S., Ma, W., Jing, K., Wang, H., Multi-tube energy-absorbing structures with different matching patterns of heights and diaphragm spacings. *Alexandria Eng. J.*, 61, 11111–11127, 2022.
11. Gu, H., Taghipour, J., Shaw, A.D., Amoozgar, M., Zhang, J., Wang, C., Friswell, M.I., Tailored twist morphing achieved using graded bend – twist metamaterials. *Compos. Struct.*, *300*, 116151, 2022.

12. Guthi, S. and Damera, V., High gain and broadband circularly polarized antenna using metasurface and CPW fed L-shaped aperture. *AEU Int. J. Electron. Commun.*, *146*, 154109, 2022.
13. Huang, Q., Wang, G., Zhou, M., Zheng, J., Tang, S., Ji, G., Metamaterial electromagnetic wave absorbers and devices: Design and 3D microarchitecture. *J. Mater. Sci. Technol.*, *108*, 90–101, 2022.
14. Huang, Q. and Xu, Y., Active cancellation stealth analysis based on interrupted-sampling and convolution modulation. *Opt.*, *127*, 3499–3503, 2016.
15. *Radar electronics*, 2022, https://www.britannica.com/technology/radar/Pulse-radar.
16. Ibrahim, T., Ahmed, O., Mirghani, M., Estimation of radar cross sectional area of target using simulation algorithm. *Int. J. Res. Stud. Electr. Electron. Eng.*, *4*, 20–24, 2018.
17. Iniguez-Rabago, A. and Overvelde, J.T.B., From rigid to amorphous folding behavior in origami-inspired metamaterials with bistable hinges. *Extreme Mech. Lett.*, *56*, 101881, 2022.
18. *Introduction to polarization*, https://www.edmundoptics.com/knowledge-center/application-notes/optics/introduction-to-polarization.
19. Karathanasopoulos, N. and Al-ketan, O., Towards biomimetic, lattice-based, tendon and ligament metamaterial designs. *J. Mech. Behav. Biomed. Mater.*, *134*, 105412, 2022.
20. Kibble, R., The electromagnetic spectrum, in: *Making Use of Physics for GCSE*, R. Kibble (Ed.), pp. 135–140, 1989.
21. Kim, C. and Kim, M., Intrinsically conducting polymer (ICP) coated aramid fiber reinforced composites for broadband radar absorbing structures (RAS). *Compos. Sci. Technol.*, *211*, 108827, 2021.
22. Knott, E., Shaeffer, J., Tuley, M., *Radar cross section*, p. 231, Artech House, Inc., Canton Street Norwood, MA, USA, 1993.
23. Kulikov, L. and Lavidas, N., Between passive and middle: Evidence from Greek and beyond, in: *Contrastive Studies in Verbal Valency*, L. Hallen, A. Malchukov, M. Cennamo (Eds.), pp. 298–325, 2017.
24. Li, C., Cao, Q., Zhou, G., Shao, M., Chai, P., Cai, D., Design and study of a new broadband RCS carbon-glass fiber hybrid metamaterial. *Compos. Struct.*, *301*, 116207, 2022.
25. Li, W., Li, C., Lin, L., Wang, Y., Zhang, J., All-dielectric radar absorbing array metamaterial based on silicon carbide/carbon foam material. *J. Alloys Compd.*, *781*, 883–891, 2019.
26. Li, W., Lin, L., Li, C., Wang, Y., Zhang, J., Radar absorbing combinatorial metamaterial based on silicon carbide/carbon foam material embedded with split square ring metal. *Results Phys.*, *12*, 278–286, 2019.
27. Li, W., Zhang, Y., Wu, T., Cao, J., Chen, Z., Guan, J., Broadband radar cross section reduction by in-plane integration of scattering metasurfaces and magnetic absorbing materials. *Results Phys.*, *12*, 1964–1970, 2019.

28. Li, W., Zhang, Y., Wu, T., Cao, J., Chen, Z., Guan, J., Broadband radar cross section reduction by in-plane integration of scattering metasurfaces and magnetic absorbing materials. *Results Phys.*, *12*, 1964–1970, 2019b.
29. Liang, M.A.N. and Haochuan, D., Study on absorbing characteristics of salisbury screen filled with plasma. *Proc. Comput. Sci.*, *187*, 241–245, 2021.
30. Mehboob, A., Mehboob, H., Nawab, Y., Hwan, S., Three-dimensional printable metamaterial intramedullary nails with tunable strain for the treatment of long bone fractures. *Mater. Des.*, *221*, 110942, 2022.
31. Mercer, C., Speck, T., Lee, J., Balint, D.S., Thielen, M., Effects of geometry and boundary constraint on the stiffness and negative poisson's ratio behaviour of auxetic metamaterials under quasi-static and impact loading. *Int. J. Impact Eng.*, *169*, 104315, 2022.
32. Norgard, J. and Best, G.L., The electromagnetic spectrum, in: *National Association of Broadcasters Engineering Handbook*, G.C. Cavell (Ed.), pp. 3–10, 2020.
33. *Radartutorial, polarization of electromagnetic waves*, Christian Wolff, Josefine-Clouth-Strabe, Cologne, Germany, retrieved 23rd/09/2022, https://www.radartutorial.eu/06.antennas/Polarization.en.html.
34. Britannica, *Pulse radar*, https://www.britannica.com/technology/radar/Pulse-radar.
35. Wikipedia, *Radar cross-section*, https://en.wikipedia.org/wiki/Radar_cross-section#Purpose_shaping.
36. Ramachandran, T., Faruque, M.R.I., Islam, M.T., Khandaker, M.U., Radar cross-section reduction using polarisation-dependent passive metamaterial for satellite communication. *Chin. J. Phys.*, *76*, 251–268, 2022.
37. Sengupta, S., Council, H., Jackson, D.R., Onofrei, D., Active radar cross section reduction of an object using microstrip antennas. *Radio Sci.*, *55*, 0–1, 2020.
38. Sharma, A., Gangwar, D., Kanaujia, B., Dwari, S., RCS reduction and gain enhancement of SRR inspired circularly polarized slot antenna using metasurface. *AEU-Int. J. Electron. Commun.*, *91*, 132–142, 2018.
39. Shen, L., Pang, Y., Yan, L., Shen, Y., Xu, Z., Qu, S., Broadband radar absorbing sandwich structures with enhanced mechanical properties. *Results Phys.*, *11*, 253–258, 2018.
40. Singh, H. and Jha, R.M., Introduction to radar cross section reduction, in: *Active Radar Cross Section Reduction, Theory and Applications*, H. Singh, and R.M. Jha, (Eds.), pp. 1–14, Cambridge Univ. Press, Cambridge, England, UK, 2015.
41. Tau, A., Bahtiar, S., Eko, R., Yuliantika, D., Hidayat, A., Sunaryono, S., Hidayat, N., Samian, S., Soontaranon, S., Fabrication of Mn1-xZnxFe2O4 ferro fluids from natural sand for magnetic sensors and radar absorbing materials. *Heliyon*, *6*, e04577, 2020.
42. Hernandez, R.V., Briones, F.C., Estrada, S.M.L., Perez, F.R., Carbon-based radar absorbing materials: A critical review. *J. Sci. Adv. Mater. Dev.*, *7*, 100454, 2022.

43. Wang, Y., Kong, Y., Li, J., Xu, S., Zhang, X., Dong, Z., Dual-band polarization conversions and optical diode based on bilayer T-shaped metamaterial. *Results Phys.*, 42, 105981, 2022.
44. Wei, J., Qi, Y., Zhang, B., Ding, J., Liu, W., Wang, X., Research on beam manipulate and RCS reduction based on terahertz ultra-wideband polarization conversion metasurface. *Opt. Commun.*, *502*, 127425, 2022.
45. Xu, C., Ren, Z., Wei, J., Lee, C., Reconfigurable terahertz metamaterials: From fundamental principles to advanced 6G applications. *iScience*, *25*, 103799, 2022.
46. Xu, S. and Xu, Y., Research on active cancelation stealth technique. *Opt.*, *125*, 6219–6222, 2014.
47. Yang, J.J., Cheng, Y.Z., Qi, D., Gong, R.Z., Study of energy scattering relation and RCS reduction characteristic of matrix-type coding metasurface. *Appl. Sci.*, *8*, 1231, 2018.
48. Yang, J., Liao, Z., Li, Y., Li, D., Wang, Z., Wang, T., Wang, X., Gong, R., Phase modulation of metasurfaces for polarization conversion and RCS reduction. *Opt. Mater.*, *119*, 111374, 2021.
49. Yi, M., Huang, J., Meng, Z., A cancelling method based on Nth-order SSC algorithm for solving active cancellation problems. *Aerosp. Sci. Technol.*, *71*, 264–271, 2017.
50. Zhou, Q., Xue, B., Gu, S., Ye, F., Fan, X., Duan, W., Ultra broadband electromagnetic wave absorbing and scattering properties of flexible sandwich cylindrical water-based metamaterials. *Results Phys.*, *38*, 105587, 2022.

Index

Printed in the USA/Agawam, MA
September 27, 2023

852255.019